T0176499

SCIENCE IN BLACK AND WHITE

How Biology and Environment Shape Our Racial Divide

ALONDRA OUBRÉ

Prometheus Books

Guilford, Connecticut

Prometheus Books

An imprint of The Rowman & Littlefield Publishing Group, Inc.
4501 Forbes Blvd., Ste. 200
Lanham, MD 20706
www.rowman.com

Distributed by NATIONAL BOOK NETWORK

British Library Cataloguing in Publication Information available

Library of Congress Control Number: 2019955277
ISBN 9781633886001 (cloth)
ISBN 9781633886018 (ebook)

For my parents and for theirs

CONTENTS

Prologue

For as long as I can remember, I have been fascinated with almost any discussion, formal or informal, aimed at clarifying the scope and causes of the "race problem"—specifically, the "American Negro" or "colored" problem, which later became known as the Afro-American problem and today persists under various names, depending upon the audience. The "Negro problem" has had multiple connotations since its usage became commonplace in the nineteenth century. Even in its current renditions, however, the concept is still invoked in some circles to suggest there are limitations to the mental and moral characteristics of blacks, based on the abiding belief that human races each have their own proclivities and destiny.[1] For individuals who espouse this view, the underlying assumption is that race-based tendencies are innate, fixed, and unlikely to be significantly transformed, even in the face of environmental change designed to improve the standard of living of a given racial group.

Whether in scholarly studies, popular media, or colloquial conversation, the "condition of African Americans," or, on a global scale, the "plight of blacks"—Sub-Saharan African blacks and their descendants—can still evoke a sense of conflicting emotions and attitudes. For many, although certainly not all, Americans, the mere mention of "black people" elicits a simultaneous sense of fear, superiority, condescension, misunderstanding, guilt, paternalism, anger, frustration, vengeance, sometimes empathy and compassion, and occasionally reserved reverence, as well as a stream of opposing feelings and perceptions too long to present here. Clearly, interethnic communication and interaction between blacks and whites, and virtually every ethnic group in the United States, have improved remarkably during the last 50 years.

Still, debate, often conducted in private settings outside public spaces, continues about the underlying reasons for ethnic or racial disparities in life outcomes. This debate revolves around a perennial question: Does nature—

meaning biology—or nurture—basically, environment—dominate in shaping human cognitive and personality traits that, according to some people, contribute to racial disparities in life outcomes? The traits of interest here include disease risks and, more provocatively, such life outcomes of intelligence performance as IQ scores, and a wide array of social behaviors. For some onlookers unabashed in their bluntness, the pivotal issue is whether biology or environment best explains not only the overall ethnic achievement gap, but also, in particular, the divide in social outcomes between whites and blacks—African Americans and other ethnic populations of African black descent.

Traditionally, the scientifically informed elements of this debate have revolved around racial—or as I prefer, ethnic population—differences, particularly black/white/Asian variance, in IQ as a measure of intelligence performance. Today, however, the nature–nurture debate has entered a new era as some researchers turn to genetics and the brain to explain group disparities in such social outcomes as marriage, parenting, educational achievement, employment, and law enforcement encounters. In the current social climate, only a relatively small number of scientific studies tackling nature versus nurture focus on race. But of those that do, the ramifications may be immensely influential for lay consumers eagerly awaiting seemingly scientific explanations for race-based differences in a wide range of social outcomes. Convinced that some nonwhite ethnic groups in the United States have squandered unprecedented educational and economic opportunities, some people point to nature—race-based genetics—as a major, if not the primary, reason for why African Americans are still disproportionately overrepresented among the imprisoned, unwed single mothers, high school dropouts, the unemployed, and low wage earners.

By contrast, other people attribute the ethnic gap in life outcomes mainly, and in some cases exclusively, to nurture—in this case, adverse environmental circumstances. Overall, public condemnation has been forceful and swift against any sentiments or claims reminiscent of racial superiority, whether a resurrection of Nazi German anti-Semitism or recent renditions of modern racial science tethered to white supremacy. Despite pervasive outcries against these ideologies, the idea that population genetics—racial genetics—might contribute, even partially, to "black social underperformance" (in fields other than sports and music) remains etched in the collective psyche of Americans of every ethnic ancestry. In some cases, blacks themselves buy into this view.

After all, in the United States, we have only to look at the divide, or perceived divide, in social achievement between whites, along with people of Northeastern Asian descent, versus people of color, especially African Americans, to realize that the gap between these ethnic populations, in general, in indicators of success is substantial, intergenerational, and unremitting. In some quarters, for example, online venues, increased chatter about race-based biology fueling race-based differences in social behaviors is on full display. These views are often intertwined with the "scientific arm" of more vocal fringe social movements such as white supremacy and the alt-right. These perspectives are also wittingly or unwittingly creeping into the consciousness of some Americans and, in fact, numerous, sometimes concealed, sectors of the global community. And yet, in the mainstream public domain, there seems to be an implicit resistance to addressing this topic in a scientific, evidence-based manner, accessible to the masses, that transcends political squabbling.

No reasonable person denies the impact of environmental forces, including unfavorable social conditions and physical toxins, on individuals and communities. The public is well aware of the devastating effects that poverty, crime-ridden neighborhoods, poor schooling, and other adverse social circumstances can have on any individual or a family, let alone on some segments of an ethnic population. But environmental hazards encompass much more than conspicuously harmful influences. In the United States, for example, the African American community has become increasingly heterogeneous, with growing numbers of people exiting the underclass and successfully entering the middle class. But despite the notable progress African Americans have achieved in the last 50 years, the American and perhaps international public still regard African Americans, in general, as a "less than" ethnic group. During this time frame, society, in general, has momentously changed it attitudes about blacks and, to its credit, learned to tolerate and increasingly accept them—at least certain blacks, more than ever before in modern history. Still the harsh reality remains that outside Sub-Saharan Africa, blacks, as a whole, are unfavorably stereotyped as heirs of inborn intelligence deficits, greater law-breaking tendencies compared to other groups, fragmented family structure, and a litany of other undesirable social traits.

If race is America's oldest dilemma, in the present day that dilemma may reflect mainstream society's hesitation to think that people of African black descent possess the innate capacity to coexist harmoniously in the modern industrialized world, particularly Western society, as benevolent, considerate, and productive members of society. This issue is not limited to blacks, as it obviously affects various non-African derived ethnic populations whose ancestors originated outside of Europe. But it holds special import for blacks in the United States in light of their unique contributions to the making of the United States, juxtaposed with their enduring overall marginalization in this society.

Today, the unutterable dilemma concerning racial differences that seemingly involve blacks may be less about physical traits such as skin color, and instead, amid emerging discoveries of personality-linked genes, more about ethnic population variation in the genetic underpinnings of social behavior. Rapid advances in neuroscience, human genetics, behavioral genetics, and psychiatric genetics, coupled with knowledge from psychology, anthropology, and sociology, have created a groundbreaking intellectual landscape for exploring the intersection between nature and nurture. Researchers from a broad range of disciplines can now probe possible interconnections between an array of biological traits—including various brain features and genetics—and environmental factors—for example, family income, level of educational attainment, aggression, violent crime, and parenting skills.

As part of the quest to understand the dynamic complexities of nature-*plus*-nurture in shaping human life, some experts are focusing on the relationship between social status and social outcomes. But only some of these studies assess race-related differences in behavioral patterns in relation to biological and environmental influences. This approach, revolving largely around socioeconomic factors, is distinctly different from, and in some cases even diametrically opposed to, the traditional racial science of the nineteenth and twentieth centuries that prided itself on patently denigrating certain ethnic populations. Racial science, historically, refers to research that generates empirical evidence intended to demonstrate the superiority of certain ethnic populations and the inferiority of others. During the 1980s through the early 2000s, several leading racial scholars in the United States and Europe proposed that race-based variation in biological traits, for instance, brain size and levels of the male sex hormone, testosterone, could account for racial differences in cognitive performance and some social outcomes.

During the last decade in the conventional scientific community, only a few studies examining the interplay between nature and nurture on such social traits as IQ and educational attainment have factored race, or genetic ancestry, into their analyses. Instead, some of the most impressive investigations in this field correlate socioeconomic indices, for example, poverty versus middle-class economic status, with various brain features or with genetics. To some degree, this direction is encouraging, for it may reveal a shift away from an emphasis on race-based genetics as a key determinant of social outcomes. At the same time, research on participants from low-income backgrounds or with limited education can be used to implicitly draw conclusions about ethnic minorities, including African Americans. And yet, as some experts note, while socioeconomic status and race are, at times, leveraged in research studies as a proxy for one another, there are also clear distinctions between them, especially regarding health outcomes.[2] The effects of early life adversity, including poverty, abuse, and traumatic stress, vary across both socioeconomic groups and ethnic populations.[3]

Although the national poverty rate has declined since 2010, as of 2015, 36 percent of African American children and adolescents versus 12 percent of non-Hispanic whites lived in poverty. The reported child poverty rate during this timeframe was 34 percent for Native Americans and 31 percent for Hispanics.[4] This may be why a chorus of onlookers, and not just highly vocal social activists, continually call attention to the detrimental ramifications of multiple environmental stressors that interact synergistically with poverty—including poor access to healthcare, adequate nutrition, and good schooling—to unfavorably impact people of color. The heavy tolls posed by these environmental afflictions are exacerbated in people of color, particularly blacks, who are also more likely to be simultaneously affected by ethnic bias (both implicit and explicit), police brutality, and encounters with the criminal justice system.

But some pronature writers, skeptical of what they dub the "egalitarian doctrine,"[5] see the situation differently, asserting that behavioral differences across ethnic populations can indeed be traced partly to genetic ancestry. And here the notion of "partly" matters because it suggests that even a minute number of ethnic population genetic differences can contribute significantly to disparities in, for example, educational attainment, child rearing practices, and rates of violent crime and incarceration. Several recent studies have sparked curiosity by examining the association between specific gene versions or brain features and genetic ancestry, or race. Studies that attempt to correlate population genetics with social outcomes like parental disciplining of children, aggression, and crime signal a new wave of contemporary research on the role of nature versus nurture in triggering the ethnic achievement divide.

These studies do not necessarily have the same ideological orientation, and some may lean toward a pronature vantage, whereas others may reflect a pronurture stance; however, as any objective researcher knows, the integrity of the scientific methods used to produce data (along with an impartial interpretation of that data), rather than an investigator's built-in biases for or against a particular outcome, defines the quality of the research. Despite this caveat, however, some, though not all, contemporary pronature studies are apparently attempting to reinvent the rationale for race-based genetic determinism, softening otherwise abrasive and vitriolic claims in a manner that seemingly befits today's more tolerant and diverse society. Most pronature advocates recognize the power of the environment on shaping social traits in individuals and ethnic populations, just as proenvironmental supporters acknowledge the influence of population genetic variation on risks for ethnically linked diseases.

The research highlighted in this book is not exhaustive, but, instead, provides a diverse sampling of findings generated from studies reflecting divergent ideologies, research orientations, and conclusions related to nature, nurture, and race. Chapters 1, 2, and 3 present an overview of the history and philosophy of the nature versus nurture debate on the origins of human social life, as well as the basic tenets of Darwinian evolution. Chapter 4 introduces aspects of biology, in addition to genetics—namely, epigenetics, gene expression, and plasticity—which play a role in determining social traits. Chapter 5 provides a brief introduction to key features of the human brain, while chapter 6 gives a snapshot of recent scientific breakthroughs on male–female differences in the human brain. Chapters 7 through 10 collectively explore a range of reported evidence on brain traits and functions, and, in some cases, genetics, with respect to differences in either racial background or socioeconomic status. Controversies surrounding ethnic population variance in the prevalence of genes linked with aggression, violence, and other severe, societal transgressions are explored in chapter 11. Chapter 12 examines the nature–nurture debate about reported differences in rates of sexual maturation in ethnic populations, focusing on female physical development.

Chapters 13 and 14 examine theories about the role of testosterone, and a key gene linked with it, in male aggression and violence. These theories have paramount implications for males of African black descent.

It is hoped that this book can offer new perspectives on the continually expanding body of research on how nature and nurture interact to influence cognitive ability and social behaviors. Be warned, however, that amid this rapidly evolving field, experts have revamped some of the conventional concepts of biological inheritance, just as they have refined classic constructs of the environment. This book is a primer—an introductory overview—of the recently charted and ever-changing knowledge about the interplay between nature and nurture that shapes behavioral tendencies across ethnic populations.

Confronting the tacit taboo of talking about ethnic population differences in social outcomes—what they purportedly mean and do not mean—potentially can, I believe, be beneficial on several levels. At a time when misunderstanding and miscommunication abound with respect to race, a candid, national conversation about nature, nurture, and ethnic population social patterns, when guided by reason and mutual respect, can potentially raise the public's scientific literacy about the contribution, or lack thereof, of genetic ancestry to the racial achievement gap. In the process, it can also improve individuals' understanding of how biology and environment interact to shape their own lives and the lives of their children and others in their world, no matter their own genetic heritage. A frank public dialogue of this topic can perhaps equip us, as a collective society, with novel cognitive tools for reenvisioning both "the other" and ourselves. Appreciating and embracing the diversity of our ancestral heritages, as a species, may be one of the most crucial endeavors we can consciously perform to ensure our self-preservation. In any case, we are long overdue for an update to the empirical scientific evidence that, despite our racial differences, demonstrates humankind's overarching shared foundations as biological, cultural, and social beings. Awareness of the fascinating research unfolding in the arena of nature–nurture and the human condition promises to be a step in that direction.

ACKNOWLEDGMENTS

IN THE PAST 25 YEARS, MANY INDIVIDUALS—FAMILY, FRIENDS, COL-leagues, and mentors—have encouraged me to write about race, even its most difficult and sensitive aspects. I am grateful for their support and inspiration. I am deeply indebted to Dr. Audrey Smedley, the late Dr. Leonard Lieberman, and the late Dr. O'Neil Ray Collins for their professorial guidance as internationally renowned scholars of "race." Dr. Smedley and Dr. Lieberman, as well as Alvin Poussaint, MD, provided valuable feedback on some of my earlier publications on the nature versus nurture debate about ethnic population differences in social behavior. Special thanks go to Irwin Zucker, Jennifer Ferro, Michael Lynton, Earl Ofari Hutchinson, Ph.D., and Rabbi Mordecai Finley, Ph.D., for their professional insights and for appreciating the need to candidly discuss race in society. I also am especially grateful to Vernon and Antionette Hickman, Ann Collins, Zoltan Soos, and Ty Stoller for their stimulating conversation, which helped make this book a reality. Finally, I want to thank the production team at Rowman & Littlefield for their diligent attention to this project. Nicole Carty provided valuable copyediting support, and Jessica McCleary assisted in managing the phases of book production.

Nature Versus Nurture

Redrawing the Battle Lines

THE ORIGIN OF NATURE VERSUS NURTURE

THE AGE-OLD DEBATE ABOUT NATURE VERSUS NURTURE—WHETHER INTEL-
ligence, personality traits, and behavioral tendencies are mainly innate or
learned—can be traced back to antiquity. In the ancient world, this debate
was birthed long before nineteenth-century Austrian monk and scientist
Gregor Mendel conducted his famous "pea experiments" demonstrating that
dominant and recessive inherited traits are passed from parents to offspring.[1]
Almost 5,000 years ago, animal breeders in Egypt and the Near East believed
that both physical and behavioral characteristics are transmitted from parent
to progeny. And by the time dog breeding became a common practice for
the Greeks, Romans, and Hebrews 3,000 to 1,500 years ago, people were
well aware that a dog's physical features were transferred to its offspring. If
physical traits of both the canine and human species were handed down from
one generation to the next, it stood to reason that human mental qualities
might also be inherited.[2]

   Some of the great thinkers of the ancient world also paid close attention to the impact of experience, loosely known as nurture, on the psychological disposition and qualities that shape the human mind. Greek
philosopher Plato claimed that the "Guardians"—the ruling elite of ancient
Greek society—required both "good ancestry" and instruction, or learning,
to achieve the ideal state. Plato acknowledged heredity, but his advocacy
of universal education for the citizens of his state, starting at an early age,

became an endorsement of the environment in shaping well-rounded individuals, no matter their background.[3]

In the modern era, the roots of the nature versus nurture debate—sometimes framed as genetics versus environment—are linked with Charles Darwin and John Locke. For Darwin, the famed nineteenth-century English naturalist whose theory of biological evolution was broadened to include social behavior, nature prevailed, but for John Locke, the seventeenth-century English philosopher and physician, nurture triumphed. Locke was an influential, antiauthoritarian thinker of the Enlightenment, and his writings on human equality had a profound effect on politics, notably the American and French revolutions. As an architect of the politics of liberalism—the belief in individual freedom and liberty—Locke had a major impact on such historic figures as Jefferson, Voltaire, and even Rousseau, who supported self-determination and self-government rather than aristocracy.[4]

Befitting the empiricist tradition, Locke envisioned the human mind as a *tabula rasa*, or blank slate, that at birth was empty of any thoughts or ideas. He believed that each person acquired knowledge through sensory perception of his exposures to the outside world, or his personal experiences in life, rather than through innate imprinting. Yet, despite Locke's liberal philosophy of natural rights and human freedom, he was not, say some sources, a pure nurture advocate, at least in a contemporary sense. Locke doubted that characteristics of the mind were ingrained at birth, but at the same time he proposed that humans have "native propensities," including innate capacities and temperaments, toward power, for example, that vary from one individual to another.[5]

Historically, Locke has often been portrayed for his patent opposition to slavery, expressed in many of his writings. And yet as an investor in the Royal African Company, one of the premier African slave mercantile companies of his era, he earned appreciable income as a merchant adventurer in the early transatlantic slave trade. Locke even helped draft state legislation that upheld the rights of Carolina freemen, essentially white men, to have absolute power and authority over their slaves. For some historians, Locke's support for black enslavement and a feudal aristocracy was the epitome of hypocrisy.[6] But Holly Brewer, chair of American history at the University of Maryland, questions critics on the "left" who believe Locke was a philosopher who advanced radical ideas while supporting slavery and colonialism. According to Brewer, Locke's contributions to drafting the *Fundamental Constitutions of Carolina*, a document published in 1669, endorsing hereditary nobility and slavery, was simply secretarial. As an agent for the Lords Proprietors of the Carolinas, Locke's role was no different than that of an attorney who prepares a will or another legal paper for his client, Brewer explained. She argued that Locke's ties to Royal African Company stock are misinterpreted, noting that he sold his investment in the Royal African Company in 1675. As a member of Virginia's Board of Trade in the 1690s, Locke opposed royal land grants that Britain had rewarded to colonists who purchased blacks as "servants," but more likely as slaves.[7] Still, Locke's published views of African blacks were usually confined to their role as slaves rather than as a particular racial or social group. This stance contrasted with his explicit opinions about Native American culture, which he called disorderly and uncivilized.[8]

In the seventeenth-century British Empire, including the American colonies, Native Americans were considered "natural men" inhabiting primitive surroundings, whereas African "negroes" were seen as "subnatural" or subhuman. The prevailing thinking of the day was that Native Americans potentially could be educated but blacks could not. Regardless of his exact perspective on African blacks, Locke was not alone in his contradictory stances regarding slavery. Other Enlightenment thinkers too have been accused, and rightly so, of inconsistencies between their personal approval of black enslavement and their public appeal for human (or least white male) freedom.[9]

These contradictions were part and parcel of the political, intellectual, and economic elite of the day. Virtually everyone in Victorian England and its empire accepted an uncontested view of the innate superiority of whites—specifically those of Northern and Western European descent—and with that belief, yielded to a blind allegiance to notions of innate race-based differences in intelligence and social patterns. This may be why some scholars have questioned if Locke truly were hypocritical with regard to African black enslavement. Yet, Locke stood apart from many of his contemporaries and even some of his intellectual successors in how he conceived of racial differences. He believed that individuals could be differentiated by not only their education, but also their race, temperament, mental capacity, and habit. Steeped in the traditional European worldview of the 1700s, he suggested that a person's racial background could influence how he was affected by life experiences.[10]

Locke crafted an argument that, although embryonic at the time, may well have laid the groundwork for the concept of nature *plus* nurture, or the *interaction* between biology and environment that was to emerge decades, even centuries, later. Despite the conflicts between Locke's public sentiments about freedom and his private actions regarding slavery, he had a momentous impact on future liberal thinkers. Ironically, some abolitionists in the United States interpreted, perhaps naively, perhaps fancifully, Locke's words to mean that environmental circumstances rather than biological inheritance best explained racial differences in the social accomplishments of whites and "Negro slaves."[11] This may be one reason why even today, John Locke remains an iconic champion of the idea that nurture, or life experience, is decisive in shaping a person's character and productivity in society.

A century later, in the mid-1800s, John Stuart Mill, an English liberal political philosopher and intellectual protégé of Locke, became an even stronger proponent of nurture than Locke. Mill challenged aristocratic privilege, arguing that opportunity was more important than birth in determining life outcomes. Mill was a contemporary of Darwin and, more importantly, Darwin's cousin, Sir Francis Galton. As an early pioneer of intelligence testing, Galton was a staunch supporter of nature over nurture. Known for his mastery of multiple scholarly fields, Galton gained recognition as a polymath with widespread expertise. When he coined the phrase, "nature versus nurture," intellectuals took note, just as they did when he announced his research on individual psychological differences and the "comparative worth of different races."[12]

Galton was a founding architect of eugenics—a set of scientific practices intended to improve physical and psychological traits of the human race by controlling who is, and who is not, allowed to breed. The goal of eugenics, in the past, as in the present, was to increase favorable heritable traits, while diminishing unfavorable ones in humankind. To ensure this goal was met, Galton proposed that society's more useful members of society birth more children, while the less useful have fewer offspring.[13] Today, Galton is still frequently vilified for his views (which were typical for his era), but he also has been commended for many of his scientific achievements. To his credit, he was the mastermind behind twin studies in psychology that set the stage for empirical research in contemporary behavioral genetics. Galton, then, indirectly at least, was the intellectual forebear of such modern bridging sciences as sociobiology and evolutionary psychology, which integrate elements of biology and psychology—heredity and environment or experience—in an attempt to scientifically explain the foundations of social behavior.

In 1859, Charles Darwin's seminal work *On the Origin of Species* sparked a media firestorm, with science going head to head with religion about the origin of humankind, indeed, the origin of life.[14] During the 1860s, Darwin's theory of evolution, the central theme of this groundbreaking book, gained traction as the educated elite, first in Britain and then in various other regions of Europe and the United States, conceded that humans are part of the animal kingdom. During this period, Galton and Mill represented the most prominent proponents of heredity and experience, respectively. By 1870, these two intellectual giants had become uncompromising symbols of the diametrical poles of a then still-nascent nature-versus-nurture controversy concerning the origin of human intelligence and behavior.[15] Darwinian perspectives became pivotal in framing the debate about nature, or inherent characteristics, versus experience, or acquired traits, as determinants of the human intellect and social life, and Darwin's ideas about the human condition remained virtually unchallenged for the next 50 years. Building on ideas about biological evolution, learned men of the day attributed human thoughts, actions, and even moral character to nativism, or inborn qualities, rather than outward circumstances.[16]

Two decades after Darwin's death at the turn of the twentieth century, *On the Origin of Species* paved the way for the modern theory of biological evolution. Darwin's legacy left an indelible mark on biology that would forever change the course of science and medicine. Leading thinkers of the nineteenth century, including Galton, Darwin, American philosopher and psychologist William James, and English philosopher Herbert Spencer, argued that heredity and individual innate abilities were crucial for intellectual, scientific, and artistic talent and achievement.[17]

Ernst Haeckel, the German naturalist and philosopher, carried this idea one step further when he attributed inborn aptitudes to racial differences. Haeckel's concept of superior and inferior races—and his division of the human species into ten races, with European whites at the helm—laid the foundation for a platform that, in the 1900s, was eventually used to promote Nazi eugenics.[18] The written works of Haeckel and other Victorian philosophers of the late 1800s conveniently coincided with Europe's nineteenth-century deepening colonial expansion into the non-Western world, especially Africa, India, and South America.[19]

For many of the physicians-turned-anthropologists and philosophers-turned-naturalists of that period, the non-Western world provided an incomparable living laboratory of diverse, if not "peculiar," non-European human specimens to probe, prod, and inevitably rank on a hierarchy ranging from inferior to superior. Most of the naturalists and philosophers of the Victorian era were not scientists, not even remotely so by present day standards, although some were professionally trained physicians. Their classifications of human racial groups revolved around popular conceptions (and quite often misconceptions) of supposed links between physical characteristics and mental or behavioral traits. African blacks were inevitably placed on an evolutionary ladder, reminiscent of the Renaissance "Great Chain of Being," above the great apes but well below the pinnacle of fully human types epitomized by European whites.[20]

The experiments of American physician and craniologist Samuel Morton to measure the size of the skulls from diverse peoples of the world and place blacks at the bottom rungs of a hierarchical ladder of human racial groups have been well documented.[21] So too have the assertions of nineteenth-century phrenologists—specialists in measuring of the human skull—such as Charles Caldwell. A physician and slave-owner from Kentucky, Caldwell described the brains of blacks as "small intellectual organs," a designation that reinforced his support for black enslavement in the United States.[22] Efforts to differentiate black Africans from other human populations were not limited to brain volume, considered a barometer of intelligence, but also other bodily parts, for instance, the limbs and pelvis. Some medical men of the nineteenth century who compared apes and human types, or racial groups, proclaimed that blacks had an ape- or monkey-like spinal column and limbs.[23]

By the end of the late Victorian age, psychology had etched its way into emergent scientific paradigms of human biological evolution. Perceptions of race—notions about the intrinsic characteristics of different racial types—did not change, but discourses on human nature took on a new tone. Motivated by a growing need to justify colonialism, black enslavement in the American South, and eventually black inferiority in the post–Civil War years, the majority of Victorian philosophers and naturalists of the 1800s were preoccupied with elevating the social worth of white Europeans and their descendants above all other racial groups.

Social Darwinism: The Foundations
of Modern Racial Science

It was not Charles Darwin but Herbert Spencer, the English philosopher of the 1800s, who created Social Darwinism, a social ideology with broad political implications. Social Darwinists proclaimed that society's elite possessed wealth and power because of the inherent biological superiority of the upper class. Spencer, who reportedly coined the term "survival of the fittest," extended Darwin's theory to the social arena by focusing on intergroup competition rather than reproductive success, the keystone of Darwin's theory of natural selection.[24] Darwin's writings centered on the struggle for survival between individuals—captured in the phrase "survival of the fittest"—but his prime interest was in the biological variation that favored reproductive success. Well before the chromosome theory of heredity became a bedrock of modern biology and long before the discovery of DNA, Darwin recognized that certain innate changes were passed on from parent to offspring.[25]

To some extent, Spencer's model of social hierarchies paralleled Darwin's theory of natural selection. Spencer's goal was to show why the "lower races," which he and his followers deemed unqualified for full membership in humankind, could never advance beyond a primitive state. By borrowing metaphors from biological evolution, Social Darwinists of the late 1800s sought to explain why human races were positioned so unequally in the world's political and economic landscapes. They introduced the idea of "racial potential"—a concept that supposedly could explain why some races succeeded in society and life while others did not.[26]

At its core, then, Social Darwinism was a social philosophy, but it was closely intertwined with nineteenth-century biology. It afforded naturalists of the late Victorian age a new context in which to discuss what they called the biological and moral inequities of human racial groups. While Africans were predictably relegated to the lowest tier of Spencer's racial hierarchy, the unconventional arguments of some Social Darwinists turned ideas about black inferiority on their heads. Spencer's followers considered blacks as degenerate, but they also thought African people in the New World benefited from slavery. In the words of one Social Darwinist, blacks were so overprotected from the fierce struggles of nature that they ought to be freed. Only then, said like-minded Social Darwinists, could blacks compete openly with whites. Paradoxically, as noted earlier, this argument was used by some Social Darwinists as a justification for abolishing slavery.[27]

Regardless of their political posture, Social Darwinists had as their ultimate objective halting the reproduction of those they stigmatized as inferior, including the "feeble-minded" and criminals, and perhaps above all, those they designated as subhuman racial groups. One key to achieve this endgame was to totally annihilate blacks, or at the very least banish people with African black ancestry from the Western world. Like their intellectual descendants of today—white supremacists, racial scientists, and some followers of the alt-right—nineteenth-century Social Darwinists labeled blacks as intellectually deficient and criminally inclined. In keeping with what would later be known as the "one drop rule," some Social Darwinists, including esteemed physicians of the day, were intent on accelerating the "extinction," essentially the genocide, of anyone with even negligible amounts of African black ancestry.[28]

For Social Darwinists, any assistance provided to "inferior races" would only advantage these "lower groups" with an undeserving edge by artificially improving their life circumstances. To offer a helping hand, Social Darwinists asserted, would be to interfere with biological evolution and the natural order of things. After all, they reasoned, evolution had already made its mark during the formation of vastly different human races. In the eyes of Social Darwinists, some racial groups, especially those of European stock, were better endowed with traits apropos for creating and maintaining civilized societies.[29]

But other races, Spencer's followers admonished, were destined to stagger behind, confined to a permanent state of mental and moral lowliness that could not be overcome by manipulating the environment. This became a grave concern for Social Darwinists because they were convinced that the lower classes bred more quickly than the middle and upper classes. If left unchecked, Social Darwinists complained, uncontrolled breeding of subhuman types would surely mean that society's misfits would (through no fault of their own) eventually outnumber and possibly overpower society's good folk. The line between Social Darwinism and eugenics was thin in the nineteenth century, just as the division between eugenics and racial science is blurred today.[30]

In 1883, Francis Galton coined the term *eugenics*, meaning "truly born," and defined it as the science of improving the "human race by better breeding." Following in the footsteps of the Social Darwinists, by the early 1900s, Galton's movement had gathered momentum in England and the United States.[31] This was the era when the term *gene* was identified as the unit of heredity, ushering in a new science that soon came to be known as genetics.[32] While conventional scientists of the day focused on uncovering genes linked with physical traits, a small but influential group of researchers called hereditarians embraced genetics for another purpose. The hereditarian pioneers, like their present-day successors, proposed that a person's intelligence and behavior are molded primarily by heredity, or biological qualities inherited from his parents.

Armed with a new set of principles designed to explain the science of inherited traits, eugenicists attempted to put hereditarian ideas into practice by weeding out from the human condition what they deemed undesirable innate traits, particularly criminal tendencies, poverty, and "over-breeding." During the early twentieth century, highly vocal eugenicist Henry Herbert Goddard, along with several other American hereditarians, established an IQ testing program on Ellis Island aimed at immigrants from Eastern and Southern Europe. According to Stephen Jay Gould, author of *The Mismeasure of Man*, Goddard concluded that more than 80 percent of Jewish immigrants were feebleminded (a notion that would be considered absurd today).[33] Some scholars alleged that Goddard never actually made this claim, suggesting that his intent was simply to compare the rate of mental deficiency in a subset of immigrants entering the United States with that of native-born white Americans. But as critics point out, attempts to reinterpret Goddard's research on intelligence testing—and its impact, even if unwitting—on U.S. immigration policies are suspect. These revisions have generally been made by hereditarians, if not outright eugenicists.[34]

By the close of the Victorian era, which coincided with the turn of the nineteenth century, Social Darwinism had gradually faded into oblivion, but its legacy lived on under new guises of hereditarian research devised to support a growing demand for eugenics social policies. By the mid-1920s, American eugenicists had launched social action programs to control the "breeding" of individuals and groups they rationalized should not be allowed to reproduce. The eugenicist platform called for sterilizing those categorized as "hereditary defectives," namely the poor, blacks, Native Americans, Jews, and various swarthy-skinned immigrants from Southern and Eastern Europe.[35] (Some of those categorized as swarthy-skinned immigrants from Europe may well have included mixed-race people of African descent who seamlessly crossed the American color line by "passing as white.")

During World War II, however, the eugenics movement backfired on itself, since by this time it had become too much of a haunting reminder of Nazi racialism to survive, let alone thrive, in the United States. And yet inevitably concerned with its own self-preservation, a few decades later eugenics managed to reinvent itself as a justifiable campaign for fair-minded scientific inquiry. Closely aligned with contemporary racial science, eugenics persists in contemporary societies of the developed world under various guises as both an ideology and a social action agenda.[36]

By 1910, behaviorism, a movement within psychology, had become increasingly more appealing to British and North American intellectuals trying to fathom the scientific basis of human thought, emotions, and action. During the next half-century, behaviorist ideas about the mind as a *tabula rasa*, or "blank slate," dominated Western intellectual culture, with the human condition envisioned as the product rather than the cause of society. As the twentieth century moved forward, leading thinkers endeavored to bridge Freudian theories of psychoanalysis with ideas of such psychologists as Ivan Pavlov and leading behaviorists like John B. Watson and B. F. Skinner.[37]

Psychoanalysis revolved around psychological therapies for treating mental illness by discovering connections between the conscious and unconscious mind. Behavioral psychology focused on explaining how behavior, or action, is conditioned by the environment. The new synthesis of these contrasting approaches within psychology elevated the importance of nurture, which at that time still prevailed in scholarly and popular narratives about human behavior. John B. Watson, who rose out of poverty and achieved international acclaim, reportedly invented the phrase "nature versus nurture" in its modern context. Watson was a leading proponent of the nurture perspective, as he had experienced firsthand how an individual's responses to environmental conditioning shapes his behavior.[38]

Watson famously boasted that with appropriate environmental conditioning, he could turn a "dozen healthy infants, well-formed" into any type of specialist—"doctor, lawyer, artist, merchant-chief, and yes, even beggar-man and thief," based simply on the upbringing and training he gave them. He insisted he could do this with each individual regardless of his "talents, penchants, tendencies, abilities, vocations, and race of his ancestors."[39]

Like his astute contemporary, anthropologist Franz Boas, Watson declared that racial disparities in behavior and personality stemmed from environmental circumstances and cultural differences.[40] For many scholars, Watson is a mere historical footnote in the annals of the history of science. But in an age when race and patriarchy informed social views about human worth and decades before it was fashionable, Watson recognized the potential power of the environment on shaping human thought and behavior.

DEMISE OF THE BLANK SLATE

Until the 1960s, many experts still considered the human mind to be a blank slate in which knowledge is not genetically preprogrammed but, instead, derived from either experience or the environment. The "blank slate" or pro-nurture era endured for more than a half-century, and not simply because of the rise of behaviorism in psychology. The "blank slate," in the way Watson conceptualized it, offered a rationale for why racial groups differed in such abilities as IQ tests and school performance. It influenced post–World War II politics, fostering public sentiments that moved closer to egalitarian ideas about human races. In the 1950s, the antiracist attitudes that fueled the Civil Rights Movement emerged at a time when the masses were still recovering from the shock of the extreme anti-Semitic ideology, inhumane policies, and brutal social practices of Nazi Germany.

By the late 1960s, Americans were more receptive to racial equality than ever before in the history of this nation (or the world, for that matter). Anyone concerned about black/white differences, although often dramatic, in social achievements was reassured that everything would work out once blacks achieved social and economic parity with whites and Asians. African American leaders and white liberals of the day attempted to persuade the masses that the solution was doable: Give blacks equality in education, employment opportunities, and access to good housing and healthcare, and surely the color divide in achievement will evaporate. The prevailing view, then, echoed by many experts in the ivory tower, was that modern psychology already had proven that the mind is a blank slate. Education is the key to knowledge, some avowed, and knowledge opens the door to equality in society, regardless of racial ancestry. Notions of the blank slate seemed to scientifically reinforce the idea that racial equality could gradually become reality if civil rights legislation were passed and implemented.

But this view had its detractors. Resistance persisted. By the mid-1900s the field of IQ testing, known as psychometrics, was almost a century old. Sir Francis Galton's vision of a science devoted to the genetic inheritance of behavioral traits was heralded by a small group of researchers who walked in his footsteps. Yet, many experts tread cautiously on the newly resurrected field of psychogenetics because it appeared to be aligned with eugenics. They had not forgotten that in the late 1800s, Galton had launched an active eugenics campaign to selectively weed out society's "lesser types" and promote "race improvement." According to some sources, Galton's views may have influenced Adolf Hitler's decision to exterminate the Jews of Europe, as well as other groups the Nazis considered inferior.[41]

Despite Galton's tarnished public image in the immediate post–World War II years, research on the heredity of behavior gradually accelerated during this period as new actors entered the arena, vividly altering its landscape. The term *psychogenetics* was introduced by Calvin S. Hall, a psychologist best known for his dream work analysis, but it was later changed to behavior genetics (and then to behavioral genetics). Like other branches of biology, including genetics, in the mid-twentieth century behavioral genetics was heavily influenced by Darwinism. The field gained legitimacy when it was endorsed by Theodosius Dobzhansky, a prominent Ukrainian American evolutionary biologist and geneticist considered one of the most important scientists of the twentieth century. Dobzhansky helped to establish evolutionary genetics as a field in its own right, based partly on his studies of genetic inheritance in vinegar flies.[42]

Dobzhansky's research did not go unnoticed. Many of the early behavioral geneticists of his day studied the heredity of social traits, for example, mating success and foraging in insects and other nonhuman animal species. The reinvention of behavioral genetics was not a contrived effort, but rather an organic synthesis of the prevailing biological disciplines of the times. It seemingly afforded biologists a morally acceptable framework for exploring the genetic underpinnings of social behaviors, for instance, intelligence, regardless of the specific behaviors and regardless of the species. For some scientists, a universal approach that supposedly could be leveraged to examine diverse species provided a solid rationale for comparing behavioral genetics in different human races.[43]

During the post–World War II period, experts in intelligence testing might have stayed out of conspicuous sight, but they never entirely vanished during the blank slate's last stand in the early to mid-twentieth century. When the doors of the new behavioral genetics finally swung open briskly and widely in the 1960s and 1970s to welcome genetic determinists, Darwinians, and a handful of race-oriented medical researchers, the pundits of intelligence testing were there, too. They were fully present, ready to stake out their claim to the largely uncharted and vaguely defined territory of genes, behavior, and evolution. These psychometricians, most of them self-described hereditarians, were the intellectual heirs to Galton's legacy. They were, after all, the one group in the entire cadre of racially if not racialist motivated scientists and scholars who, for almost a century, had chimed a consistent message about the innate superiority of certain races and the inborn inferiority of others.

Psychometricians, on the whole, never ceased conducting research that potentially could support eugenics (not that their research necessarily had to achieve this end), even if it meant repeatedly attributing ethnic differences in IQ to racial genetics (or what would later be termed population genetics). The public may not have been privy to or appreciated the potential implications of the new behavioral genetics for race relations. But shortly after World War II, conversations about racial and ethnic differences generally appeared unattractive to the masses of democratic nations.

Yet, from the vantage point of a pristine academic endeavor free of racial undercurrents, the intelligentsia in the United States and Europe praised psychometrics and its ally, behavioral genetics, as a commendable sign of scientific progress in the twentieth century. What could be more intriguing than the genetic transmission of mating behaviors to maximize survival, manifested in the Darwinian fitness in fruit flies? Better yet, innovative studies like these would surely allow the United States to advance its scientific acumen—even in such potentially sensitive areas as the genetics of social patterns, without threatening the social morass of a democratic and free society. With Dobzhansky leading the foray into uncharted scientific ventures, American and European intellectuals showed little, if any, resistance to the then-burgeoning field of behavioral genetics.[44]

During this period, a small number of hereditarian psychologists, one of them University of California, Berkeley, professor Arthur Jensen, were vigorously merging their embryonic, if not crude, empirical knowledge of human genetics with research on IQ and memorization. In 1969, Jensen, at the time a leading expert in intelligence testing, published his forever-contentious article "How Much Can We Boost IQ and Scholastic Achievement?" in the *Harvard Educational Review*. Jensen's message was clear and unwavering: Compensatory education programs, the hallmark of free schooling in the United States, had failed to meet expectations of closing the social gap between blacks and whites.[45]

The widely noted IQ gap between African Americans and white Americans was evident in the late 1960s. When Jensen's article first appeared in print, the average score of 85 points usually cited for blacks was almost a full standard deviation, nearly 15 points, below the overall score of white Americans. In a carefully worded essay standing in stark opposition to the popular sentiment of the day, Jensen implied that the ethnic divide in IQ might be due, partly at least, to inborn differences between these two ethnic groups. He encouraged his fellow psychologists and biologists to use the latest research tools available to them: the rapidly evolving techniques for studying human genetics. Only then, Jensen advised, could scholars properly evaluate the role of race-based heredity in the black/white gap in IQ and scholastic success.[46]

Jensen's article signaled a turning point in the history of the nature–nurture debate. Historically, this controversy had revolved around a concept of the mind involving nativism—innate mental qualities hardwired into the brain—versus the blank state.[47] With political overtones and pervasive social implications, Jensen's provocative writings reframed the largely unspoken controversy concerning underlying causes of racial friction in the United States. Almost single-handedly and almost overnight, Arthur Jensen supplied the momentum that reversed the direction of the nature–nurture pendulum, at least in the United States.[48]

Little by little, the consciousness of many educated Americans edged its way closer to notions about nature as the most likely explanation for racial differences in social outcomes. The shift was felt far and wide throughout society's elite, including myriad scholars and scientists. It even affected some psychologists who had once unconditionally upheld the blank slate, giving them reason to perceive that inborn race-based differences are the true culprit of "black underachievement" and the "black lag." For scholars and policymakers who jumped on Jensen's bandwagon, racial genetics came to be seen as a crucial, if not paramount, factor underlying the black/white IQ divide. During this period, novel discoveries emerged both in human genetics and the brain and behavior sciences. The academic setting was ripe for crystallizing a new interdisciplinary field of behavior genetics that blended these scientific pursuits.

The rising popularity of behavior genetics went well beyond the ivory tower, coinciding, perhaps not accidentally, with the peak of the Civil Rights Movement. Although American society had grown increasingly more democratic during the 1970s and 1980s, interethnic tensions persisted, at times turning into visible confrontations. Racial conflict, particularly between blacks and whites, continued even though ethnic minorities seemingly had greater opportunities for educational and economic advancement than in eras past. Overall, the public accepted affirmative action and other "Great Society" programs as vital for promoting social and economic parity between the races. But civil rights policies, especially those linked with public policies labeled as problack, had their share of detractors, and some critics doubted from the start that these programs had merit. They complained that only a small percentage of the many "black beneficiaries" of Affirmative Action measures could perform academically, professionally, and economically on par with whites.

In the face of an ethnic gap in IQ scores and scholastic accomplishment that rarely narrowed and more often appeared to widen, some cynics questioned if the human mind were truly a blank slate. These naysayers to Affirmative Action—to the very idea of racial parity itself—asserted that even though black Americans had been offered decent, if not excellent, educational and job development training opportunities, as a group they still fell substantially behind whites and Asians in life outcomes. Was this not proof that nurture had run its course as a possible remedy for closing the ethnic achievement gap?

Skeptics of the egalitarian doctrine became ever more suspicious of the "nurture assumption," particularly with regard to the ethnic divide across numerous life outcomes. For them, the disparities between blacks and whites, and blacks and Asians, was a reflection of Nature—the natural order of things that allowed people to distinguish "us from them." Some of these skeptics redefined themselves not as insensitive, let alone vicious bigots, but as scientifically neutral "race-realists" who claimed the empirical evidence surrounding this issue spoke for itself. Their self-description grew out of their deliberate focus on what they called the reality of overall black underachievement in school, the workplace, personal and family wealth, and obedience to the rule of law.[49]

Some intelligence specialists rejected Jensen's assertion that genetics played a significant role in both individual and group differences in IQ scores. For many people across the demographic spectrum in the United States, the rhetoric of environmental interventions and socially engineered solutions still promised hope for closing the achievement divide. But during this time, the tide of scholarly sentiment began to turn, slowly but surely. By the 1980s and certainly the 1990s, most biologists and a growing number of experts in the social sciences and behavioral sciences, notably prolific maverick scholar Steven Pinker, expressed their weariness with the "blank slate dogma."[50]

In the post–World War II era, many of these scholars were gradually swayed by what they declared was robust evidence for strong links, if not irrefutable cause-and-effect relationships, between genes and personality, and genes and IQ. Academics, intellectuals, and countless individuals from diverse backgrounds, frustrated with what they feared was a pronurture bias in science and education, could, in their own words, finally be liberated from a pretentious allegiance to an unrealistic ideology of equality. Some psychologists, once firm advocates of the blank slate and environmental primacy, sought refuge from the authority of behaviorism—for them, an invalidated idea that experience, learning, and nurture shape intelligence, personality, and social behavior.[51] To some extent, then, the embrace of "Nature," particularly from unexpected quarters of academia during the 1980s and 1990s, signaled a backlash against public pleas for racial parity in American society and beyond.

During the last two decades of the twentieth century, new discoveries about DNA revolutionized the life sciences, not simply biological science, but also applied biological research in biotechnology. With access to such sophisticated tools as more efficient DNA sequencers, researchers were finally able to sequence the genes of an entire living creature and even an entire species, including the genome of the entire human species, which was completed in 2003. The entry into the age of genomic technologies starting in the early 1990s occurred so precipitously that the line between the pregenomic and genomic periods is still blurred for some of the researchers who witnessed firsthand the transition from one stage to the next.

As the arsenal of biological data expanded, the scientific community at large began to question the long-held conventional wisdom about the delicate balance between genetics and environment in shaping intellect, personality, and social behavior. Informed by brave new scientific theories (with or without data), some academics outside of biology—in philosophy, psychology, and social science, for example—repositioned themselves in relation to changing social and scholarly currents regarding the origin of human cognition and behavior. They too wanted in on the prestigious "Nature" bandwagon. Who would not choose to be part of a fast-paced scientific movement that assured, for once and for all, a tangible and quantifiable explanation, rooted in genetics, for why some individuals and, perhaps more importantly, some ethnic populations, succeed in life, whereas others do not?

What's in a Name: Race, Racial Science, Racialism, and Race-Realism

That discomforting four-letter word—*race*—describes not only a contentious social and political issue, but also the source of some of the most vehement controversies among scientists, scholars, and, in recent years, the general public. The popular notion of a human race is a group with a common genealogy and physical traits, including skin color, facial features, and possibly hair texture.[52] The concept of a racial group in the social sciences is similar, but many experts in this broad field view racial classifications as social constructs— artificially invented categories of nature and reality that correspond to society's shared assumptions and beliefs. Many anthropologists, sociologists, and economists consider racial groups to be social constructs because they are based on society's consensus view that people can be divided into such discrete, physically distinct groups as whites, blacks, Asians, or East Indians (or South Asians), along with additional smaller racial groups worldwide that exhibit different combinations of physical features. For some members of the lay public, racial groups seem to intrinsically differ from one another not only physically, but also in their cultural beliefs, social behaviors, and, at times, social status within a given society.[53]

In sharp contrast to these definitions, in biology, "race" refers to a more technical but inclusive term that extends beyond humans and encompasses virtually every living species, both plant and animal. A race represents a subdivision of a species and is usually called a subspecies in the scientific literature. Traditionally, biological races are categories of geographically isolated breeding populations that vary in the frequencies of some of their genes. For many experts, a race is not a biological race unless it has specific genes or a specific cluster of genes that differentiate it from other populations within the same species.[54]

From the antirace or "race denial" perspective, the problem is that human populations cannot be divided into discrete categories that meet the same standard used to define biological races in nonhuman species. This is one of the main reasons, although not the only one, for why many scientists oppose the existence of separate human races. Instead, they claim there is only one race: the human race. Supporters of the biological reality of human races disagree, citing research suggesting that genetic ancestry can be used to divide humans, as well as chimpanzees, into biological groups that correspond to subspecies.[55]

But for antirace naysayers who challenge this position, racial divisions within the human species are erroneous because they do not meet the statistical criteria on which scientists rely to define biological races in other species. One method to determine whether races exist within a given species is to measure the degree of genetic variation between different subdivisions, or groups, within that species. Experts have calculated the amount of genetic variation needed to demonstrate the presence of biological races within a species, but the extent of genetic variation in humans is below this threshold.[56] As evolutionary biologist Joseph Graves points out, genetic structure, or genetic variation, is evident in the various populations of the human species, but there are no natural divisions that correspond to scientific categories of races.

The fixation index, or F_{st}, is commonly used to measure the degree of genetic differences between populations, with values of .05 to 0.15 showing moderate differentiation, 0.15 to 0.25 indicating a large degree of differentiation, and 0.25 revealing a very large amount of genetic differentiation. Most experts designate a F_{st} value of 0.25 as the minimum cutoff for genetically distinct races within species.[57] Nonetheless, researchers do not always agree on the extent of genetic variation within the human genome needed to divide people into biological races. Nor is there consensus on the best methods for measuring this variation. Still, the average range reported between various sets of populations is a F_{st} of 0.05 to 0.13, a value considerably beneath the threshold for well-demarcated races in nonhuman species. In short, then, population subdivisions within the human species have less than the minimum statistical amount of genetic variation required for the existence of biological races within a living species.[58] Not surprisingly, hereditarians who embrace the concept of human races contest the accuracy of this threshold. But a close inspection of the data on human genetic variability confirms geneticist Alan Templeton's observation that the bulk of genetic diversity in humans reflects individual uniqueness rather than race.[59]

Literally thousands of pages published in textbooks, peer reviewed scientific journals, and an assortment of print media during the last six decades have been devoted to formal arguments for and against the biological existence of human races. Today, the debate is far from settled. Although the terms *race* and *ethnic group* are sometimes used interchangeably, there are distinct differences between the two. Ethnicity pertains to a group's shared social and cultural heritage, whereas race, as a biological concept, connotes a sense of common genetic ancestry. Some researchers, particularly medical scientists, use the hybrid term *ethnic population*, because even though it is imprecise, it embodies a sense of biological ancestry, as well as the historical experience of most, if not all, members of a group.

At the end of the day, it may matter less what human groups loosely categorized on the basis of hereditary traits are called. Perceptions of human populations that vary in their common ancestry are, to some extent, arbitrary and difficult to separate from views about a group based on its known or perceived history and cultural patterns. Overall, ethnic populations are smaller units than traditional racial groups, but they can be combined into larger continental populations that, for practical purposes, sometimes appear to correspond to the conventional major divisions of human races. Still, this is not a precise or valid way to biologically partition human groups. On the other hand, the term *ethnic population* offers a realistic twofold starting point—one that melds elements of both genetic ancestry and cultural affinity—for rethinking racial classifications.

Racial science, also called scientific racism, is often used to describe naturalists' systematic investigations of various racial groups throughout the world beginning in the eighteenth century. Historically, racial science was characterized by notions, for example, "pure racial types" and fixed "racial hierarchies" of intellectual and social worth, allegedly based on scientific theory and evidence.[60] Throughout the decades and centuries, the language used by racial scientists or, more accurately, racialist scholars has changed (since the majority of them were and still are not professionally trained biologists), but the intent of racial science has remained unaltered. Its goal is to demonstrate, using scientific methods from such fields as genetics and neuroscience, that racial groups, ancestral populations, ethnic populations, or whatever term is in vogue, differ fundamentally in their capacity for intelligence, civilization, and morality.[61]

Racialism refers to the idea that a person's race (or in current scientific terminology, population) determines his physical and psychological traits, as well as his social and creative capacities. While this definition may seem to be synonymous with racism, there are differences between the two terms. Speaking about the reactions of society toward blacks, one psychologist defined racism as hostility toward a particular racial group based on the supposed inferiority of its members. By contrast, he described racialism as a heightened consciousness of the race of others. Elaborating further, this writer viewed racialism as rational, morally neutral, and inevitable in a society like the United States, with its history of enslavement, racial discrimination, and ethnic differences in many areas of society.[62]

Yet, racialists are not typically unbiased and objective in recognizing what they perceive as inherent differences in race-based traits. In keeping with the mindset of twenty-first-century racial science, white racialists generally think that certain nonwhite populations, particularly blacks, have genetic deficits that diminish their capabilities for high intelligence, discipline, and behavioral restraint. This is hardly a neutral assessment of race-based traits. In the white racialist worldview, these traits are essential for building "advanced" civilizations equipped with inventions ranging from writing to computers to robotic devices to nuclear weapons. In the context of the alt-right and white nationalism, which have gained momentum in recent years, racialism may in theory be neutral in its judgment of racial differences. But in practice it is virtually indistinguishable from racist attitudes and behaviors toward others whom a person despises, dislikes, or, at the very least, tends to avoid. These same tendencies and beliefs also describe a race-realist—a person who believes human races are a biological reality and not simply a social invention.

Race-realism is based on the notion that racial groups overall differ innately and often dramatically in such attributes as intelligence, physical stamina, temperament, and social behaviors, including criminal proclivities. Race-realists are convinced there is sufficient empirical scientific evidence to support their worldview. Although race-realism can be applied to a full gamut of human populations, the focus of most race-realists—who once again overlap considerably with the alt-right, white nationalists, and white supremacists—is on the aptitudes, mores, and social patterns of African blacks and their descendants worldwide.

The Rise of Behavioral Genetics

Whether behavioral science research of the late twentieth century constituted bona fide racial science conducted by racialist and race-realists may be a matter of perspective and perhaps historiography, or the manner in which historical events of the past are reinterpreted in the future. The decline of the blank slate in the nature versus nurture conundrum in the mid-twentieth century did not happen instantaneously. Its demise was a gradual process, an incremental fading of an idea born in the nineteenth century, if not farther back in time starting in the first century B.C.E.

At the close of the twentieth century, ideas about nature and nurture were drastically revamped, and nurture was no longer seen as a force standing solely in stark opposition to nature. Instead, nature and nurture became intertwined as inseparable partners, forever locked in an interminable tango of war and peace. No one denied the import of nurture, but with the advent of the genetics information age, nature once again unabashedly reared its head. Amid growing public interest in the origins of human behavior, nature resurrected itself, at least for genetic determinists and their like-minded adherents, as a more powerful force than nurture in the ongoing debate about what makes humans human. It was only a small step from this debate to the quarrel about the underlying causes of ethnic and racial differences in a gamut of personality and social traits.

The battle regarding innate qualities versus experience in determining cognition, temperament, and social behavior performance is not new, but the intellectual and scientific insights that have fueled this controversy have undergone radical transformations in the last two centuries. Many of the core controversies have revolved and continue to revolve around the impact of nature versus nurture on not only IQ, but also other psychological traits. Hereditarians, including prominent behavioral geneticists of the 1970s, were among the first custodians of innovative investigations of how genetics affects personality. In that era, only a few genes, usually referred to as candidate gene variants, were implicated in intelligence. Some bench scientists who worked in genetics laboratories soon began collaborating with psychologists, and together they formulated theories of human behavior in an intensive search to identify precise genes, or specific sections of DNA, associated with cognition. By today's standards, their laboratory tools were crude and their research methods unsophisticated and imprecise.

By the 1990s, when scientists pioneered techniques to track genetic markers called SNPs (single nucleotide polymorphisms), pronounced "snips"—short repeating sections of DNA that correspond to different versions or variants of a gene—the foundation was laid for formal studies on the heredity of human social behavior. Although still in its infancy, behavioral genetics opened the door to a data-driven discussion about nature versus nurture in personality traits, as well as intelligence.[63]

Some behavioral geneticists initially specialized in the genetics of IQ test scores, often comparing the IQs of identical twins with those of nonidentical twins or siblings to estimate the amount of genetic influence on IQ. Other researchers focused on racial differences in IQ, as well as elementary cognitive performance, including the speed of mental reaction time and movement time. Following Arthur Jensen's lead, the race-oriented researchers of the 1970s tackled controversial issues that penetrated the core questions about racial equality in the United States and beyond.[64]

It took little time for the new behavioral genetics of the 1970s and 1980s to draw harsh criticism as the reconfigured face of scientific racism. While it may well have served that function for some researchers, most behavioral geneticists fought hard for a reputation of objectivity, fair-mindedness, and respect among their peers in the broader scientific community. When sociobiology emerged as a distinct field in its own right in the 1970s, some racial scientists, as many in this group came to be known, aligned their research with its theories. Sociobiology is based on the notion that evolution has had a direct impact on shaping human social behaviors, including efficient strategies for feeding and reproducing. For sociobiologists, the origin of social life in social animals, just like the origin of physical traits in any species, is tethered to ecological pressures and competition for resources. Although socio-biology is not limited to the study of humans, some researchers in this field explore human social traits like mating patterns and parental investment.[65]

The central tenet of E. O. Wilson, the esteemed biologist and cofounder of sociobiology, was that the social behavior of any species, including the human species and its many genetically varied populations, could be explained by biological evolution. There are key differences between racial science and sociobiology, although some areas of these two fields have common ground. Most racial scientists borrow heavily from sociobiology, but it would be mis-guided to suggest that all sociobiologists are necessarily racialists. Yet, many pro-nurture critics consider the distinction, if there is a distinction at all, cosmetic, calling the self-proclaimed "open-mindedness" of various "socio-biologists" disingenuous, at best. Some sociobiologists, along with other Darwinian experts in social behavior, for instance, behavioral geneticists and evolutionary psychologists, have protested these labels, claiming they are unbiased scientists who are simply conducting objective research.[66]

Other experts in behavioral Darwinism seem immune to the onslaught of public disapproval against them for turning to genetics in an endeavor to explain ethnic population differences in social traits. And those who wittingly align themselves with the scientific arm of "race-realism" or the alt-right proudly wear their nonconformist attitudes about race as a badge of honor. Some have remained indifferent to accusations of not only uncon-scious implicit bias, but also labels of deliberated bigotry thrust upon them, proudly displaying their patent prejudices. Still, others have fought back, intent on stirring controversy related to racial differences by openly prop-agating, if not distorting, controversial research findings, seemingly as if to intentionally sway public opinion against racial harmony and instead toward racial polarization, segregation, and interethnic discord.

Behavioral genetics has had more than its share of critics, but regardless of public perception, it carved out a place for itself in the geography of the new biology of the 1980s and 1990s. It was the pivot of a novel synthesis of established academic fields, one in which DNA could help unlock the mysteries of the human mind and personality, and from all appearances, could possibly even create a biological blueprint of human social behavior. Behavioral genetics was not a mere theoretical academic exercise to discover causal relationships between certain genes and personality traits. Indeed, much of the research in this field is relevant to health, particularly mental health, offering a profusion of new insights into medical conditions ranging from neurotransmitter, or brain chemical, imbalances in psychiatric disorders to Alzheimer's disease.

Even before the advent of gene therapy, some scientists had correctly predicted that once the genes associated with specific psychiatric conditions had been identified, those genes could be targeted to potentially treat a host of mental illnesses. The unveiling of various drug classes of antidepressants, including SSRIs like fluoxetine (commonly known as Prozac), designed to inhibit the reuptake of the neurotransmitter serotonin, occurred before scientists understood the genetics of such psychiatric conditions as depression, anxiety, and panic attacks. With the rise of pharmacogenomics during the last 10 years, some patients have benefitted from individual testing to determine how well they respond to specific psychiatric medications.[67] And despite the stall in big pharmaceutical companies' active development of novel psychiatric drugs in the last decade, novel genetic targets for treating mental health have been identified.[68]

In the 1980s and 1990s, in close parallel with medical scientists mapping the genetic underpinnings of psychiatric disorders, behavioral geneticists identified new genes that contribute to a wide range of personality traits—novelty seeking, sensation seeking, addictive behavior, sexual orientation, and antisocial tendencies.[69] This research generated keen interest among both professionals and the public, helping to drive the expansion of behavioral genetics as a socially acceptable and scientifically legitimate field. Some investigators turned their attention to the DNA of personality, deviating slightly but never completely losing sight of their core interest in cognitive performance. During this period, behavioral geneticists were poised to influence, and be influenced by, their peers from a parallel field developing within medicine known as molecular psychiatry.

Also called psychiatric genetics, molecular psychiatry commenced in the late twentieth century as a field devoted to the study of genes that increase susceptibility to psychiatric disorders. Capitalizing on the rapid accumulation of new genetics data of this period, some research was aimed at manufacturing new drugs that could target candidate genes linked with mental disorders. As molecular psychiatry expanded, cross-disciplinary experts—geneticists, psychiatrists, pharmacologists, and other medical scientists—soon realized that rarely does a single gene cause a person to exhibit a particular personality trait, let alone a psychiatric condition, nor does it even by itself predispose him to it. Instead, multiple genes are almost always involved in the genesis of mental illness or disorders characterized by distinct personality and behavioral features. Most importantly, study after study strongly suggested that genetic proclivities toward a psychological imbalance were not likely to manifest unless distinct environmental conditions were also present.[70]

With the rise of computerized computational technologies to analyze large and complex amounts of data, by the early 2000s researchers finally had the innovative laboratory tools needed to start quantifying the effects of genetic and environmental factors, separately and combined, on several key psychiatric disorders.[71] Today, the findings from these studies are increasingly leveraged to assist healthcare providers in predicting which individuals are most susceptible to developing a psychiatric illness and which therapies are most likely to benefit specific patients.[72]

It was a logical step from this type of medical research to studies on links between genes and cognition, and, soon thereafter, on links between genes and personality traits. Just as medical scientists examined genes and disease risks across different ethnic populations, by the 1990s behavioral geneticists had started comparing genes associated with personality and social behaviors along population lines. As momentum grew for an unprecedented scientific revolution linking biology and human social life, science-savvy proponents chimed in that the benefits of research on connections between genetics, social traits, and race far outweighed any potential costs, notably the risk of uncovering hard-core scientific evidence that might just prove racial superiority and racial inferiority.

As pronature advocates put it, the research of race-oriented behavioral geneticists might have been contentious, but it was mandatory for society to "know the truth."[73] During this era, the majority of scientists were prudent and strived ardently to avoid both the appearance and the reality of personal and ethnic bias. But for other researchers torn between, on the one hand, the safe, status quo of "political correctness" in matters of race and, on the other hand, a defiant, if not brazen, stance to challenge the blank slate—indeed to reject the very essence of egalitarianism that dominated public opinion at the time—there was another beckoning. A deep dive, some said, into possible relationships between population-linked genes and social behaviors might finally tell us if some ethnic groups—certain racial groups—are genetically predisposed, more than other ethnicities, to lower intelligence, academic failure, poor parenting, aggressiveness, violent crime, and a litany of other "social pathologies." Wouldn't this information, sensitive that it is, gleaned from onward marching science, in the long run benefit society as a whole? Who could ask for more?

CHAPTER TWO

Darwin's Legacy

Selection and Adaptation

CHARLES DARWIN: FROM NATURALIST TO BIOLOGIST

THEODOSIUS DOBZHANSKY, ONE OF THE TWENTIETH CENTURY'S MOST BRIL-
liant scientists, authored an essay famously titled, "Nothing in Biology
Makes Sense Except in the Light of Evolution."[1] One hundred years before
Dobzhansky penned this article, nineteenth-century English naturalist
Charles Darwin introduced the theory of biological evolution to the educated masses of his day. Like his contemporaries in affluent Victorian society,
Darwin learned about the evolution of living things while in school. In the
1820s, as a young medical student in Edinburgh, Scotland, he was taught the
prevailing wisdom surrounding biological evolution, specifically the concept
of Lamarckian inheritance, of his day. During that era, ideas about evolution had been debated since the seventeenth century, if not since antiquity:
invented, discarded, and reinvented again to produce what was then called
the theory of the transmutation of species.[2]

Popular in Darwin's youth, this theory was the brainchild of Jean-
Baptiste Lamarck, the eighteenth-century biologist whose term *Lamarckian
inheritance* has become immortalized in the history of science, although not
always admirably. By the early 1800s, Lamarck proposed that any physical
changes or new physical traits an organism acquires through adapting to its
environment are incorporated into its own physical makeup. These acquired
changes are then passed on to its offspring. To illustrate how traits not present at birth can be obtained and transmitted to an individual's offspring,
Lamarck told the story of how the giraffe developed its elongated neck. In

retelling a fanciful scenario concocted by another French naturalist, Lamarck speculated that throughout time, the giraffe's neck lengthened as it habitually stretched it, generation after generation, so that it could reach the leaves of trees without having to stand up on its hind legs.[3] The tale of the giraffe's neck is a "just-so" fable and hardly a reflection of bona fide modern scientific inquiry. For Lamarck and his followers, the giraffe's extended neck represented an acquired trait that emerged from a need for nutrients and water. They believed giraffes' modified anatomy was eventually transferred to their offspring as an inborn characteristic. This was the cornerstone of Lamarckian evolution: Acquired traits can and are passed on from parents to progeny.[4]

Lamarck's hypothesis of the inheritance of acquired characteristics was accepted by most naturalists—the predecessors of biologists—and philosophers of that era. However, by the late 1800s, when the academic world embraced Darwinian natural selection as the foundation of biological evolution, the notion of Lamarckian inheritance fell by the wayside, only to be eventually discredited, rejected, and finally discarded. For decades, starting well before the discovery of DNA in the 1950s, the concept of Lamarckian inheritance has been mocked, ridiculed, and invalidated as an outmoded, scientific-sounding tale of the implausible. After all, one does not have to be biologically savvy to know that genetic inheritance involves the transmission of genes from one generation to the next.[5] The idea that acquired characteristics, whether physical, psychological, or behavioral, could be passed on through successive generations has, until recently, been considered the antithesis of conventional biology. Ironically, however, emergent discoveries in epigenetics—the inheritance of traits by way of genes but not involving DNA—has since turned some elements of the almost 120-year-old science of genetics on its head.

Growing evidence, initially from animal investigations but more recently from human studies, has shown that some disease risks and behavioral tendencies may indeed be passed on from parents to offspring. Epigenetics seems to offer a mechanism for this intergenerational transmission of traits. Epigenetics has not validated Lamarckian inheritance. But it has given newfound credibility to some components of Lamarck's hypothesis and, in the process, provided a reason to not entirely reject his ideas as totally implausible.[6] The theory of biological evolution states that physical change transmitted from one generation to the next occurs in living creatures throughout time. In Darwin's day, the concept was neither novel nor unimaginable. But scientific understanding of evolution remained largely unknown until Darwin identified and articulated the biological mechanisms that drove some species to evolve into other species. In an age that predated the discovery of chromosomes in the twentieth century, Darwin relied on his astute observations of nature to pen a coherent narrative of the origin of the Earth's diverse species.[7] Well-versed in both natural history and geology, he fit the classic definition of a nineteenth-century naturalist—a biologist who studies plants, animals, and insects, focusing on the relationship between living creatures and their environment.

Darwin's famous voyage that transpired between 1831 and 1836, with the crew of the HMS *Beagle*, a vessel of the Royal British Navy, laid the backdrop for his groundbreaking theory of biological evolution. In 1835, the HMS *Beagle* docked at the Galápagos Islands, a volcanic archipelago in the Pacific Ocean situated about 600 miles off the coast of Ecuador in South America. Known for its richly diverse wildlife, this wonderland of nature has been called an "ideal living laboratory," a "museum that showcases evolution."[8] The islands, islets, and rocks comprising the Galápagos were formed from volcanic upheavals that erupted from the bottom of the ocean between 500,000 and 3 million years ago.[9]

During the five-week period that the HMS *Beagle* was stationed in the Galápagos, Darwin visited four islands, where he carefully inspected a variety of species that were apparently unique to the archipelago. This setting provided a geological and ecological wellspring for the curious 26-year-old Charles Darwin to track the biological evolution of species confined to habitats isolated from mainland environments. The geographical space of the Galápagos Islands as a whole afforded the future father of evolutionary thought the opportunity to examine why, for instance, approximately 15 closely related species of finches, or seed-eating songbirds, in this archipelago had beaks of widely varying sizes and shapes. Darwin traced the origin of these birds, now fondly dubbed "Darwin's finches," to a common ancestral species in South America. After monitoring their behavior in intricate detail, he concluded that each species of finches that had migrated from its original homeland eventually spread to a different environment once it reached the Galápagos Islands.[10]

Throughout time, the birds acclimated to the conditions of their new ecological niches so they could obtain sufficient food to survive and reproduce. To optimally adapt to different surroundings, Darwin's finches evolved novel features in the size and configuration of their beaks. These anatomical transformations occurred during the span of many generations, eventually endowing each species of finches with physical traits that afforded them an edge for surviving and living longer in their own unique ecological niches.

This process of change, or acclimation, is known as biological adaptation, and it is central to the theory of evolution, regardless of the species. An adaptation is an advantageous biological change that improves a species' overall chances of survival, or the survival of a population within a species. Natural selection refers to the process whereby organisms that have favorably adapted to their environment survive and produce more viable offspring, which themselves survive and bear offspring that also successfully reach adulthood. This pattern allows the cycle of survival and reproduction to continue. Often linked with the cliché "survival of the fittest," natural selection embodies mechanisms that enable a group of organisms, whether a population or a species of living things, to achieve reproductive success in a particular niche.[11]

In biology, a species is a group of individuals that can breed with one another. Populations are subgroups or subdivisions within a species that have varying amounts of genetic differences among them—differences, in humans, for instance, in such physical characteristics as skin color and disease risks. Populations within the same species can mate with one another. This is one of the cardinal conditions that defines a species. In scientific terms, a population contrasts with a race because members of a particular population are genetically closely related to one other, much more so than are the members of a biological race. If we accept these criteria, the human race is a single species made up of an arbitrary number of populations, a few or many, depending upon how they are grouped. Overall, human populations are genetically so similar that many scientists today agree that the human species is a single race—the human race.[12] At the same time, the human species is characterized by genetic structure, or subpopulations that differ in the frequencies of various gene variants and display varying degrees of genetic isolation from one another.[13]

Today, the mechanisms of natural selection can be explained largely in terms of genetics, but the science of DNA was nonexistent in the late 1800s,

when Darwin formulated his theory of biological evolution. Forced to articulate the origin of adaptations without any knowledge of genetics, Darwin
examined the links between a habitat and noticeable physical traits in different groups of organisms. He deliberately studied various subdivisions of
finches that seemingly belonged to the same species. (Scientists have yet to
reach a consensus on whether "Darwin's finches" belonged to a single species
or different species. Today, some experts surmise that the finch populations
Darwin observed may have been slowly undergoing a process of biological
differentiation sufficient to transform them into new species.)[14]

Regardless of how the different subdivisions of finches are categorized,
some of their distinctive features, or physical traits, were unique to their
specific habitat. These traits emerged in each subdivision of finches as they
adapted to environmental pressures when first settling into their new ecological niches. Darwin eventually realized that the diet of each group of birds
was related to one key adaptive trait—the shape and size of its beak. These
two interrelated anatomical characteristics are essential for finches being able
to successfully exploit potential sources of food in their surroundings.[15]

Scientists who have since marched in Darwin's footsteps have elaborated
on his original fieldwork, refining the relationship he reported between the
physical form of the finches' beaks and the practical benefits, or survival-
related functions, of those beaks. For example, the fine bill of the warbler
finch is suitable for preying on small insects and spiders, whereas finches
occupying environments with cactus flowering plants have sharply pointed,
long beaks. Variations in the size, shape, and function of the beaks of the
Galápagos finches are highly specific for each ecological niche. Two species
of cactus finches have elongated and decurved beaks for extracting nectar
and pollen from cactus flowers.[16] Two other species, one of them being the
woodpecker finch, use their beaks to break off twigs or leaf petioles, which
they can then use to dig for insect larvae and termites underneath bark and
within the cavities of trees.[17] The iconic large ground finch has a noticeably
deep, broad beak for cracking open hard seeds, so hard that apparently pliers
would be needed for humans to perform the same feat.[18] The sharp beak of
the vampire finch can cut wounds on such marine life forms as sea lions and
iguanas, allowing these birds to literally suck the blood out of their prey.

The Galápagos finches had to survive to mate, and adequate nutrition
was crucial for their survival. The birds that birthed healthy offspring were

the most fit. Those with an anatomical form well suited for efficiently obtaining food in a particular niche became the beneficiaries of natural selection. They were the ones that displayed "survival of the fittest" —successful adaptation to their environment that helped ensure the continued existence of their population and, by extension, their species. Finches that found new homes in locales where they could cash in on a readily available food source without serious competition from other birds or wildlife were more likely to reach reproductive age, enabling them to bear offspring. Although all of the finches inhabiting the Galápagos Islands were close genetic relatives, their adaptations to their respective environmental niches required them to evolve specialized features. For Darwin's finches, those features inevitably took the form of specialized beaks.[19]

Natural Selection: The Basics

Gregor Mendel, an Austrian monk and botanist, laid the foundation for the science of genetics when his research on heredity was published in the mid-1860s. By cross-breeding pea plants with conspicuously different traits, Mendel famously demonstrated that physical traits are passed on from parents to offspring in a random but overall statistically predictable manner. His investigations of more than 10,000 pea plants belonging to 34 different strains of peas spanned eight years.[20]

Mendel fused the pollen, or "male component," of a common garden pea plant with the part of the flowers, analogous to the "ovaries," of a closely related garden pea plant. He identified seven contrasting traits that distinguished the two types of pea plants in his experiments. For example, one plant was dwarfed, whereas the other was tall, with the two plants producing either green or yellow seeds, each of which had an exterior characterized by either a smooth or wrinkled texture. Mendel discovered that each trait had an underlying factor, or unit, responsible for the physical appearance of the trait. Contrary to the predominant theory of the time, the results of his experiments clearly demonstrated that traits inherited by offspring were not blended characteristics of their parents.[21]

Mendel's discovery included several core principles that were incorporated into what has become known as the classic theory of *Mendelian inheritance.* One of his paramount tenets was that every living plant and animal inherits a unit or particle for each trait from its parents. The units are passed from one generation to the next, even though the traits linked with the units do not manifest in every generation, sometimes skipping a generation. Mendel soon realized that any given trait depended on the random combination of the units or "factors" that it inherited. After three generations, some of the observable physical traits of the pea plants were dominant, whereas others were recessive, meaning they were less likely than the dominant traits to appear.[22]

When Mendel cross-fertilized a tall pea plant with a short one, the offspring were tall. None were of medium height, the likely outcome if the traits were blended. The height of pea plants, then, was not an intermediate characteristic, but instead depended on the dominant trait. When Mendel cross-bred second-generation plants, 75 percent were tall, while 25 percent were short. These were the outward, or observable, physical characteristics. But the inheritance pattern—what is now called the genotype—underlying these traits was a different matter. In any given population, each gene can have one, two, or even more alternate genetic variations, known as alleles. Alternate gene forms, often called gene variants, exist as pairs, with one gene variant located on each chromosome.[23] Each individual pea plant inherits a gene variant from each parent for a particular trait. In Mendel's experiment, the inheritance pattern for each plant was 2 tall (TT) or 2 short (SS), or one gene variant of each (TS). Based on the possible combinations, Mendel calculated the ratios of the inheritance patterns of second-generation plants. (These ratios represent the genotype.) The percentages of each potential combination were 25 percent for the dominant pattern of tall (TT), 25 percent for the recessive pattern for short (SS), and 50 percent for the hybrid pattern (TS), where one T and one S gene variant were inherited from each parent, respectively. Since T was the dominant trait, plants that had the hybrid genetic makeup grew to a tall height.[24]

Although Mendel lived during Darwin's era, contemporaneous biologists of this period did not learn of Mendel's experiments until 1900. For almost 30 years, Mendel and his research on plant hybridization and inheritance patterns were virtually ignored by experts. Even Darwin was unaware of the existence of "inheritance factors," or genes, when he penned *On the Origin of Species* in 1859. Ironically, Mendel had read this book— a treatise in which Darwin correctly predicted that the source of anatomical changes in adaptation underlying biological evolution lies within individuals and not in their environment. Still, Darwin could not explain the mechanism of inheritance—how physical traits are transmitted from parents to offspring. Nor could he clarify why certain traits skip a generation. Without detailed evidence to fill these voids in knowledge, Darwin's concept of natural selection was rightly criticized as partially flawed.[25]

Despite their individual brilliance as men of science, neither Darwin nor Mendel had any awareness of genetics, particularly DNA as the core substance of each gene. Mendel, considered the father of genetics, possessed only a rudimentary understanding of chromosomes, although he accurately conceptualized the existence of genes lying along chromosomes.[26] Other researchers eventually realized that the seemingly disparate ideas of Darwin and Mendel were highly compatible and complementary. By the early 1900s, Darwin's theory of natural selection was gradually merged with Mendel's theory of chromosomal genetics, eventually leading to the modern synthesis in evolution formulated in the 1920s. The pivotal idea of this new theory, sometimes termed neo-Darwinian evolution, is that biological adaptation results from genetic change in a population or a species and that this change occurs in one or more genes.[27]

Today, we know that genes represent the units within each living thing that contains the hereditary material, DNA. As the transporter of genetic information, DNA is a threadlike chain of small molecules—two linear strands that form a ladder-like structure called a double helix. The strands are composed of four chemical base pairs, commonly called the four-letter code of the DNA alphabet. Each gene is a sequence of DNA—a chemical blueprint for manufacturing larger molecules that in turn combine to produce proteins. In technical terms, genes are the functional units or sections of DNA located along a pair of chromosomes that regulate the inheritance of outer or observable traits. Humans, like most other higher living species in the animal kingdom, inherit two chromosomes, one from each parent, for every gene. Genes range in size from a few hundred to more than 2 million DNA chemical letters. Chromosomes are larger threadlike structures made of DNA and proteins that reside in the nucleus or center of most living cells. In short, chromosomes contain genes, and genes are composed of DNA.

The human genome is the complete set of human genetic material for all humans, comprising almost 3 billion DNA base pairs. Amid ever-increasing advances in technology, the number of genes in the human genome has

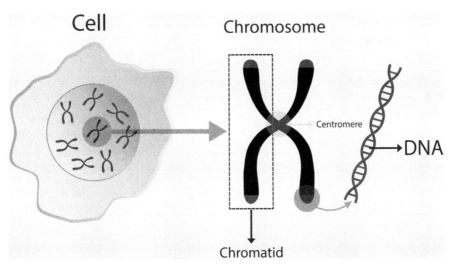

DNA inside the cell.
SHUTTERSTOCK

been constantly shrinking the last few years. According to a recent estimate, 20,000 to 25,000 genes lie within the 23 pairs of chromosomes of each individual. Most genes do not directly produce a trait, but, instead, each gene carries a code for manufacturing specific amino acids. Each amino acid in turn represents a single building block of a chain that ultimately makes up a larger molecule called a protein.[28] Inside the body, proteins form a variety of structures, for example, skin, hair, and nails, as well as hemoglobin, the compound that transports oxygen in the blood to tissues and organs.[29] And yet, less than 2 percent of the genes in the human genome codes for proteins.[30]

Genetic variation is the driver behind adaptation and evolution, and natural selection signals one force among several that results in genetic change. Other forces—mutation, genetic drift, migration, and breeding behaviors—can also lead to the genetic variation that underlies Darwinian evolution. A mutation—a variation in a gene that can make it defective—usually arises as a spontaneous and random change in the genetic code—the DNA sequence that translates into amino acids, or the building blocks of proteins. Mutations sometimes yield altered traits, but in other cases, they have no effect on a trait. Many mutations are harmful, whereas others are beneficial for survival or reproduction, or both.[31] During gene flow, genes are passed back and forth between two or more populations as the members of these populations mate with one another. The genetic distance between these populations decreases, bringing them genetically closer together. In genetic drift, specific genes are confined to a single population as it becomes increasingly more isolated from other populations throughout time. Of all of the mechanisms involved in genetic variation, however, only natural selection is inextricably intertwined with biological adaptation.[32]

Eloquently articulated by Charles Darwin, natural selection is the process that endows individuals in a population or a species with outward physical traits that allow them to adapt to their environment. During this process, organisms improve their odds of surviving and reproducing, and, as a result, so too do their offspring and their offspring's offspring, and so on down the line. The traits that the offspring and subsequent generations inherit ideally enable them to function more efficiently within their environment. And while physical traits are crucial for adaptation, the most pivotal force underlying natural selection is genetic alteration, namely mutations, which gives rise to new or different traits in the first place. Nature, as a metaphor for evolution, acts on genetic variation in the form of mutations.[33]

Both the nonhuman animal kingdom and the plant kingdom are replete with countless illustrations of natural selection. During the last 6 million years, natural selection laid the groundwork for the gradual anatomical changes in our ape-like ancestors that influenced the evolution of the human species today. Upright walking, hands with opposing thumbs for grasping and carrying objects, and a bigger brain with an expanded prefrontal region for governing executive thought are among the most important physical adaptations that have occurred in the human lineage in the last 2.5 million years.[34]

Human populations, or the groups we traditionally call races, can be differentiated by numerous physical traits that evolved as adaptations to different physical environments. Local or regional environments may vary in climate, temperature and solar radiation, plant and animal life, food sources, predators, altitude, and disease risks. Some of the physical traits that distinguish human populations from one another emerged as small groups of people adapted to different geographical niches. Just as Darwin's finches evolved customized beaks to meet their survival demands in a variety of habitats, human populations have evolved various physical traits in response to their own unique environmental selection pressures. Population differences in body build, body proportions (including leg length to body length or trunk length), body surface area, and skin color evolved as physical adaptations to contrasting ecological conditions.[35]

Populations also sometimes vary in such physiological traits as proteins in the blood, the mineral content of bone, vitamin D synthesis, and the frequencies of different blood groups. These traits represent adaptations to geographical factors, for instance, diet and climate. In some cases, they are related to risks for various genetically linked diseases. Certain gene variants that code for distinct blood types—type A, B, AB, or O—may protect people from specific ailments, but they also may increase their susceptibility to other diseases like plague (the bacterial disease that dramatically reduced the population of Europe during the Middle Ages), bronchial pneumonia, and infantile diarrhea. People with blood type O may be more resistant to anemia and malaria, whereas those with type A, B, or AB may have an increased risk for anemia.[36]

Natural selection is an ongoing process that equips some individuals in a population with an inherited advantage, or an edge in fitness, in competing with other members of their population. This advantage manifests as a biological adaptation, but the changes that define the adaptation are sometimes difficult to detect because they occur at the molecular level rather than the anatomical or physiological level.[37] Approximately 9% of the genes in living humans are still undergoing rapid evolution through natural selection. These are the genes that vary the most between different individuals. Natural selection acts on the most variable genes, including those involved with the immune system, sensory perception, and sexual reproduction. And it continually influences human evolution.[38] When natural selection results in an adaptation that benefits the members of a population, it is called positive selection. By contrast, negative selection works hand in hand with positive selection to remove, or select against, detrimental gene variants that may arise in a population. Almost 13 percent of mildly harmful genes may undergo weak negative selection. Today, most manifestations of natural selection in humans are associated with diseases, usually in the form of protection against a disease or as a buffer to diminish its severity.[39]

Sickle cell anemia develops in individuals who inherit two copies of the mutant gene that causes this potentially fatal hereditary blood disease. But carriers—those who inherit only one copy of this mutant gene—do not develop sickle cell disease. Instead, they have sickle cell trait, which protects them against malaria. The evolutionary origin of this disease illustrates how an otherwise injurious gene variant can increase the probability of surviving a second disease and, in so doing, potentially enhance the overall survival of a population. Sickle cell trait evolved in several regions of Sub-Saharan Africa where malaria is endemic.[40] Natural selection also may have played a preeminent role in the origin of various other hereditary conditions, for instance, lactose intolerance and Tay-Sachs disease.[41] Lactose intolerance refers to a reduced ability to digest milk due to lactase deficiency—an inadequate amount of lactase, the enzyme that breaks down milk sugar. Lactose intolerance is not a pathological condition. But Tay-Sachs disease is a serious medical condition—an inherited metabolic disorder caused by the buildup of lipids, or fats, that usually results in spasticity and eventually childhood death. Although rare in the general population of the United States, Tay-Sachs disease is more likely to occur in Ashkenazi Jews and, to some extent, French Canadians, Cajuns of Louisiana, and people of Irish descent. Like sickle cell anemia, Tay-Sachs disease is a genetic disorder caused by a potentially fatal mutation.[42]

An inherited disease can remain fixed in a population and affect only a few individuals rather than the majority, helping to ward off a second disease that could harm many more people. But if the detrimental mutant gene is not eliminated from the population through negative selection, it can persist and be passed on generation after generation. Most inherited diseases seem to fall in this category. Hereditary diseases are by definition typically triggered by deleterious gene variants that might have originated as mutations. But not all hereditary diseases are necessarily adaptations. If a disease does not confer an adaptive benefit, then, strictly speaking, it cannot be called an adaptation—a trait that evolved through natural selection.[43] This is an important distinction for not only identifying a disease-linked trait, but also appreciating potential genetic contributions to personality and behavioral tendencies.

Certain genes that appear to be influenced by negative selection are linked with hereditary diseases. In some cases, a disease will not manifest without the necessary environmental stimulus. Today, in developed nations like the United States, for instance, negative selection may be reducing the frequency of congenital blindness and deafness. In this case, environmental conditions of improved sanitation in modern society appear to be counteracting potentially damaging tendencies of mutant genes related to Usher syndrome. This inherited disorder is characterized by partial or total hearing or vision loss that is exacerbated throughout time.[44]

DNA: THE GENETIC SPIRAL LADDER

In 1953, famed American biologist James Watson and English physicist Francis Crick forever altered the course of genetics and medicine when they discovered the double helical structure of the DNA molecule.[45] At that time, scientists had already known for almost a century that DNA was the hereditary material residing in humans and most living things. The genomic code is a DNA sequence comprised of paired chemical bases, or building blocks. The four bases are adenine (A), guanine (G), thymine (T), and cytosine (C), with each base represented by the first letter of its name. DNA consists of paired bases located on opposite strands of the double helix, which code for specific proteins.[46]

In genetics, the term *code* refers to a set of rules relayed via information within DNA or other genetic material that tells living cells to synthesize certain proteins. Genes do not produce traits, however, nor do they directly create a trait like eye or hair color. Instead, each gene instructs cells to build amino acids—the units that make up a protein. The sequence of the DNA base pairs, or building blocks, determines the types and sequence of the amino acids. Once a protein is formed, it regulates various biochemical processes that eventually influence the specific characteristics of a physical trait.[47] The paired bases from the opposite spiraling strands in DNA are arranged like rungs on a ladder, such that A pairs with T and C with G. The molecules attached to each base pair form the vertical sidepiece bars of the spiraling ladder or double helix. A nucleotide is a DNA single base connected to two other molecules. The sequence of the bases serves as an instruction code for building proteins, with every three base pairs directing the production of a specific amino acid. Genes, then, are simply segments of DNA used to make a functional product like an amino acid sequence. Amino acids are the building blocks of proteins, which in turn function by themselves or combined in the form of enzymes, hormones, and various body chemicals. These substances are used to build and repair key molecules that guide a host of biological processes inside the body.[48]

A DNA segment of three base pairs is called a codon, and codons are sometimes popularly dubbed "DNA words." The lineup of bases is analogous to the way letters of the alphabet are ordered in a specific fashion to form words and sensible sentences. In humans, the size of a gene ranges from a few thousand to more than 2 million pairs of nucleotides (or "base pairs"). The ordered sequence of base pairs along a strip of DNA functions as a code or blueprint for producing amino acids, which in turn combine to make proteins. If a deviation or change occurs in this sequence code, the gene is considered a mutant variant. Mutations are transformations to DNA that alter a gene's set of instructions or messaging for manufacturing amino acids. A mutation in the base pair sequence shifts the amino acid sequence of the protein encoded by the gene, possibly altering the action of the gene.

These alterations constitute the genetic variation needed for evolutionary change at a broader group level—a population or species. Each mutation, or genetic change, can be as small as a single DNA base or as large as an entire segment of a chromosome. Humans have 23 pairs of chromosomes, or strands of DNA encoded with genes, for a total of 46 chromosomes. (In both males and females, 22 pairs of chromosomes are the same, but the 23rd pair constitutes the sex chromosomes that differ for males and females.)[49]

Mutations are often triggered by environmental events or stressors that in some cases can turn a normally functioning gene into a faulty gene with destructive effects. Not all mutations are detrimental, however, as some are beneficial, whereas others are neutral, neither improving upon nor deceasing the function of the gene. Most mutations are small, and good and bad mutations can sometimes interfere with one another, essentially canceling out each other's effects. Despite their potentially adverse actions, mutations are the raw genetic material that ultimately drives the biological evolution of living species, including improvements in both plants and animals. Mutations can occur as a single phenomenon or collectively, such that different mutations accumulate and then influence the direction of evolution in a population or, throughout an extended period of time, a species.[50]

Whether the majority of mutations as a whole are considered good, bad, or indifferent may depend upon the species. Neutral mutations are more likely to emerge in small populations, but in a larger population beneficial mutations have an increased probability of being selected—acted on positively by natural selection—as adaptive traits.[51] Overall, the vast majority of mutations in living things on the planet are beneficial or at least indifferent, but the situation may be less optimistic for humans. One study reported that most mutations in humans are neutral, with only three out of 175 mutations posing harm in each generation.[52] But as one research team recently warned, the human genome is now overburdened with millions of rare gene mutations, making us literally a mutant species.[53]

In most species, mutations occur at a slow rate and through a variety of changes to DNA. Natural selection is fueled by multiple types of mutations, including movement of one part of a DNA strand to another section, deletion or insertion of a section of DNA, chromosome alterations that affect a large enough portion of DNA to be observed through a microscope, and point mutations that change only a single letter of a DNA sequence.[54] The accelerated expansion of the human species has been accompanied by an abundance of rare mutations. In the last 5,000 to 10,000 years, mutations have increased so rapidly that nature has not had time to purge all of the faulty changes to DNA in our species. Most mutations in the human race are considered harmful, even though individual mutations that change genetic activity may not cause serious defects.[55]

SINGLE NUCLEOTIDE POLYMORPHISMS (SNPs)

Many of the mutations that severely diminish fitness—survival of the fittest—tend not to accrue in a population, lest the population die off. But the cumulative effects of single small changes in DNA can be maladaptive, such that many of the rare mutations in humans are linked with increased disease risk.[56] Research in human genetics is increasingly aimed at identifying point mutations, or "single nucleotide polymorphisms"—SNPs—within single genes that sometimes differ in their frequencies across populations.[57] A single point mutation does not usually lead to either a boost or decrease in fitness. And yet, the effects of mutations are additive, perhaps even synergistic, and shaped by a complex web of interactions with other mutations and even environmental factors.[58]

Mutations present in more than 1 percent of the population are called polymorphisms, a term that literally means different biological forms or types in the same population.[59] A SNP—a small variation affecting a single DNA base pair in the genome—represents the most common type of genetic variation in humans. A SNP occurs when a single nucleotide—a single base or letter in a stretch of DNA—is substituted with another nucleotide. If less than 1 percent of a population has the same single base mutation, and

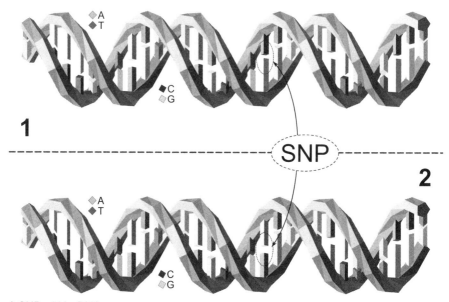

A SNP within DNA.
WIKIPEDIA. SNP MODEL BY DAVID ECCLES

essentially the same alteration at a specific location in the DNA sequence, the mutation is called a SNP.[60]

SNPs are responsible for numerous traits, including hair color, vulnerability to developing certain diseases, individual reactions to viruses, metabolism of various drugs, and such environmental factors as toxins.[61] Usually located between genes on a strip of DNA, SNPs are not necessarily disease-causing mutations, even though they are sometimes inherited along with disease-linked genes.[62] Some SNPs are associated with cancer, infectious diseases like AIDS and hepatitis, sickle cell anemia, cystic fibrosis, and such autoimmune conditions as systemic lupus erythematosus. In complex multifactorial conditions—diseases caused by multiple factors, for example, osteoporosis—SNPs may interact with one another rather than act separately to alter the body's normal function by changing lipid, or fat, levels.[63]

SNPs are generated through copying errors that occur on average about once per every 300 DNA bases when a cell divides. Based on this rate, the human genome contains approximately 10 million SNPs. Most SNPs do not result in visible differences between people, but they are increasingly used as biological markers to help scientists identify specific genes associated with particular diseases.[64] These minuscule chemical markers allow researchers to estimate the association between genes and various diseases, or even psychological traits, based on the distribution of SNPs throughout the human genome.

The International HapMap Project was launched in 2002, with the goal of discovering gene variants affecting health, disease drug responses, and environmental triggers of illness. As its namesake suggests, the original mission of the International HapMap Project was to develop a map of the haplotypes, or common patterns of variation in the DNA sequence of the human genome. The search for haplotypes in humans is centered on correlational studies, also called genome-wide association studies, or GWAS, of diverse populations worldwide. Although the International HapMap Project reportedly has lost its momentum in recent years, it successfully laid the groundwork for the presently thriving 1000 Genomes Project. Started in 2008, this international collaboration among research groups is actively creating an extensive catalog of human genetic variation to advance medical research on genetically linked diseases throughout the world.[65]

By identifying genes linked with both diseases and normal traits, GWAS can provide new insights into the potential role of SNPs in the evolution of human populations. In people of European descent, for example, a SNP in a gene affecting the coloration of the skin, hair, and eyes determines whether a person has blue eye color.[66] Other SNPs play a pivotal role in various versions of dopamine receptor genes that regulate the activity of the dopamine brain chemical, or neurotransmitter. Dopamine is often labeled the reward-seeking brain substance because it is released during pleasurable experiences, providing the chemical basis for subjective experiences of highs or euphoria.[67] Medical investigators are also analyzing SNPs to determine a person's risk of developing specific diseases. And single base variations in DNA offer important clues about the timing and trajectories of human evolution, including the microevolution of continental populations, or racial groups. The previously mentioned well-known genetically triggered conditions—sickle cell anemia, lactase deficiency, and Tay-Sachs disease—evolved as unique, ecologically driven adaptations that affected certain populations but not others.

In addition, SNPs represent the type of mutation most likely to affect personality traits and social behaviors, as well as such psychiatric conditions as schizophrenia, bipolar disorder, major depressive disorder, attention-deficit hyperactivity disorder (ADHD), and autism spectrum disorder. Some SNPs in the dopamine-related DRD4 gene appear to raise the risk of developing ADHD and bipolar affective disorder. DRD4 has a far-reaching impact on multiple types of social behavior, including behaviors as wide ranging as cigarette smoking and age of first intercourse.[68]

Variants of DRD4 are linked with such a broad array of personality traits that even a small number of SNP differences in various versions of this gene could potentially have a profound influence on a person's psychological and behavioral tendencies.[69] Certain regions of the DRD4 gene associated with psychological traits seem to be especially sensitive to the effects of natural selection.[70] As possible targets of natural selection, these are the sections of DNA that are more likely to have base substitutions—DNA sequences that vary due to different SNPs. If, for example, a particular psychiatric disorder, such as ADHD or bipolar affective disorder, provides some benefit to a population overall, it might mean that the condition evolved as an adaptive response from some individuals in that population to environmental selective pressures. At the same time, individuals with psychiatric conditions like ADHD or bipolar affective disorder might suffer personal setbacks. And yet, their unconventional psychological makeup may have emerged as a tradeoff between, on the one hand, an increased risk for a psychiatric condition and, on the other hand, protection of their population against the ever-present threat of starvation.

Consider a hypothetical scenario regarding how the potential benefits of ADHD, or a precursor, might have evolved in a prehistoric population of early humans. Compared to most people in their social group, hyperactive individuals, like those with ADHD, would have been "revved up" and ready and willing to go beyond the normal call to duty in their daily activities. Some of them might have been endowed with a risk-taking personality, anxious to engage in full-fledged hunting or gathering, even in extreme duress or during harsh climatic conditions, which the nonhyperactive members of their social group likely would have avoided. With unrelenting motor activity, some individuals with ADHD would have had behavioral tendencies that might have helped ensure that their community had an adequate food supply.[71] A handful of hyperactive individuals in a social group or population that sometimes faced food shortages or had other urgent needs requiring immediate, even impulsive, action could have benefitted from the hyperactivity of individuals with ADHD-like traits.

Since ADHD appears to have genetic underpinnings, thousands or even tens of thousands of years ago, some of the gene variants, notably SNPs, contributing to this condition may have become stabilized in a small percentage of people in various populations throughout the world. SNPs have the ability to transform the function or expression of a gene. If a genetic alteration increases the fitness of a population—if it improves the population's capacity to survive or ability to reproduce in its own environment—that change tends to be favored by natural selection.[72] Alternatively, a gene variant present in a population could have instead arisen through evolutionary forces other than natural selection. In some cases, neutral selective forces—migration, random genetic drift, or mutation—might have been responsible for the origin of a particular gene variant.[73]

Geneticists have traditionally studied physical traits, but with the rise of interdisciplinary approaches that meld biology, psychology, and the social sciences, researchers are increasingly turning their attention to the genetic underpinnings of personality and behavioral tendencies.[74] This is the terrain of some of the most fascinating but also some of the most contentious pursuits in current explorations of the role of nature versus nurture in the ethnic achievement gap.

Darwinian Medicine

NATURAL SELECTION: THE DISEASE CONNECTION

HUMANS ARE OVERWHELMINGLY GENETICALLY CONNECTED AND SHARE 99.9 percent of their DNA, an indisputable fact that explains why many experts agree that humankind constitutes a single biological race. At the same time, during the last 200,000 years, the ancestors of diverse ethnic populations inhabiting various parts of the world today evolved different physical features and, in some cases, slightly different anatomical traits as they adapted to their ecological surroundings. Throughout this span of time, natural selection acted on genes linked with traits that improved the chances for surviving in a particular environment. If a trait were undesirable and likely to harm a population, it was usually weeded out, lest the population face the risk of extinction.[1]

In this manner, natural selection gives rise to adaptive or beneficial traits in a population by favoring particular gene forms, called gene variants, over other versions of the same genes. Multiple variants of myriad human genes are the source of population differences in a host of physical traits, including skin pigmentation, the structure of hair follicles, the speed of body metabolism, the development of the nervous system, and response to various medications.[2] During the human lineage's multimillennia-long evolutionary history, biological adaptations to climate and geography resulted in not only diverse physical attributes, but also numerous distinct disease-related traits. Natural selection has been an indispensable partner in the making of the human species. One of its most profound manifestations is its momentous impact on ethnic populations' susceptibility to, or resistance against, various diseases.

Darwinian medicine, as the name suggests, is a field within biology that focuses on evolutionary explanations for why some genetically related populations are more prone than others to certain diseases. It applies evolutionary principles and research methods in an effort to understand, prevent, and treat human disease.[3] Even before the advent of Darwinian medicine, scientists knew that some diseases evolved thousands of years ago as environmental adaptations in certain populations, while sparing others. As noted in chapter 2, sickle cell anemia, Tay-Sachs disease, and lactase deficiency (a shortfall in the enzyme needed to digest milk) are classic examples of biological adaptations that potentially involve physical disorders and affect some, but not all, ethnic populations.

Modern Darwinian medicine has expanded the terrain of evolutionary medicine beyond traditional disease-related adaptations and pushed the envelope in terms of the kinds of questions researchers now ask. As Randolph Nesse, one of its most prominent proponents, puts it, Darwinian medicine uses an evolutionary lens to explain why the body is not better designed and why diseases exist in the first place.[4] This field has made important contributions to understanding the origins of obesity, bodily defenses like fever and cough, and even such psychological conditions as anxiety.[5] Once experts can decipher why, in Darwinian terms, these conditions evolved in the first place, they may be better prepared to engineer effective treatments for these disorders. One way to accomplish this goal is to compare the frequencies of a multitude of genetic markers across multiple populations worldwide.

Recall that the International HapMap Project, commencing in 2002, was designed to produce a map detailing DNA patterns within the human genome. Its purpose was to identify gene variants, including a wide array of SNPs, that influence health and disease, as well as responses to drugs and other environmental factors in people whose ancestors came from Africa, Asia, and Europe.[6] The HapMap database, along with its successor, the 1000 Genomes Project, offers new clues about the possible impact of natural selection on the genesis of various diseases in populations worldwide.[7] In some evolutionary scenarios in which natural selection played a pivotal role in the origin of a disease, positive selection of a mutant gene variant may have been responsible for a favorable trait—a beneficial trait—that was transferred from parents to children. As in the case of sickle cell anemia, however, the same gene variant that may have conferred a protective advantage against one potential disease also may have carried a risk for harm caused by another disease. In this situation, the notion that positive selection yields a favorable trait is somewhat misleading. Favorable, here, is a relative term because the benefit of being genetically shielded from malaria came at a cost of increased genetic risk for sickle cell anemia. We will return to this topic shortly.

In negative selection, a potentially threatening trait, usually in the form of a disease, is eliminated or bred out of a population; however, mutation and natural selection are not the only Darwinian forces that determine the prevalence of gene variants in a population. Other processes—migration (synonymous with gene flow) and genetic drift—can also influence the rate of a physical trait or genetic disease in a population. Since migration and genetic drift do not involve natural selection, they are considered neutral evolutionary forces. But it can be difficult to determine if a disease (or, for that matter, other types of traits as well) originated in a population as an adaptation through natural selection or, instead, a nonselective force other than positive or negative selection.[8]

Genetic drift refers to the evolutionary changes that arise in a population during a span of many generations due to chance. It occurs when a population is isolated or cut off from contact with other populations or breeding groups. If the members of the population have no mates outside their own community, their genes cannot intermix with different DNA from people outside their own group. By contrast, admixture occurs when two or more otherwise genetically distinct populations mate with one another, resulting in an exchange of DNA commonly known as gene flow. If a population's genetic composition is not altered by an external influence, or another group, then throughout time, the frequency of one or more of its gene variants will tend to either increase or decrease. Just as mutations sometimes work hand in hand with natural selection, mutations may also be the source of the initial change in DNA that eventually leads to a shift in the frequency of one or more specific genes. This process is called genetic drift. In this case, the shifting prevalence of a particular gene variant is not due to either positive natural selection or negative natural selection. Instead, it comes from a neutral genetic effect that neither selects for nor against a gene variant. For this reason, genetic drift is also called random drift or neutral selection.[9]

SNPs, Natural Selection, and Disease

During the last 20 years, Darwinian medicine has advanced momentously with the discovery of SNPs (single nucleotide polymorphisms), genetic mutations involving only a difference in a single base or chemical letter of a gene—a stretch of DNA. As noted in the last chapter, SNPs represent the most common type of mutation found in living things, serving as the source of most of the genetic variation observed between people, both within a population and across divergent populations. SNPs are also the simplest type of genetic marker that can be detected. These small variations targeting a single DNA base pair in the genome are associated with a wide range of traits, including a host of diseases. SNPs not only help to identify a person's risk of developing particular illnesses, but also provide pointers for mapping the evolution of inherited disorders across diverse continental populations. Today, SNP mapping is vital for accelerating research on diagnosing and treating a wide gamut of genetic diseases, including conditions that tragically target disproportionately high numbers of individuals in some ethnic groups.[10]

During the 1980s, SNPs transformed the landscape of genetics, only to be replaced in the 1990s by another type of genetic marker—the repeat of a DNA strip, commonly called a repeated sequence, or simply a repeat. By the late 1990s, the field of genetics witnessed a reversal when the tools for mapping genetic changes in the form of SNPs were once again in vogue. In the current research climate, SNPs signal the mainstay of one of the most extensively used tools in genetics research: genome-wide association studies, or GWAS.[11]

Recall that these studies are not cause-and-effect investigations, but observational explorations designed to demonstrate if specific genetic variants are connected with regions of the human genome involved in the onset of a disease.[12] Recent studies indicate there are fewer population differences in immune-related genes—genes that help protect people against infectious diseases—than in genes correlating with positive selection. A high percentage of disease-linked genes contain SNPs, a finding that was virtually unknown just four decades ago when researchers first began comparing immune genes across populations. Some experts now think that many disease-related genes tend to be eradicated through negative selection—the elimination mechanism that selects against the presence and persistence of a harmful gene variant. Negative selection reduces the genetic variation within a population, especially in disease-related gene variants involving SNPs. At the same time, gene variants, including some SNP's that evolved as adaptations, may contribute to both diseases and physical traits.[13]

Consider the highly publicized p53 gene. Sometimes called the "guardian of the genome," p53 produces a protein that switches on cancer-suppressing genes to protect the body from forming tumors.[14] If p53 is abnormal, the body's ability to suppress tumor development may malfunction. This tumor-suppressing gene has a potentially formidable impact on a person's risk for cancer, in general, with almost 50 percent of all human cancers containing mutant versions of it.[15] The vast majority of SNP mutations in the p53 gene are not selected for but, in fact, against. They are not adaptations, meaning they are not the outcome of positive selection. As potentially detrimental triggers to human health, they do not flourish but, instead, simply die off. In short, these SNPs largely undergo negative selection.[16]

SNPs overall often have neutral effects on selection that are neither good nor bad, although there are exceptions. After meticulously mining databases of SNPs in the human genome associated with p53, scientists discovered one SNP that stood out from the others. This marker, called the KITLG SNP, is associated with cancer and is also attached to p53. The KITLG SNP is located in a DNA sequence where p53 binds, specifically a site where the attachment between p53 and DNA is stronger than in other areas along the DNA strand.[17]

As geneticists probed further, they soon realized that the KITLG SNP dramatically increases the risk of testicular cancer, even though it also seemingly has a favorable evolutionary benefit. The most reasonable explanation for this paradox is that the KITLG SNP gene variant has undergone positive selection to help protect people with fair skin from the sun's damaging ultraviolet, or UV, rays, thereby decreasing their risk for skin cancer.[18] On the other hand, the KITLG SNP bound with p53 acts as a double-edged sword, diminishing fitness by increasing the risk for one disease while simultaneously enhancing fitness through protection against a second disease. The KITLG SNP is a rare example of a functional p53 SNP mutation that, despite its drawbacks, holds selective benefits for males who carry this gene variant.[19]

Scientists arrived at this conclusion after examining the widely varying prevalence rates of the KITLG SNP-p53 gene across populations. The following findings related to cancer susceptibility in males from different ethnic populations are revealing. Testicular cancer occurs four to five times as frequently in European white men compared to African black men, a statistic suggesting that the KITLG SNP-p53 gene evolved through natural selection. In addition, this SNP is more common in men at increased risk for testicular cancer, with frequencies of about 80 percent in European Caucasians but only 24 percent in people of African black descent. The variable rates of this mutant gene in different populations may mean that it evolved as an adaptation, especially during the lengthy period when the immediate ancestors of European whites migrated out of Africa approximately 80,000 years ago and entered Europe almost 40,000 years ago.[20]

During a span of tens of thousands of years, the skin pigmentation of these migrants faded during their gradual northern bound journey out of Africa. Without substantially melanized, or dark, skin to protect them from sunlight, fairer-skinned humans heading north from Africa toward Europe would have become increasingly vulnerable to skin damage induced by solar rays. The hazards of solar radiation in stimulating rapid and abnormal cell growth are well documented. For the fairer-skinned males participating in this mega multimillennia-year odyssey, the harsh radiant energy emitted by the sun would have inevitably put them at heightened risk for testicular cancer.

A biochemical mechanism to help repair ultraviolet, or UV, injured skin would have been an expedient adaptation for the men in populations whose skin coloration gradually faded during the course of tens of thousands of years. And such a mechanism indeed did evolve in the form of an enhanced tanning response, but the benefits came at a price. The tradeoff was a mutation in the p53 gene that resulted in increased stimulation of cell production—in this case, abnormal cell proliferation, a cardinal sign of cancer. With simultaneously favorable and unfavorable effects, the KITLG SNP-p53 gene endowed the males who inherited it with both increased protection from UV radiation and an elevated risk for testicular cancer.[21]

SICKLE CELL DISEASE

Sickle cell anemia is an inherited blood disorder that remains one of the most well-documented genetic diseases in the human species, if not *the* most well-documented genetic disease. There are several types of sickle cell anemia, commonly grouped together under the umbrella of sickle cell disease, and despite some variation in the mutations that cause them, all of them are medically grueling conditions. If left untreated, every form of this potentially deadly disorder can cause such life-altering symptoms as pain, arthritic symptoms, anemia, blood clots, and possibly fatal infections.[22] Sickle cell anemia evolved through natural selection as an adaptation to buffer the extreme decline of populations occupying malaria-prone environments. The sickle cell gene mutation may have emerged independently at least five different times in four different regions of the world: four in Africa and one in either the Arabian Gulf or Central and South India.[23] The time line for these mutational events is illusive, but the evolution of the earliest sickle cell mutation that protects against malaria reportedly evolved approximately 2,000 to 3,000 years ago; however, the "natural selection" of the sickle cell mutant gene as an evolutionary adaptation in various populations may have occurred much earlier, perhaps 6,000 years ago, corresponding to the dawn of agriculture. This was the period when humans throughout many regions of Sub-Saharan Africa began practicing swidden agriculture, also known as slash and burn, shifting cultivation, or rotational farming.[24]

Experts are still debating the exact time line for the origin of the mutant gene that underlies the sickle cell trait, but most of them concur that it was closely aligned with the spread of shifting cultivation. Prior to 10,000 years ago, humans throughout the world subsisted on foods they obtained through hunting-gathering or foraging, or both.[25] Early agriculturalists who practiced rotational farming would first clear the land for cultivation, usually with fire, plant their crops, and then leave the land to regenerate before returning to plant new crops after a few years. As this cultural innovation of transforming the tropical rain forest into areas amenable for cultivating yam crops spread throughout tropical Africa, the newly created fields became fertile breeding grounds for mosquitoes that transmit the malaria-causing parasite.[26] The slash-and-burn practice, followed by incipient farming methods, caused the soil to become compacted, which in turn allowed stagnant water pools to form. These changes, introduced by human intervention, were vital for the successful growth and expansion of humans, banded together in small social groups. But as their communities expanded and became denser, so too did the Anopheles mosquito populations, eventually finding new homes in still pools of water. Amid the heightened risk for malaria in these early agricul-tural populations, the genetic mutation that gave rise to the sickle cell gene became favored, or selected. This was natural selection in action. Sickle cell disease, then, emerged as an adaptation.[27]

A person bitten by an Anopheles female mosquito that harbors the single-celled parasite *Plasmodium falciparum* is at heightened risk for developing malaria unless he has at least one of two forms of protection. He can either take a prophylactic drug or rely on an inherited predisposition—genetic protection against this parasitic disease in the form of sickle cell trait.[28] As previously noted, people who inherit two copies of the mutant sickle cell gene, one from each parent, are invariably afflicted with sickle cell anemia. But sickle cell trait carriers—otherwise healthy individuals who inherit one mutant gene variant from only one parent—have a built-in genetic protection against malaria.

Falcifarum malaria is the most severe form of this disease, as it is linked with high rates of complications and death. It is the predominant form of malaria found in Sub-Saharan Africa, where it accounts for more than 90 percent of the global burden of the *P. falciparum* parasite.[29] Today, sickle cell trait affects millions of blacks in Sub-Saharan Africa, as well as some of the descendants of enslaved blacks transported to the New World, including North America, South America, Central America, and some regions of the Caribbean. Different variants of sickle cell disease caused by a related parasite, *Plasmodium vivax*, can be found in Mediterranean populations of Greece, Turkey, and Italy, as well as some populations of the Arabian Peninsula and India.[30] Until recently, most experts assumed that the varieties of sickle cell disease originated from independent mutations, but recent research suggests this was not the case. In 2018, a team of geneticists described how they recently traced the first appearance of the sickle cell mutation to a singular event thousands of years ago. They analyzed the genomes of 2,932 people and identified 156 sickle cell carriers. Using sophisticated genetics simulations, they calculated the time line of the sickle cell mutation and found that it arose only once: 259 generations, or about 7,300 years, ago.[31]

In every form of sickle cell anemia, the hallmark of clinical disease is sickled, or abnormally shaped, hemoglobin—the protein inside red blood cells that transports oxygen through the bloodstream and on to tissues and organs. Sickle cell disease, then, is characterized by defective hemoglobin triggered by a genetic mutation. When oxygen levels are low, the faulty hemoglobin changes the typically smooth doughnut-like shape of the red blood cells into a sickled or crescent form with spikes. This is a departure from normal red blood cells, whose disc-like configuration is flexible enough to navigate through both tiny and large blood vessels. But in sickle cell disease, the red blood cells become rigid and distorted, rendering them so fragile that they can collapse within the network of tiny blood vessels, called capillaries. If the capillaries are blocked, blood clots can develop, reducing the blood supply to the heart and brain.[32]

Carriers of sickle cell trait are individuals who inherit one normal hemo-globin gene variant, or "hemoglobin A," from one parent and one abnormal gene variant, or "hemoglobin S," linked with sickle cell disease, from the other parent. Their genotype for this trait is "AS." With two different variants of the hemoglobin gene, these individuals are said to have sickle cell trait. They do not develop the full-blown disease of sickle cell anemia because only some of their red blood cells are misshaped. Rarely do they develop any of the serious complications associated with sickle cell disease unless they are deprived of oxygen. But those who inherit two hemoglobin S gene variants—two identical variants for the mutant hemoglobin gene, with one from each parent—have the full genetic makeup for sickle cell condition. Their genotype is designated as "SS." Individuals born with a double dose of the defective sickle cell–linked hemoglobin genes face the potentially dire consequences of living with full-blown sickle cell anemia. And yet, they also have a unique biological advantage: an increased likelihood of surviving in heavily mosquito-infested tropical regions where malaria is prevalent.[33]

Until the advent of modern Western medicine, sickle cell disease was inevitably fatal. For humans inhabiting regions with dense mosquito pop-ulations, the mutant hemoglobin gene that gives rise to sickle cell trait also had a benefit. This disorder evolved thousands of years ago as an evolutionary trade-off to help ensure the survival of a population. It took a toll on the loss of life in the individuals who inherited full-blown sickle cell disease. But the mutant S hemoglobin gene provided a crucial countermeasure—protection against malaria in those who possessed the sickle cell trait or only one abnor-mal hemoglobin gene variant, along with one normal hemoglobin gene vari-ant.[34] Almost 70 years after scientists like E. A. Beet and James Neel, as well as Anthony Allison, independently published their discoveries of the genetics of sickle cell anemia, the evolution of sickle cell disease remains the prime and, perhaps, the only definitive illustration of natural selection in humans.[35]

LACTOSE INTOLERANCE

With ever-expanding consumer choices for alternatives to milk that is easy to digest, the public is well aware of lactose intolerance. This condition is technically not a disease but an inability to digest lactose, a sugar found in milk. It is a genetic trait that affects individuals after the weaning period, or whenever they stop drinking breast milk and start consuming a bottled milk replacement, along with an adult diet. An estimated 65 percent of people from various ethnic populations throughout the world cannot fully metabolize lactose after infancy. Weaning is the gradual process when a baby starts ingesting food other than its own mother's breast milk. When weaning ends, usually between six months to two years of age, the genes controlling lactose production are switched off because milk ceases to be an essential food for humans past infancy. By five years of age, lactase activity has declined in most populations, with levels decreasing from this point onward; however, the amount by which this enzyme is reduced can vary across populations.[36]

The breakdown of lactose is regulated by an enzyme called lactase. Crucial for newborns and infants, who require milk for normal growth, lactase is under the control of the gene that keeps the enzyme turned on, instructing the body to digest milk through the weaning period. But after infancy, this gene is usually turned off, inhibiting a person's ability to completely digest milk. People who lack the fully functioning lactase enzyme after weaning are considered lactose intolerant, even though in some cases they may be able to partially digest milk sugar.[37]

The loss of the totally intact enzyme for digesting milk sugar is a genetic condition called lactase nonpersistent trait, a term that reflects the discontinued activity of a once-viable enzyme. Lactase nonpersistent trait is not a medical disorder because it is quite common, and, in fact, it affects the majority of humans worldwide. It manifests in a wide spectrum of activity, as many people with this trait can digest milk in varying degrees without experiencing such symptoms as bloating or abdominal cramps.[38] Just as lactose intolerant people have inherited lactase nonpersistent trait, lactose tolerant individuals—adults who are able to digest milk sugar—are born with lactase persistent trait. In lactose tolerance, the genes controlling the ability to break down milk protein after weaning were never deactivated, allowing the digestion of milk in childhood to continue on into adulthood. Lactose tolerance appears to be the evolutionary by-product of multiple mutations that emerged in Middle Eastern and North African populations that domesticated cattle 7,500 to 9,000 years ago.[39] For small-scale societies that domesticated cows or other dairy-producing animals, a high-protein food source like milk would have given them a potential edge for staving off starvation and surviving during harsh cycles of famine. But this strategy would have been viable only for the older children and adults who had a genetic capacity to break down milk past infancy. Cattle would have provided a rich food source for humans who were genetically equipped to break down milk protein and harness its nutrients for growth and reproduction.

In most areas of the world where cattle herding was absent thousands of years ago, the lactase gene was inactivated after weaning. The geographical regions where the lactase persistent genes remained functional through adulthood correlate with dairy herding societies that have existed for centuries, if not millennia. Lactase persistence trait most likely arose independently through natural selection in several different parts of the world. Regardless of its geographical origins, it was a favorable adaptation in populations with access to milk from cows and goats. In the present-day living descendants of early cattle herding populations, lactase persistence trait has been passed down as the dominant inherited trait, with the lactase nonpersistent trait representing the recessive, or less common, inherited condition. Many contemporary West African and Central African populations are characteristically lactose intolerant (like their genetic relatives, African Americans), but some East African populations derived from dairy herding ancestors can readily metabolize milk and milk products.[40]

Lactose tolerance seems to be similar, at least in its outward manifestations, in living populations whose ancestors practiced dairy herding. Still, the genes controlling lactase persistence vary across these populations, depending on their geographical origin. At least five closely linked genes, or haplotypes, associated with lactase persistence worldwide have been identified in various human populations.[41] Each mutation arose independently as a SNP variant in a specific section of the lactase gene that governs the ability to digest milk sugar.[42] Haplotypes are gene clusters that provide additional clues about the evolutionary origins of human populations throughout the planet. Not only do the lactase persistence haplotypes vary among European, Arabian, and African populations, but also the precise evolutionary mechanism underlying each lactase persistence gene variant may differ.

Lactase persistence trait may have evolved initially through genetic drift in a limited number of populations when it first appeared in certain non-African populations about 9,000 years ago.[43] As humans migrated north toward Europe from Sub-Saharan Africa and the Middle East, cattle herding spread widely throughout these areas. Dairy-herding populations whose adult members could digest milk and milk products like yogurt and cheese would have had increased access to dietary protein. This would have given them a nutritional edge, one that ultimately may have helped to improve the overall fitness of the population. After the original mutational event in which a lactase persistence gene variant appeared, the capacity for milk sugar digestion may have become solidified as an adaptation in later populations. At present, the evidence is still unclear regarding whether natural selection or genetic drift was most responsible for the presence of lactase persistence in European populations. It may be that both evolutionary mechanisms were involved during different periods of time. Natural selection may have contributed to the eventual stabilization of the lactase persistence trait in Northern European populations.[44]

TAY-SACHS DISEASE

Tay-Sachs disease is a rare, life-threatening inherited disorder in which lipid, or fat, storage goes awry. There is no cure or effective therapy for this disease, and people afflicted with it usually die before they reach four years of age. Tay-Sachs disease is caused by a genetic mutation that occurs 10 times more frequently in Eastern European Ashkenazi Jews, Cajuns of Louisiana, and French Canadians, especially people of Cajun ancestry, compared with the global population. Tay-Sachs disease is also more common in people with Irish ancestry. If both parents are carriers of the mutant gene, their child has a 25 percent chance of developing this disorder. This debilitating condition causes an excessive buildup of fatty substance in the brain and spinal cord, eventually destroying nerve cells. In the classic infantile childhood form of Tay-Sachs, mental and physical symptoms start in infancy. As children born with this disorder grow, this inevitably fatal disease advances, eventually leaving them blind, deaf, and unable to swallow. Progressive muscle wasting and paralysis follow the premature destruction of nerve cells, causing death in young childhood. Tay-Sachs disease can affect people of any age, but the types that begin in adulthood do not lead to early death.[45]

The current conventional wisdom is that Tay-Sachs is not an adaptation due to natural selection. Rather, evidence increasingly points to genetic drift—a selectively neutral evolutionary mechanism—as the underlying Darwinian force behind this devastating illness. Sophisticated DNA sequencing techniques developed since the 1970s reveal that Ashkenazi Jews probably descended from a small founder population. Mutant genes for Tay-Sachs disease apparently originated in the population that founded the Ashkenaz settlement in the early Middle Ages.[46] In the last few years, more than 100 mutations, including some SNPs, have been linked with Tay-Sachs. Multiple mutant genes underlying Tay-Sachs could have spread in concert with the geographic expansion of Ashkenazi Jews throughout Central Europe starting as early as the seventh century but certainly by the twelfth century CE.[47] Although Tay-Sachs disease probably did not evolve as an adaptation, it is nevertheless a prime example of a heritable trait (in this case, a disease-related trait) that exists in low frequencies in specific populations.

Today, innovative genetics techniques have advanced our understanding of molecular evolution, fueling new debate about the role of natural selection, whether positive or negative, versus neutral selection, for example, genetic drift, in the origin of numerous traits. With an increased ability to identify precise genetic changes like SNPs, scientists are poised to unveil the genetic underpinnings of physical and disease-related traits. Using this same approach, some scientists are exploring a similar line of inquiry: whether genetic traits linked to personality characteristics and social behaviors represent adaptations that evolved through natural selection.

GENE–ENVIRONMENT INTERACTIONS

Human physical traits are always influenced by genetics, but in the final analysis, many of them are also shaped by complex interactions between genes and environmental factors. For example, blood type and iris color are determined solely by genetics, whereas weight, height, and skin color are shaped by a mixture of genetics and environmental conditions.[48] Certain environmental factors—including diet, sunlight, stress, hormones, and possibly toxins and pollutants—interact with genes to influence the degree to which an inherited predisposition toward a physical trait like obesity or a short stature is expressed, or not expressed.[49] The keynote here is "expressed"—gene expression, or the concept of whether a genetic tendency actually manifests as an overt trait, be it a physical, personality, or behavioral trait, or a disease.

Gene–environmental interactions can also determine if a clinical illness will develop in individuals who have a genetic proclivity toward a specific disease. Gene variants can assume multiple forms, and their expression may depend on factors other than DNA. In conditions like type 2 diabetes, genetic proclivities do not typically lead to the full-blown disease unless environmental triggers are also present.[50] This scenario contrasts with the onset of heritable diseases triggered exclusively by mutated DNA inherited from both parents, as in the cases of sickle cell anemia, Tay-Sachs disease, phenylketonuria, and cystic fibrosis. Lactose intolerance and color blindness are not diseases, but they too are associated with mutated genes transmitted from parents to offspring. Classic inherited disorders induced by a genetic defect will almost certainly manifest, even though environmental interventions, including modern medicine and lifestyle changes, can at times mitigate the complications of these conditions.[51]

Most people are well aware that excessive food intake, lack of physical activity, and genetic tendencies contribute to obesity, but the exact gene–environment interactions underlying this condition are complicated. More than 40 gene variants have been linked with common forms of obesity and fat distribution. And almost 80 percent of obese people have parents who were obese, suggesting that the tendency to store excess body fat may be partially in the genes. Obesity has a broad heritability of 6 to 85 percent, with a range of 40 to 70 percent for most populations. This also means that the amount of environmental influence on excess body weight varies widely. After all, heritability is not a measure of the degree of genetic influence of a trait in an individual. It refers to the proportion of variance of a trait in a population that is due to genetic rather than environmental differences.[52]

Solving the riddle of obesity and finding its cure could substantially reduce the burden of many of the most serious chronic diseases in the United States and the developed world. A decade ago, scientists began examining obesity through the lens of genome-wide association studies, or GWAS. Much to their surprise, they soon learned that gene variants that influence body mass, metabolic rate (the rate at which the body burns calories), insulin resistance, and perhaps even hunger account for only a modest amount of risk for obesity. As research progressed, it soon became apparent that there is another force at play. Obesity is shaped by not only genetics and environment, but also epigenetics—a form of heritable but reversible imprinting that impacts gene expression without altering the DNA sequence.[53] Epigenetic marks are chemicals that lie above the DNA molecule and instruct a gene to turn on or off. And while these marks are not an intrinsic part of the DNA transmitted from parents to offspring, they can be inherited and passed on from one generation to the next. On the other hand, epigenetic tags, or signals, can also be reset such that, unlike DNA, they are not inevitably passed on to other generations.[54]

An epigenome refers to a network of epigenetic signals—the chemical compounds surrounding DNA—that can affect the activity of a particular gene. Each person has multiple epigenomes that are capable of modifying his genome, or complete set of genes, without disturbing the sequence of his DNA.[55] A single epigenome, then, can alter a specific gene through its epigenetic activity or tags. Growing evidence suggests that epigenetics can

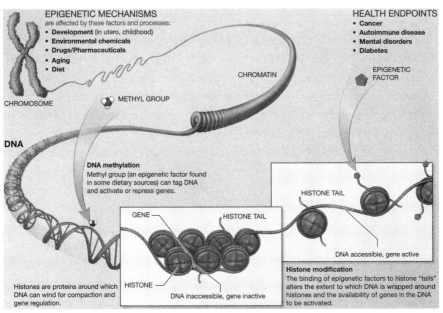

A scientific illustration of how epigenetic mechanisms can affect health.
NATIONAL INSTITUTES OF HEALTH (NIH)

increase a person's susceptibility to obesity by changing the expression of genes that regulate the body's metabolism and energy balance.[56] The key point here is that although epigenetic marks are "innate" and can be passed along with DNA, they can also be changed by environmental exposures during crucial developmental periods, particularly in infancy and early childhood.

Epigenetic marks can be transformed through the response of a person, especially a young child, infant, or even a fetus, to such external factors as malnutrition, tobacco smoke, air pollutants, infections, metals, organic chemicals, and other physical stressors. Any epigenetic alteration, in turn, can potentially shift gene expression, in some cases blocking an otherwise protective action of certain genes. Modified gene expression occurring at developmental milestones in childhood might make a person vulnerable to developing obesity, type 2 diabetes, or heart disease later in life. Amid the expanding obesity epidemic, these conditions are now seen at alarming rates in adolescents and, in recent years, even young children.[57]

Converging research on metabolic diseases, genetics, and epigenetics has identified the epigenome as one of the prime triggers for the childhood obesity epidemic in the United States. Epigenetic traits may interact with various other conditions, namely an inherited genetic tendency toward obesity, environmental risk factors for excess body fat and impaired insulin function, or any and all of these factors. As a "malleable interface of gene–environment interactions," the epigenome is pivotal for gene expression. It may even serve as a plasticity master control center that either enables or inhibits the intrinsic potential of a specific gene variant.[58] To some extent, gene expression signals the missing link between the interplay of nature and nurture in causing obesity. Not unlike physical traits, for example, height, which are not genetically fixed at birth, gene expression appears to influence how genes and environment interact to either promote or mitigate against obesity.

The sphere of influence of gene–environmental interactions ventures far beyond a host of physical and disease-related traits. These interactions underlie a broad gamut of cognitive and behavioral traits, ranging from brain size to brain wiring, from IQ to school achievement, from child rearing to law abiding behavior. In humans, the role of epigenetics in the making of intelligence, personality traits, social behaviors, and personality traits is less clear, but it is currently under intense scrutiny. We will revisit this issue in later chapters.

Tracking Genes: From Populations to Clans

With the use of GWAS, experts can rapidly scan SNP markers across the genomes of a large number of people to locate gene variants linked with a particular disease. Their findings have unveiled numerous clues about how genetic variations impact individual responses to certain medications, as well as how these variations influence interactions between a person's genes and the environment.[59] GWAS are successfully aiding experts in pinpointing the genetic blueprints associated with Parkinson's disease, heart disorders, obesity, prostate cancer, and Crohn's disease, as well as type 2 diabetes.[60]

These studies illustrate a convenient way to identify SNPs linked with specific diseases, but they are not the only tools in the geneticist's armamentarium. Scientists are also exploring the genetics of a person's ancestral group, or racial group, to better assess his susceptibility to heritable diseases. But this approach has its shortcomings. Instead, the heredity of an individual's own recent ancestors—in essence, his clan—may reveal more about his risk for certain diseases than can be gleaned from population genetics, or the overall genetic makeup of his ancestral group. Investigating both the genetic variation inherited from an individual's parents and the genetic risks of his recent relatives is called clan genomics.[61]

In the future, clan genomics may be beneficial in complementing traditional population genetics, given its practical value in identifying gene variants that either protect against or increase susceptibility to a disease in a small group of closely related kin. It remains to be seen if this emerging field, when combined with research on environmental forces, can provide new insight into why individuals in the same population sometimes differ greatly in their social behaviors. Environment and experience reportedly exert a profound effect on a person's social tendencies. However, to the extent that genetics is involved, the most relevant gene variants impacting personality and behavior may be tied more to clan genomics—family genetic patterns—than population genomics, per se. While we wait for evidence to resolve this conundrum, the potential role of clan genomics in explaining not only inherited risks for physical diseases, but also the intergenerational transmission of social traits, including traits that may contribute to the ethnic achievement gap, is worth pondering.

From Bodies to Behavior

Population Genetics and Social Life

Serotonin and the Biology of Behavior

For decades, serotonin, a chemical found in the gastrointestinal tract and the brain, has captured the public imagination as a possible panacea for a wide range of psychological and social ailments. Inside the brain, serotonin acts as a neurotransmitter—a brain chemical that stimulates nerve cells to fire and trigger a wide range of physiological events. Serotonin helps regulate such diverse functions as appetite, sleep, sex drive, and digestion, while also affecting mood and even playing a key role in boosting emotional well-being and happiness. Some antidepressants and antianxiety medications achieve their pharmacological effects by raising levels of this chemical in the brain. In the 1930s, Vittorio Erspamer, an Italian pharmacologist, discovered serotonin inside the intestines, originally giving this substance a different name, enteramine.[1]

From the start, the search for this once-enigmatic compound promised intrigue. Researchers knew that the unidentified compound was chemically similar to adrenaline and ergot alkaloids, a group of medicinal substances with broad therapeutic benefits for migraines, Parkinson's disease, dementia, and excessive bleeding after vaginal childbirth.[2] By the early 1950s, the illusive substance was isolated in the laboratory and named serotonin. Paradoxically, it excited the heart muscle of lower organisms like mollusks but relaxed other muscles in these same shellfish.[3]

In 1953, famed American biochemist Betty Twarog succeeded in isolating serotonin from the mammalian brain, a discovery that many of her colleagues had once thought was impossible. At that time, perhaps no one fully grasped the powerful impact this compound would later have on scientific thought, particularly its role in the genetics of human social behaviors.[4] Early studies launched in the 1980s suggested that serotonin activity in the brain might contribute to such psychiatric disorders as obsessive compulsive disorder (OCD), aggressive behavior, lack of impulse control, alcoholism, bulimia, panic disorders, depression, and suicide.[5] As genetics research progressed during the 1990s, scientists eventually learned that serotonin levels in the brain are controlled by a protein called the serotonin transporter. This protein is essentially a reuptake pump, regulating the level of serotonin after it is released into the synapse—the junction between two neurons, or nerve cells, in the brain. When neurotransmitters like serotonin flow into the synapse, they create a chemical message system that allows neurons to communicate with one another. Nerve impulses are then transmitted from one nerve to another in various parts of the brain and body. The size, or length, of the transporter protein determines the amount of serotonin removed from a small space or gap in the synapse. Any excess transporter protein remaining in the synapses that is not successfully recycled causes neurons to continue firing, potentially triggering psychological imbalances that, throughout time, can contribute to psychiatric disorders.[6] In this manner, the serotonin transporter protein determines the degree to which nerve cells bathed in serotonin will fire nerve impulses, resulting in numerous physiological processes.[7]

This was a novel finding, but the genetics underlying the activity of the serotonin transporter—the function of the serotonin transporter, SERT, or gene—may have been the most fascinating discovery of all. In the 1990s, scientists identified multiple versions of the genes regulating the reuptake, or recycling, of serotonin in the brain. As knowledge about the genetic underpinnings of this neurotransmitter increased, eventually several psychiatric drugs, one of which was fluoxetine, were designed to target the serotonin transporter.[8] Fluoxetine (commonly known by the trade name Prozac) is an antidepressant medication that blocks serotonin transporter reuptake, leaving more serotonin in the synapse to keep nerve cells firing and help stabilize mood.[9] Serotonin transporter biology opened the door to a then-emerging field dedicated to the genetics of personality and social behavior, a field outside of mental health, per se. Experts soon uncovered another set of genes—each with their own alternative forms or variants—associated with the activity of yet another neurotransmitter, dopamine. Like serotonin, dopamine is a key neurotransmitter in the brain. Along with some of the "risk genes" that regulate its activity, dopamine plays a preeminent role in several psychiatric traits and social behaviors, including such substance dependencies as cocaine addiction, attention deficit hyperactivity disorder (ADHD), and certain personality traits.[10]

Even before the discoveries of serotonin activity in the 1980s, population genetics had gained recognition for identifying disease risk in different ethnic groups. Soon thereafter, the related but distinct field of biological psychiatry, sometimes called molecular psychiatry, sought to explain mental illnesses in terms of the underlying activity of the nervous system, including the brain. Some researchers in this field began investigating population variance in genes possibly linked with mental illnesses. Their focus on the genetic foundations of psychiatric conditions soon catapulted molecular psychiatry into the scientific spotlight. A few experts in this field turned their attention away from the search for individual genetic differences and, instead, toward a quest to map racial variation in the genes linked with such brain chemicals as the serotonin transporter. (At that time, "race" rather than "population" was the common terminology.)

Both laboratory scientists and armchair biologists debated the proper placement of serotonin-linked genes in assessing risks for mental illness, based on links between specific serotonin gene variants and either psychiatric disorders or psychiatric drug responses across ethnic populations.[11] Aware of the potentially dangerous implications of tethering genes to social behaviors, ethically minded observers demanded exercising caution before rendering judgments about a person's genetic makeup as a supposedly inevitable pathway to a predetermined personality and social life. These critics were never completely silenced, but their message played largely to deaf ears. At the close of the twentieth century, the correlations between certain gene variants and a variety of personality tendencies were still viewed as preliminary. And yet, a shift was already under way in how scientists perceived the association between genetics and nonphysical traits, namely social attributes. The nurture perspective of the human mind as a blank slate that had once engulfed mainstream psychology and, to some extent, even the neurosciences was passé. A new model of human cognition, personality, and behavior was unfolding—a model that this time around was grounded not in philosophy or psychology, but the empirical evidence of biology.

The overall findings from early investigations in behavioral genetics at times revealed population differences in the frequencies of specific gene variants. As researchers penetrated more deeply into the activity of personality-related genes, they discovered that environmental influences often determine whether a genetic predisposition toward a particular personality trait actually manifests.[12] In some cases, individual genes interact with one or more other genes that individually or collectively further interact with various environmental forces. As more layers surrounding genetic activity were lifted, it soon became apparent that the same gene variant might have a slightly, or perhaps dramatically, different impact on a psychiatric or personality trait like depression, alcoholism and substance dependency, or antisocial tendencies.[13]

In some cases, these discrepancies seemed to be due to specific variable sections within the serotonin transporter gene, which sometimes differed across populations.[14] Research conducted throughout the ensuing decades revealed that gene expression ultimately might depend on the synergistic activity of a serotonin transporter gene variant with other genes and the social environment, notably culture.[15] Apparently, gene expression can vary, depending on the life experiences of the individuals or populations being studied.[16]

NATURE MEETS NATURE *PLUS* NURTURE: THE NEW INTERACTIONISM

Today, almost a decade after the sequencing of the full human genome, scientists have been forced to reconsider classic pronurture views about how personalities and behavioral impulses are formed. No one can ignore the compelling association between numerous mutant genes and personality predispositions or behavioral tendencies. The once-fashionable opinion that life outcomes of wealth, marriage, childhood experiences, family structure, and social relationships are mainly, if not exclusively, the province of environment has also transformed. This idea has been replaced, to some extent, by the notion that social outcomes too are influenced by genetics. Amid this shifting tide within the scientific community, many researchers from diverse disciplines are now examining whether genetic and environmental influences vary for different psychological and social traits within the same individual, within families, and, in some cases, across populations.

Today, the widespread view holds that intelligence, personality, and social behaviors are not determined by either nature or nurture independent of one another. Instead, it is the interaction of both forces—genetics and environment, represented by G × E—that exerts the strongest influences on these nonphysical traits. Gene–environment interactions signal an unavoidable reality that makes practical sense when considering individual differences. Mapping the interface between genetic and environmental forces is a sound, evidence-based approach to explaining how personality develops within an individual, in people within the same family, or in individuals randomly selected from society. Molecular genetics takes this logic one step further by making gene–environment interactions a central theme in comparative studies of ethnic populations—racial groups.

Today, molecular psychiatry, commonly known as psychiatric genetics, is partly an outgrowth of behavioral genetics, which is generating new data on the genetic architecture of a wide array of mental health conditions, including major depressive disorder, panic disorder, OCD, alcohol dependence, bipolar disorder, and schizophrenia.[17] Some investigators who explore health issues tied to psychological conditions are now examining genetic variation between ethnic populations, usually within the context of gene–environment interactions, which tend to either lower or raise the risk for specific psychiatric disorders.[18] Their research is mainstream and, despite a few critics inherently suspicious of genomic psychiatry, generally accepted as a scientifically viable and potentially benevolent application of modern medical science. After all, most of the findings produced in this field relate to health outcomes. But some of their data may also hold important implications for other researchers, notably behavioral geneticists, who intentionally go beyond the boundaries of health and medicine by focusing instead on population differences in social behaviors, including crime and other antisocial activities.[19]

Charles Darwin's cousin, nineteenth-century eugenicist Francis Galton, is often considered the "father of modern behavioral genetics," but this field as it exists today did not emerge until the 1970s. With intellectual roots embedded in hereditarian science, the psychologists and other scholars who embraced behavioral genetics in the mid-twentieth century borrowed analytical methods from classic quantitative genetics. They reasoned that research techniques used in zoology laboratories could also be leveraged to study the genetic underpinnings of social traits.[20] In the last

50 years, behavioral geneticists have traditionally studied personality traits, primarily such cognitive performance as IQ test scores, in keeping with their scholarly heritage of psychometrics, or intelligence testing. They too operate within the broader realm of psychological conditions, but their emphasis historically has been on intelligence and social behaviors related to education and, to some extent, the work force.

The immediate predecessors of many behavioral geneticists were medical geneticists interested in the inborn tendencies that seemingly made some individuals—and some ethnic groups—more prone to diseases like obesity, hypertension, and type 2 diabetes.[21] During the 1980s, some, but not all, behavioral geneticists remained focused on comparing IQ test scores in individuals and, in some cases, ethnic groups. By the close of the twentieth century, some researchers in this field, no longer confined to its psychometric wing of intelligence testing, had shifted their interest to identifying genes linked with such personality traits as impulsiveness, aggressiveness, and violence.[22]

Today, behavioral geneticists use many of the same laboratory and analytical techniques used in biological psychiatry. In some academic and research institutions, there is no clear division in the work performed by specialists in each discipline. Many practitioners in both fields are currently transitioning from quantitative genetics approaches to gene mapping techniques that analyze a huge number of gene variants, into the thousands, or data on tens of thousands of research participants at the same time.[23] Still, the philosophical orientations of behavioral genetics and biological psychiatry are noticeable, even though some degree of overlap is to be expected. Investigators in biological psychiatry are, foremost, medical scientists who operate within the realm of healthcare. By contrast, behavioral geneticists span a wide range of disciplines, with expertise in fields as divergent as psychology, social science, evolutionary biology, and genetics. Hereditarian behavioral geneticists represent a core segment of the pronature advocates in the nature–nurture debate with regard to IQ, in general, and, specifically, ethnic population differences in cognition and personality traits.[24]

Behavioral Genetics and Race-Related Research

In contemporary American society, the most vehement exchanges in the nature–nurture debate continue to be waged about the likely causes of ethnic differences in intelligence and social behavior. To some degree, new evidence has reshaped this debate, but the core questions remain the same as they have for decades: Are black/white/Asian differences in IQ, school performance, material wealth, and (for those who dare ask) "temperament and civility" caused by social and economic factors? Or, instead, do these differences, in cases where they truly exist, stem from inborn "racial" differences— population genetics? Pronurture supporters argue that much of what appear to be ethnic differences in average IQ and social behaviors, including aggressiveness and violent crime, are simply a reflection of socioeconomic, or SES, disparities. (SES is a measure of class and social status that includes household income and, to some extent, level of education and type of employment.) Proenvironmental advocates claim that the gaps in social outcomes are not due to genetic variation in different races or ethnic groups. But pronature hereditarians disagree, declaring that population genetics is a crucial and possibly major cause of multigenerational ethnic gaps in scholastic achievement, IQ, and, alas, "antisocial behavior" as well.[25]

In the United States, the ethnic achievement gap in school performance has narrowed slightly during the past 15 years, based on some assessments, but overall, blacks and Hispanics still lag substantially behind Asians and whites in scholastic performance.[26] In the 1970s through the mid-1980s, the white–black divide decreased sharply, but eventually the pace of the narrowing reportedly stalled. Compared to the ethnic gaps of the 1970s, in 2012 the white–black and white–Hispanic achievement divides were 30 to 40 percent smaller.[27] According to investigators at the Brookings Institute, the white–black achievement gap in the United States has declined 0.15 standard deviations in fourth grade and 0.11 standard deviations in eighth grade.[28] Some states, including Washington, Ohio, Iowa, and California, have fared better than others in reducing the achievement divide between blacks and whites. Despite notable progress in narrowing disparaties between these two ethnic groups in the last 50 years, the white–black achievement gap remains. The largest gains in shrinking this divide have usually occurred in the states with the smallest African American and Hispanic populations.[29]

Similarly, black/white variation in IQ test scores has persisted since the time IQ test results were collected for African American school children in Philadelphia starting in 1913.[30] Ethnic gaps were reported in members of the U.S. military more than a century ago.[31] The dual achievement gaps in both IQ and academic performance correlate closely with ethnic disparities in other social outcomes. Both in the United States and elsewhere there are notable, if not statistically significant, divides between populations—particularly blacks, Latinos, whites, and descendants of Northeast Asians (e.g., the Chinese and Japanese) in several key indicators of social success.

Among the most important measures of social well-being are family stability, rates of employment, teenage unintended pregnancy, family wealth, and health status. More sensitive indicators tied to individual behavior (often perceived, rightly or wrongly, as ethnic behavioral tendencies) include interpersonal aggression, psychiatric diagnoses, crime rates (especially for violent criminal activity), guilty verdicts in criminal prosecution, and incarceration. These social traits comprise some of the predominant social behaviors that have complicated the contemporary nature–nurture debate concerning the ethnic achievement gap. To some extent, the ethnic divide in this realm is no longer confined to population differences in cognitive and academic test results. As in the traditional debate, however, the pronature and pronurture camps are distinguished by their contrasting views of the main underlying causes of ethnic differences in numerous social patterns.[32]

Most experts, no matter their perspective, agree that social behaviors are complex and shaped by multiple factors that extend well beyond a single gene, a single environmental factor, or even a linear interaction between a specific gene and a specific environmental force. Until the mid-1990s, at least on record, many knowledgeable people in this field attributed the ethnic achievement gap to nurture—a wide array of environmental forces. Even after gene–environmental interactionism became the accepted rationale for best explaining population variance in intelligence performance and academic success, some pronurture scholars have adamantly rejected this explanation. This group, sometimes called social constructionists, argues that the ethnic divide in IQ, scholastics, and, by extension, social behavior is due primarily, if not exclusively, to nurture—the social environment and experiences. Their staunch proenvironmental perspective is diametrically opposed to the hereditarian, or historical, pronature position.[33] From this perspective, ethnic variation in IQ, academics, employment, and other life outcomes is viewed as largely the result of innate differences in ancestral populations, racial groups, ethnic populations, or any other group designation based on common gene frequencies.[34] The nature–nurture debate regarding race and

IQ, and to some extent social behaviors, intensified after the resurgence of hereditarian science, usually equated with biological determinism, in the 1990s. Spearheaded by scholars like Arthur Jensen, J. Philippe Rushton, and Charles Murray, it left many scientists and scholars on both sides of this fierce intellectual conflict polarized. At times, it seemed to render them too immobilized to engage in long-term rational and solution-oriented discussions about how to close this divide.

Just as the concept of the mind as a blank slate—a mind that develops largely, although not exclusively, in response to nurture—has faded in the last two decades, so too has evidence increased, incrementally at least, for the influence of nature on social behaviors in both individuals and groups. In the last few years, some researchers have reported small but potentially significant population differences in a few of the gene variants associated with such psychological traits as aggression, violence, and other socially disruptive behaviors. Again, however, these links represent correlations, and few, if any, have proven to be causal relationships. Genes do not invariably determine a person's psychological makeup, although they may contribute to his personality traits. Few experts dispute this, as it constitutes a core axiom in nature–nurture interactionism. This idea is not inherently controversial with regard to race relations because the role of genes in personality has historically revolved around individuals rather than groups. But when the discussion shifts to allegedly tangible evidence for differences in gene variants across populations, the situation becomes exponentially more sensitive, guarded, and politicized.

We know that overall, ethnic populations vary in social outcomes like rates of high school graduation, criminal prosecution, and teenage unintended pregnancy. What, then, are we to infer if some social behaviors appear to have genetic underpinnings and some of these potential genetic contributions to behavior also vary across populations? Is the writing on the wall? Is it a rational deduction or, instead, a leap of faith to conclude that population genetics—"racial genetics"—influences the risk for various adverse social behaviors in different ethnic groups, in human races? Are new scientific findings pointing to a twenty-first-century hereditarian heyday?

Nature, Nurture, and the Other "Biologies"

For some people, the claim that there is a cause and effect relationship (even if only a partial cause) between genetic ancestry ("racial genetics") and ethnic social norms may seem like a logical assumption. But genes do not directly determine psychological or social traits in individuals, let alone in populations in which individuals may vary substantially in their genetic makeup. Some mainstream experts studying the human condition, whether biologists, psychologists, or social scientists, are creating novel interfaces between genetics studies and environmental research in relation to human behavior. They are scrutinizing the genetic underpinnings of human intelligence and personality, and, for the most part, they are doing so without allegiance to the hereditarian worldview.

These scientists are perhaps better described as nature-friendly rather than exclusively pronature. Although they are not traditional pronurture advocates, they are by no means hard-core, racialist, genetic determinists. In the emerging camp of nature *plus* nurture (rather than nature *versus* nurture), this new breed of bridging or hybrid researchers are careful not to underestimate the power of life experience and the multifaceted environment in molding the life experiences of individuals and even ethnic populations. At the same time, they are reclaiming the salience, although not necessarily the supremacy, of genetics, albeit often within the context of individual experience, in shaping human behavior, even in the contentious context of ethnic population differences in social behaviors.

As human genetics research grows ever more sophisticated, mounting evidence reveals that gene–environment interactions are not simply additive, but also sometimes synergistic. These interactions appear to be even more influential than either genetics alone or experience and environment alone as separate forces that shape social behaviors. Interactions between nature and nurture are much more complex and variable than previously thought. Genes and hormones, along with life circumstances and experience (or individuals' reactions to their social environment), do not operate in a vacuum as isolated factors. They act reciprocally in a dynamic manner to mold a host of psychological conditions and social traits—for example, personality and psychological temperament, depression, anxiety, aggressiveness, violence, and alcohol or drug dependence.[35] Yet, the relative impact of nature, as it interacts with nurture, on social behaviors may vary multigenerationally throughout time and differ from one ethnic population to another.

Investigators have identified some links between genes and personality, but rarely do these links reflect a direct causal relationship. Each person's social tendencies have the potential to be continually refined by the interplay of numerous elements of the social environment that interact with genes, with other environmental influences, and then as part of gene-environment complexes with other genes and life circumstances. In the current tussle about nature versus nurture, individual experience has assumed a new, or perhaps renewed, level of legitimacy. Whether a genetic tendency toward any trait—physical, psychological, or social—actually manifests depends partly on gene expression. Multidirectional interactions between genes and environment, and gene variants and experience, may ultimately determine if the genetic underpinnings of a personality trait or social behavior are expressed. Gene expression in turn involves other key biological phenomena, in addition to DNA—namely, a nongenetic form of biological inheritance called epigenetics.[36]

The influences of nature, nurture, and nature–nurture interactions on social traits are intricate. Biology alone cannot explain all of the ethnic variation in social behaviors, but it may play a role. Even when it does, however, the biology at stake here is not necessarily or inevitably a rigid heredity, at least in the strict sense of inherited traits governed by DNA, which cannot be easily altered. In some cases, the predominant biological forces at work in group differences in cognitive performance and social behaviors involve plasticity—the ability of an individual to biologically adapt to changes in its surroundings. Biological plasticity—also known as phenotype plasticity—is closely interwoven with gene expression and developmental biology.[37]

Epigenetics, plasticity, and gene expression are closely intertwined biological processes. These processes are shaped by genes, as well as a wide array of environmental factors that span the physical milieu, social circumstances, and interpersonal relations. In the emerging paradigm of gene–environment interactions that drive human social activity, these three processes are increasingly assuming a paramount position. Many experts now agree that social behaviors and personality traits are molded by the interplay between genes and environment, but for some researchers (and not just the proenvironmental ones), the role ascribed to genetics has diminished considerably. That role has not altogether disappeared, however, even though there is growing evidence that genes per se do not exert large effects on social behaviors.[38] In regard to social traits, genetics seemingly are more likely to influence how an individual responds to various environmental conditions, especially adverse events, than to directly govern his behavior. And genes do not necessarily orchestrate the biochemical changes that give rise to distinct personality traits. But genes sometimes play a key part in mediating the effects of a wide range of environmental forces on an individual's personality and social proclivities.[39]

Some of the most well-known examples of the moderating effects of genes on a person's response to environmental stressors involve the interaction of genes versus environment in shaping antisocial behavior in young men. In 2002, Israeli American psychologist and neuroscientist Avshalom Caspi and his associates reported their findings on adult males who had either low activity or high activity variants of the MAOA, or monomine A, gene. MAOA is a key enzyme that breaks down important neurotransmitters in the brain, for example, norepinephrine, dopamine, and serotonin.[40] Caspi's team discovered that among adult males who had been mistreated as children, those with the low activity gene variants had an increased risk for antisocial behavior, whereas those with the high activity gene variants showed no heightened risk for troublesome behavior when they grew older.[41]

In a second study published one year later, Caspi and colleagues examined another gene variant—the short version of the serotonin transporter gene, dubbed 5-HT—which is implicated in the onset of such psychological conditions as depression. The link between life stress and depression was stronger in people who carried this particular gene version. If they had at least one copy of the short serotonin transporter gene variant and also experienced life stress, they were more likely to develop depression than individuals who experienced life stress but had the long serotonin transporter gene variant instead. As in the previous example of the MAOA low-activity gene version, life stressors did not appear to alter the activity of the gene variant itself. Instead, adverse life circumstances seemed to affect the short serotonin transporter gene variant's ability to moderate the impact of an environmental experience—life stressors—in triggering clinical depression.[42]

Caspi's seminal research helped usher in a new dawn in the search for candidate genes thought to underlie mental illness; however, his research team's original findings on the association between the serotonin transporter gene, life stress, and depression were not replicated in subsequent studies.[43] In the decade that ensued after the publication of this study, evidence from other research in psychiatric genetics and later in behavioral genetics called into question the validity of uncovering the biological roots of psychiatric illness based on individual candidate gene variants. The reality was that as research continued, the data failed to demonstrate irrefutable empirical support for proven cause-and-effect relationships between candidate genes and specific psychiatric conditions. Nonetheless, this approach—exploring gene-by-environment interactions—to unlocking the genetic underpinnings of mental health and personality predispositions was not a futile pursuit. It opened the door to an entirely novel strategy for deciphering the molecular genetic code, and identifying genetic contributions to psychological and social traits.[44]

In the wake of diminishing returns from candidate gene studies, by the early 2000s many scientists had turned to genome-wide association studies (GWAS) as a more expedient endeavor for pinpointing genes tied to social traits. As we will learn in later chapters, GWAS have consistently shown that the effects of individual gene variants on social behaviors or cognitive skills, notably IQ, are exceedingly minute. At the same time, multiple gene variants can cumulatively produce a measurable, although usually still small, effect on a personality or behavioral trait.[45] Despite this revelation, some investigators have shifted their pursuits to what could be called missing links in gene–environment interactions that impact psychiatric disorders, psychological tendencies, and social behaviors.[46] One of those missing links may well be epigenetics—the biological process in which physical traits and apparently some psychosocial tendencies are transmitted from parent to child via a part of the genome without involving DNA.

At present, epigenetics and its impact on gene expression are gaining increased visibility in nature-versus-nurture explanations of the underlying causes of intelligence, personality, and social behaviors. According to some experts, the emerging empirical evidence in this field is so compelling that concepts of epigenetics are gradually, if not precipitously, supplementing and even supplanting the historically simpler model of genetic determinism.[47] This does not mean that genetics is less important than environment in conditioning how a person thinks, feels, and acts. The impact of genes on social traits may be quantitatively less than in previous paradigms of gene–environment interactionism, but genes are still considered crucial in influencing a person's reactions to his life circumstance, including life stresses. From this perspective, genetics—coupled with epigenetic processes that impact gene expression—is on equal footing with environment in determining mental health and social behavior.

In contrast to this "middle-of-the-road" position on nature–nurture interactionism is the vantage of pronurture scientists who embrace the environment rather than genetics per se as the dominant force driving social behavior. To some extent, this biology-friendly, pronurture camp also takes a centrist or hybrid stance with respect to the rift concerning nature–nurture and personality or social behavior. For pronurture advocates, environmental circumstances and personal experience are often the most decisive forces in determining social traits. But they are quick to acknowledge the interplay between an excessively stressful or adverse environment, epigenetics, and gene expression in shaping disruptive or worrisome psychological and social tendencies.[48]

Closely aligned with plasticity, developmental biology guides the formation of physical traits related to growth and maturation that directly or indirectly influence social behaviors. These traits include the brain's softwiring, or new and changing neural connections that form as a result of experience, the formation of the skeleton and muscles, and height or tallness. Developmental biology is also related to such physiological traits as the timing of sexual maturity. In this context, it is influenced by hormones, particularly during adolescence, which in turn impact social behaviors ranging from social interactions to mating, competitiveness to aggressiveness, and social productivity to delinquency.[49]

To some extent, developmental biology is an all-encompassing phenomenon of human psychological and physical growth. Like its closely allied process, plasticity, developmental biology is largely shaped by a host of environmental conditions—the broad social and economic landscape that encompasses income and education, family upbringing, a community's social attitudes and values, and the ecological surroundings. Some environmental factors are obvious, while others are not, but they are still important because they exert subtle, although sometimes forceful, effects on individuals and groups. Nutrition, exposure to toxins (in air, water, and food), and even sleep deprivation can influence if, when, and how a genetic tendency is expressed and, if so, how that manifest tendency affects a child's overall physiological and psychological health.[50]

The distinction between genetics and biological plasticity is often nuanced because genes, in their interaction with both the environment and epigenetics, are crucial in crafting the brain and central nervous system's capacity for flexible, or plastic, change in every person.[51] Determining whether environment or genetics has a greater influence on biological development and plasticity ultimately may depend on the specific trait. Consider a meta-analysis of 2,748 studies, published in 2015, which examined correlations and differences in various psychiatric traits and diseases in 14.5 million pairs of twins.[52] (A meta-analysis is a summary of the results of multiple studies conducted to answer the same research question. The summary is based on a statistical analysis of the data as a whole reported for these studies.)

The overall variation for diseases and social traits, including social interactions and social values, was almost equally attributable to nature and nurture. An estimated 49 percent of the variance was due to genetic factors, and 51 percent was due to environmental factors, even though the amount of variance varied for different traits.[53] Although epigenetics was not evaluated in this meta-analysis, a growing number of recent studies have examined the potential effects of epigenetic changes on shaping physical traits, diseases, and social traits.[54] In later chapters, we will examine emerging evidence for the intersection between genes, environment, epigenetics, gene expression, and plasticity, as well as their synergistic impact on human cognitive performance and various social behaviors.

Architecture of the Human Brain

THE THREE-POUND INNER GALAXY

BEFORE WEIGHING IN ON WHAT SOME RESEARCHERS CALL SUBSTANTIAL differences in the biology, notably the brains, of different ethnic populations, a brief introduction to the brain could be instructive. A spongy mass encased in a covering marked by ridges and crevices, the human brain remains the indisputable seat of human feelings, thought, and conscious behavior. In the simplest terms, the mind-brain—the physical structure and its associated functions—are controlled by the central nervous system, or CNS, which consists of the brain and spinal cord. These two organs are arranged in a hierarchy of centers that, when combined, control cognition, emotion, and action. At the lowest level, the spinal cord receives sensory information from the body and then sends motor messages to other neurons and muscles, and sometimes to the endocrine glands, which house and secrete hormones.[1]

The brain has three major divisions: the cerebrum, brainstem, and cerebellum. The cerebrum is partitioned into two halves called cerebral hemispheres that communicate with one another through a broad tract of nerve fibers called the corpus callosum. The cerebral hemispheres control the highest levels of mental activity. Various regions of the cerebral hemispheres are involved in sensation, voluntary muscle movement, emotions, thought, reasoning, and memory. Specific areas within the cerebral cortex, or outer covering of the cerebrum, regulate thinking, planning, emotions, speech, and learning.[2] The cerebrum contains four distinct anatomical sections, or lobes—parietal, temporal, frontal, and occipital—which are each found in both the left and right hemispheres.[3] A small area called the insula is sometimes classified as the fifth lobe of the brain.[4]

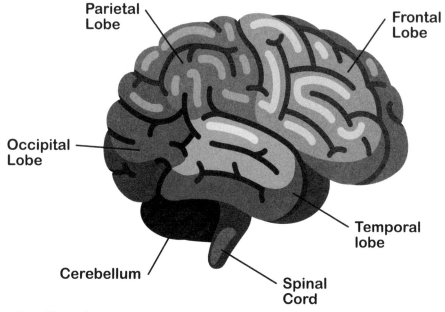

Lobes of the brain.

Each lobe is associated with general functions, except for the occipital lobe, which is solely responsible for processing visual perception and processing. Hearing, language, attention, and spatial cognition, as well as tasks of interpreting and processing sensory information, are regulated by the parietal lobe. The temporal lobe controls speech and language skills, as well as visual and auditory memories. It also contains the hippocampus, a structure crucial for forming and storing memories. As the brain's executive control center, the frontal lobe, situated in the front of the brain, serves as the control station for a wide array of functions—motor skills, attention, abstract thinking, problem solving, and expressive language. Within the frontal lobe is the prefrontal region, a highly specialized area associated with personality, decision-making, and complex, willful behavior.[5] Located behind the cerebral hemispheres is the cerebellum, the largest portion of the lower brainstem. And sitting beneath the cerebrum and atop the spinal cord is the brainstem, a part of the brain proper, which seamlessly connects the brain with the spinal cord. The brainstem provides an intermediate level of coordination, regulating such complex vital functions as breathing, heartbeat, blood pressure, and consciousness, including wakefulness and sleepiness. The limbic system, lying partially within the cortex and partially below it, governs emotions, memory, and motivations that guide behavior, ranging from appetite to sex drive. The infrastructure of the limbic system regulates autonomic and endocrine functions, usually in response to emotional stimuli. In addition to the hippocampus, pivotal structures of the limbic system include the amygdala and hypothalamus, both of which influence a person's response to emotional stimuli.[6]

The CNS works in tandem with the peripheral nervous system, an organized panoply of nerves located outside the brain and spinal cord that connects the CNS to the limbs and organs. The fundamental unit of the CNS is the neuron, or nerve cell—a unique type of cell in the brain and spinal cord that can be electrically stimulated to receive signals from other neurons and transmit these signals to different target neurons. Neurons are not the only type of cell in the brain, but they comprise major players in a complex electrical and chemical system that communicates messages to target areas throughout the nervous system. The organization of the CNS can be viewed from the most basic level of neuron connections—synapses, or junctions between single cells—to an intermediate level of groups of cells within defined anatomical areas, to a higher level of large-scale regions of the brain, along with their interconnecting pathways, or *neural* circuits. Home to an estimated 86 billion neurons, the human brain is a complex network that facilitates dynamic communication via electrical signals across neurons within the brain.[7] Electrical impulses are relayed to other neurons, as well as cells in muscles or endocrine glands, where they trigger a wide variety of physical actions.[8]

A neuron typically refers to the round or polygon-shaped body of the cell, but each neuron also has extensions called axons and dendrites. Many axons—long tentacles projecting from the nerve cell body—are encased in a fatty coat called a myelin sheath. Dendrites comprise a network of short, tree-like, narrow branches radiating from the body of a neuron and into the synapse. Axons and dendrites are part of the brain's elaborate electrical-chemical signaling system to deliver messages to nearby, intermediate, or distant target cells. Electrical impulses are often transmitted from the axon of one cell to the dendrites of another cell within a minuscule space, or gap, called a synaptic cleft. The brain's communication network reveals an intricate architecture of neurons, including their myriad axons and dendrites, and neu-rotransmitters, or chemical signals, communicated across synaptic junctions dispersed throughout the brain.[9]

Myelin is the fatty coating covering some axons that provides a layer of electrical insulation around the body of a neuron. This fatty substance assists in the transmission of electrical signals, while at the same time conserving space and energy in the brain's infrastructure. In some brains, neurons have more extensive myelin sheathing, possibly allowing them to accelerate the rate at which electrical impulses are transmitted along the axons of neurons. A nerve impulse stimulates an axon to produce neurotransmitters, which, in turn, are released into the synaptic gap. There, the neurotransmitters stimulate the transmission of an electro-chemical signal across a synapse—a message from one neuron to another.[10] Adjoining neurons in the same region and neurons from different regions of the brain communicate with one another at the synaptic gap, or cleft. With an estimated 100 trillion synapses in the human brain, primarily in the cerebral cortex, the magnitude of synaptic firing is astonishing.[11]

ORGANIZATION OF THE HUMAN BRAIN

Links between specific areas of the brain and aspects of human conscious-ness—for example, speech, hearing, spatial perception, learning, cognition, and emotion—have been the focus of myriad investigations since the time of ancient Greece. Philosopher-scientist Alcmaeon of Croton described the sensory nerves and the optic nerve, posthumously earning him the unofficial honorific as the father of neuroscience.[12] During the last 200 years, perspec-tives on the organization of the human brain have shifted dramatically in response to scientific breakthroughs. In the mid-1800s, neuroscience became a recognized field when French physician Paul Broca demonstrated that the center of spoken language resides in the frontal lobes. Known today as Bro-ca's area, this region is damaged in people who lose their ability to articulate language. A separate region of the brain called Wernicke's area, named after its discoverer, German physician Carl Wernicke, is involved in language comprehension. In the nineteenth century, neuroscience underwent another revolution when researchers began to identify key anatomical locations in the brain involved in the production and comprehension of speech, and even some mental functions and emotional states.[13]

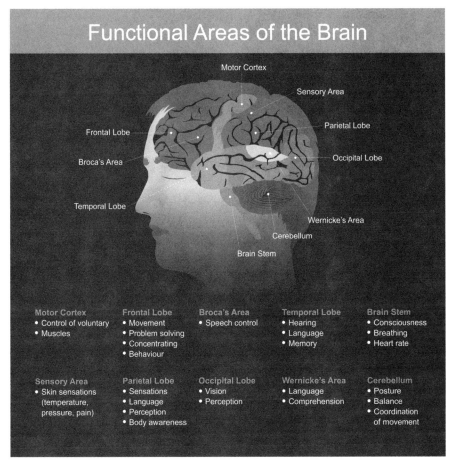

Functional areas of the brain.
SHUTTERSTOCK

In the early 1900s, German neurologist Korbinian Brodmann pioneered a novel method for classifying areas of the cerebral cortex (called cortices in the plural), based on several characteristics of nerve cells in various regions of the brain. He identified 52 discrete cortical divisions—now known as Brodmann areas—and linked some, although not all, of these sections with specific functions of the nervous system. The different regions were distinguished by the shape, size, density, and distribution of the neurons.[14] Brodmann's discoveries of structure–function relationships in the brain paved the way for scientists following in his footsteps to refine and eventually revamp his original cortical map. Even though the brain is organized by modules—structures, columns, or discrete regions—each self-contained area may perform multiple functions. Some mental functions are widely distributed throughout the brain, reflecting functional connectivity—the dynamic interplay and exchange of information between different regions of the brain.[15]

By the late nineteenth century, Spanish neuro-anatomist Santiago Ramón y Cajal used a microscope to successfully visualize neurons, along with their projecting axons and dendrites interacting with one another. Cajal correctly determined that information flows from the dendrites to the axon via the cell body.[16] He became a momentous figure in the history of neuroscience, indeed biology, in general, for not only his meticulous description of neurons, but also his introduction of the concept of plasticity. Cajal suggested that the connections between nerve cells are plastic, or pliable. Brain plasticity, or neuroplasticity, in its modern usage, refers to the brain's ability to continually undergo adaptive change, or reorganize connections between neurons. It is essentially a process of the brain rewiring itself. Plasticity allows individual neurons and nerve cell networks to alter their connections and firing patterns in response to new information from both within and outside the brain.[17] Cajal's notion of neuroplasticity may have differed from the concept used in contemporary neuroscience. But his seminal insights into the structural plasticity of the CNS laid the groundwork for contemporary studies that track the impact of the multifaceted environment on the brain and other parts of the nervous system.[18]

Today, perspectives on the functional architecture of the human brain are rapidly changing as new techniques for mapping structure–function relationships between the human brain and mind become available. Such sophisticated tools as magnetic resonance imaging (MRI) technologies now allow the brain to be visualized in two-dimensional slices. Advanced imaging can produce three-dimensional scans of the outer surface and internal structures of the living brain in vivid imagery and real time. In an earlier era, neuroscientists assumed that emotional, social, and cognitive states were generated in separate brain regions or brain networks. But recent studies leveraging state-of-the-art neuroimaging techniques suggest that emotional and social phenomena are best understood by examining large-scale networks found in different regions of the brain.[19]

With MRI scans, scientists can also visualize activation patterns—areas of electrical activity—that occur in different regions of the brain when a person experiences a particular mental or emotional state. Still, the information generated by MRI brain scans is limited because they produce only static images. Functional MRI—fMRI—is a noninvasive modality that can be used on humans to observe "live" brain activity without harming them. This technique enables scientists to localize brain activation patterns to areas as small as only one cubic millimeter. For fMRI, the technology provides a window on the networks of the brain in specific areas of a particular lobe or lobes by displaying activity on a scan (similar to a computer monitor) while a person is performing a mental task. The neocortex—the area of the cerebral cortex involved with sight and hearing—has functional modules that are often the site of highly orchestrated neuronal activity. Sophisticated fMRI images conveniently capture shifting patterns of activity in neurons and synapses. This technique has also been used to identify modules mapped in the six layers of the cortex, as well as in the subcortical areas (beneath the cortex) and limbic system.[20]

GRAY MATTER AND WHITE MATTER

The size of the brain and the number of neurons it contains are the subjects of intense scrutiny because, according to some experts, these attributes determine a person's mental speed and efficiency. Burgeoning interest in the brain's gray and white matter, rather than just its lobes and surface area, has opened up new vistas for peering into the intricate geography of the human brain. Gray matter is composed of nerve cell bodies that process information for cognition. Gray matter is found mainly on the surfaces of the cerebral cortex and the cerebellum, and in some of the structures located deep within the cerebrum and cerebellum.[21] By contrast, white matter does not contain neurons or synapses but, instead, consists of a mixture of tightly packed supporting cells, as well as axons—both with and without a fatty sheath. Myelin acts as insulation for the axons of neurons, which lie largely beneath the surface of the cerebral cortex. In this manner, white matter functions like a network of highways that distributes electrical signals between cells, facilitating communication across divergent areas of the brain. Despite their contrasting functions, however, both gray matter and white matter are essential for brain function.[22]

With the aid of ultra-sophisticated technology, the new science of brain connectivity is transforming the entire field of neuroscience. MRI is frequently the neuroimaging tool of choice for probing the brain's infrastructure and mapping patterns of connections between diverse brain regions. In medicine, MRI scans provide a valuable method for evaluating the integrity of structures and tissues in the brains of patients who have experienced head trauma or potentially have a neurological disorder. This imaging strategy is also used to determine and quantify the shape, size, and integrity of structures in the brain, including areas of gray matter and white matter. This information has helped to advance research on brain connectivity.[23]

During the last few years, scientists have been able to access data from the Human Connectome Project, an endeavor funded by the U.S. government to study the brain using powerful and innovative brain imaging technology. Connectomics, or neural mapping, is an investigative technique to compute the brain's wiring and then produce a map, or connectome, of its various connectivity patterns. Initiated by the National Institutes of Health (NIH) in 2009, the Human Connectome Project signals a monumental effort to map major pathways of nerve fibers over long distances in the human brain and identify the main circuits within it. The product of this endeavor—the human connectome—refers to a large and complex network that allows the brain to be analyzed as a whole, rather than as separate or isolated regions or tracts.[24]

Neuroscientists can now exploit multiple brain imaging devices to navigate intricate regions of the brain and eventually create detailed topographies of neural connections between divergent areas of the brain. This field relies on multiple technologies, notably specialized types of MRI that yield high-resolution images of newly discovered architectural landmarks in the brain. These include structural and myelin mapping of wide brain regions, structural connections between distant brain regions, and depictions of white matter—the myelin sheathing surrounding the axons that project from neurons. The ultimate goal of the Human Connectome Project is to produce a "network map"—a blueprint of the connectivity of neural pathways, revealing associations between the brain's anatomy and its functional activity. Probing the architecture of the human brain in both healthy individuals and those with neurological conditions is expected to shed new light on the biology of Alzheimer's disease, schizophrenia, dyslexia, and autism.[25] By correlating findings on the brain's structure with extensive behavioral and genetic data, scientists are poised to provide additional clarity on the influence of genes on brain development. With new discoveries continually unfolding, experts are also inching closer to mapping the influences of environmental and other nongenetic factors on the development of the brain and its many varied functions.[26]

CHAPTER SIX

The Gendered Brain

THE PARADOX OF SEX AND BRAIN SIZE

ONE OF THE MOST WELL-KNOWN PARADOXES IN STUDIES OF THE BRAIN AND IQ is that men and women of the same ethnic population tend to earn similar scores on IQ tests, even though females have smaller brains than those of males.[1] What are the implications of this widely cited finding in light of reports showing that when compared to whites and Asians, blacks overall have smaller brains? Can we extrapolate from the results of gendered brain studies and convincingly argue that ethnic differences in IQ scores are not due to racial differences in average brain size? Are comparisons of gross brain size between adult men and women analogous to comparisons of brain volume between diverse populations? Perhaps, but not entirely.

Despite the similarity of scores sometimes reported between the sexes, the IQ scores of age-matched males and females in a given population are not exactly the same. In 2005, British researchers Paul Irwing and Richard Lynn found that males earn scores averaging roughly four points higher than females' scores on IQ tests measuring verbal reasoning and spatial abilities.[2] For Lynn, the higher average IQ among males compared to females in the same population is predictable because of the larger male brain. But he also noted that since girls mature faster than boys, the gap in biological development, or physical maturity, between the sexes means that IQ scores for eight-year-olds to 14-year-olds might be misleading.[3] Several investigations published since Lynn made this claim in the 1990s suggest that the sex difference in IQ within a single population is less than previously thought. In some studies, females tend to outperform males on verbal and memory tasks of intelligence tests, a finding that counters the general trend observed in most studies.[4]

Male Brain, Female Brain, and Spatial Perception

Males historically have also earned higher scores than females on specialized mental rotation tests—tasks designed to assess how well a person can move and then locate a visual representation of various images within his mind. As one of the ultimate barometers of high-level intelligence, mental rotation tasks provide a means of evaluating spatial visualization skills, regardless of the test taker's verbal skills. These tasks test a person's ability to rotate and compare—or "flip, *rotate*, and piece together"—mental representations of two-dimensional and three-dimensional objects within the mind.[5] In the 1970s, when formal mental rotation tests were first administered to students of various ages, these tests were partly used to clarify how spatial ability develops in girls and boys, especially as they transition from childhood to puberty. Even then, psychologists recognized that sex differences in spatial ability increased with the onset of puberty, although they did not know why this occurred. Hormones, particularly the male sex hormone, testosterone, was, and still is, a prime candidate for the biological underpinnings of potential male–female variance in mental rotation abilities.[6]

Multiple studies published in the last four decades have found that males across the age spectrum overall score significantly higher than females on mental rotation tests.[7] Ironically, however, a higher proportion of males fall in the low-ability end of several types of cognitive impairment, for instance, mental retardation, attention disorders, dyslexia, stuttering, and delayed speech.[8] Apparently, then, mental rotation abilities do not correlate closely with *g*—general intelligence—the "thing" measured on standard IQ tests. Perspectives vary regarding the reasons for differences in male and female performance on spatial intelligence tasks. In a meta-analysis of mental rotation test results from students in various countries, the pooled findings suggested that boys had an edge over girls on general knowledge exams. But the male advantage was small, and it may have been tied to female students being more interested in the humanities than the sciences.[9] This is one of the central arguments for why girls tend to underperform on spatial cognition tests compared to boys.

Still, the brain's white matter forms at different rates in males and females, a finding that, hypothetically, could support a biological basis for sex differences in spatial intelligence. Recall that white matter is largely composed of myelin-covered axons—the projections from nerve cells that transmit electrical signals, or nerve impulses, from one neuron to another. Lynn speculated that girls from approximately eight to 12 years of age earn better IQ scores because their physiological development is more accelerated than that of boys during this stage of growth.[10] Male–female differences in the rate of white matter maturing in the brain can vary at different stages of the life cycle, possibly accounting for sex-linked discrepancies in IQ. As one researcher put it, if Lynn had conducted his experiment on the IQ gap in males and females on younger children, the white matter might have been greater in the girls, revealing that they had an IQ edge over the boys at this stage of life.[11] The potential contribution of white matter to intelligence performance might explain why, within a particular ethnic population, girls overall may score higher than age-matched boys when they are younger and why the situation tends to be reversed in adolescence, with young men averaging higher IQs.

In *IQ and the Wealth of Nations,* a book often cited to condone, if not promote, racial science and racialism, Richard Lynn and Tatu Vanhanen attributed the male–female IQ divide, which averages almost 4.4 points, to the larger human male brain.[12] Like other writers who have borrowed from Darwinian theory, Lynn and Vanhanen reasoned that males have larger brains because they faced greater selection pressures during the evolutionary history of the human lineage. Males had to hone and refine their hunting skills, if only to ward off animals and human enemies or capture prey so they could feed their mates and children. A brain with an infrastructure suitable for enhanced spatial perception, including good performance on mental rotation tasks, and motor abilities would have afforded men an upper hand in performing these feats.[13]

For critics, many evolutionary scenarios to explain the origins of present-day social behaviors are seen as merely fanciful just-so stories, pensive ponderings that have no basis in fact or empirical data. Lynn and Vanhanen's narrative about the origin of the human male brain may have been dismissed as sheer conjecture were it not for evidence supporting their position. As noted here, testosterone has been linked with the overall better performance of males on mental rotation tests. In one study, testosterone levels correlated with the amount of white matter in the brains of adolescent males as they advanced through the stages of puberty. The hormones associated with puberty, as well as the boys' increasing chronological age, seemed to contribute to an enlarged volume of white matter in their brains.[14]

In 1999, a team of University of Pennsylvania scientists, led by Ruben Gur, made a startling discovery that radically altered neuroscience research on sex differences in the human brain. They found that compared to men's brains, women's brains have a larger proportion of gray matter—areas dense in neurons—and a relatively smaller quantity of white matter.[15] While gray matter contributes to verbal abilities and computing skills, white matter is essential for the brain's communication network, where it plays a pivotal role in solving spatial problems. These different attributes of gray matter and white matter suddenly provided an explanation for why men and women in the same population often earn similar IQ scores, despite differences in their brain volume. According to Gur, women exhibit a larger enhancement of their information-processing capacity for every equal increase in brain volume seen in both men and women. A larger amount of gray matter in a woman's brain means that she has proportionately more brain cells, specifically neurons, or nerve cells. Gur's team reasoned that this difference might be related to sex variance in cognitive ability tests, including the tendency for females to perform better than males on verbal IQ tasks.[16]

In 2003, a study leveraging high-resolution MRI scans confirmed that the volume of white and gray matter varied in different regions of the brain in men and women. As expected, the male brains were larger than the female brains. But the women's brains displayed a higher ratio of gray matter to white matter volume in certain areas, including the temporal, parietal, and occipital lobes. Despite these sex variations in parts of the brain's architecture, the researchers admonished that it was premature to conclude that men and women vary significantly in their cognitive functions.[17]

In another study, the average male–female difference in IQ was not correlated with sex differences in brain size. And the relatively poorer performance of women on spatial IQ tests—mental rotation tasks—was not associated with brain size or number of neurons. This is not surprising in light of other research suggesting that the more a person practices taking a visual-spatial intelligence test, the more she can improve her performance. In one study, females increased their scores on mental rotation tests by as much as 30 to 50 percent after taking the test just one time. If women can raise their scores to this extent in such a short period of time, it is doubtful that brain volume, at least within the normal range, has a limiting effect on spatial performance.[18]

Scans using the advanced imaging technology of fMRI reveal that total brain size does not affect spatial cognition, even though specific regions of the brain are apparently activated when a person performs visual-spatial tasks.[19] But the brain areas that are activated differ somewhat between men and women, even when they perform at the same level while doing the same mental rotation exercises.[20] In one assessment, males showed significantly stronger activation in the parietal region—the lobes located on the sides of the brain—where high-order visual input and sensory input from the body converge within the cerebral cortex. This is the site of neural networks involved in visual attention. In females, the right frontal region, an area associated with nonverbal communication, was more strongly activated.[21] These findings suggest that men rely on a more automatic process when analyzing complex visual reasoning tasks. By contrast, women appear to use a more top-down process—perceptions driven by thought or cognition.[22] Again, however, spatial ability can be enhanced through training, especially if educational interventions are tailored to individual differences in ability.[23]

REIMAGINING AND REIMAGING THE BRAIN'S GENDER GAP

When the gender debate concerning the utility of science, technology, engineering, and mathematics—STEM—for girls and young women commenced in the 1980s, some experts argued that females performed below males in STEM subjects because girls were discouraged from studying science in the classroom. Some women, along with a few men willing to stand up to the status quo, challenged the then (and until recently) traditional stance that gender disparities in science-related subjects and the real-world high-tech workforce reflect, to some degree, inborn sex differences between males and female across the age spectrum. After all, some said, the striking contrasts between male and female physiology, notably reproductive functions, offered ample proof that the two sexes are not biological equals. To support their position, these pronature advocates cited emerging evidence indicating slight, but potentially important, anatomical contrasts between the male and female human brain.

For many experts, the pivotal issue was whether anatomical variation observed in male and female brains correlated with functional differences that made men and women think and behave differently. The initial findings were not conclusive, but Gur's team eventually reported that white matter is more efficient for transmitting nerve signals in the female brain than in the male brain.[24] Still, other neuroscientists proclaimed that the known sex differences in the human brain, although small, were substantial enough to make men innately smarter than women. But some scholars, including feminist academics and a minority chorus of neuroscientists, argued that environmental conditioning best accounts for different behavioral tendencies observed in males and females starting at an early age.

The mystery of how male and female brains function—and the debate about whether their brains are fundamentally different—remained at a standstill until the 2010s, when novel findings emerged on brain wiring connections in men and women. The main catalyst for this shift was the connectome—a cutting edge map detailing the organization and integration of neural networks of the human brain. During the last decade, scientific explorations of the intricate interconnections of the brain have become so crucial for treating neurological disorders that the National Institutes of Health (NIH) in the United States created a consortium to address this issue. The consortium's overarching five-year mission is to create a map of human brain circuitry in 1,200 healthy adults. At present, multiple research teams are simultaneously conducting connectome studies to compute structural and functional connections in both male and female brains.[25]

Sorting out the intricacies of sex-based differences in the size of various structures and wire patterns in the brain is an arduous task, given the conflicting findings reported in the last few years. Some studies have not shown statistically significant male–female differences in cortical connectivity—the fibers that connect different areas of the cerebral hemispheres in the structural connectome. But other research has pointed to a female brain that houses more connections in the cerebral cortex—the brain's main thinking regions—compared with the male brain.[26] With improved imaging technology capable of detecting minute and subtle variations in the physical infrastructure of the human brain, experts have recently identified several overall anatomical differences consistently seen in the male versus the female brain. Significant sex differences have been reported in the proportion of fibers in the frontal lobe, with a greater number of connections occurring between the left and right sides of the upper frontal cortex in women. The frontal lobe regulates executive function, including such mental skills as self-control and speech.[27]

And yet, the female brains in this study had fewer fibers in the right lateral, orbitofrontal region, an area linked with decision-making.[28] Located in the frontal lobes, this region is a key part of the prefrontal cortex, lying just above the orbits, or eye sockets, where the eyes are located. Some scientists have speculated that there might be gender differences in the functions associated with the frontal area, along with the temporal lobe; however, skeptics have cautioned against a rush to judgment regarding male and female differences in cognition and behavior because the sample sizes in some of these studies are too small to produce definitive results.[29]

Compelling evidence of sex differences in the structural connectome of the human brain has come from Madhura Ingalhalikar, a neuropsychiatrist at the University of Pennsylvania. Ingalhalikar's team evaluated brain images of almost 1,000 youth using diffusion tensor imaging, an MRI-based technique that traces how water molecules spread to pinpoint white matter tracts in the brain. The participants, aged eight to 22 years, in this study were divided into three age groups—eight to 13.3 years, 13.4 to 17 years, and 17.1 to 22 years—corresponding to the developmental stages of childhood, adolescence, and young adulthood. When the researchers examined the cerebrum—the largest mass of the brain, which governs voluntary movements and coordinates mental actions—they found that the male participants had more connections between nerve pathways within the lobes of each hemisphere, or greater intrahemispheric connections. The male cerebrum also displayed a larger number of coherent network modules, or interconnected areas within the brain, as well as an increased ability to form connections with neighboring brain regions. Females exhibited greater interhemispheric connectivity, or more nerve fibers between hemispheres and across modules.[30]

These sex-linked patterns were reversed in the cerebellum—the area beneath the cerebrum, which coordinates voluntary movements, posture, and balance. Compared to females, males had stronger connections between the left hemisphere of the cerebellum and the right cerebral cortex. This piece of evidence marked a key turning point in research on structural differences in the connectome between the two sexes.[31] It demonstrated that the male human brain is organized by modular and discrete structures that help promote neural connections between perception and coordinated action. This feature may contribute to the overall better performance of males on motor and spatial tasks. In the female human brain, the larger number of connections across hemispheres seems to increase communication between analytical and intuitive processing areas of the brain. According to Ingalhalikar's team, this infrastructure affords the female human brain an advantage over the male brain in two specific areas: enhanced memory and social cognition. When the researchers compared the three groups of participants, they found that sex differences in brain connectivity became more prominent as the individuals, especially females, grew older.[32]

The brains of the male children revealed strong connections within each hemisphere, with the connections persisting through adolescence and young adulthood. But in female brains, connections between the hemispheres were dominant, especially during adolescence. The connectivity between hemispheres diminishes in adulthood because interhemispheric connections are not as strong as the connections within the same hemisphere. And yet, as the males and females in this study aged, they did not differ significantly in the degree to which their connections changed, either within a particular hemisphere or between hemispheres.[33] The sex differences in the brain observed in this study may mean that gender differences in behavior complement one another. The contrasting sex-linked patterns of brain wiring connections may reflect an evolutionary adaptation resulting from male and female differences in procreation and social roles. If Ingalhalikar's team is correct, sex-based variation in brain structure may contribute to biological underpinnings for physical aggression in males, as well as increased verbal-based memory and social intelligence in females.[34] The trends reported by some experts point to a potential cause-and-effect relationship, but definitive causality has not been established. Correlations, commonly called associations, are not automatically causal relationships. The evidence to date, then, does not unequivocally prove that the different brain connectivity patterns detected in male and females necessarily account for sex or gender behavioral differences.

DETANGLING THE GENDER-BASED CONNECTOME

In the last two decades, research on brain and behavioral differences in males and females has become increasingly nuanced, partly because of the novel perspectives that women scientists themselves are bringing to the laboratory. Still, men and women (perhaps like the brains they possess) cannot inevitably be placed in discrete buckets. In response to Ingalhalikar's interpretation of the female versus male connectome, University of Melbourne psychology professor Cordelia Fine questioned whether this study generated "new insights into gendered brain wiring" or, instead, was simply a "perfect case study in 'neurosexism.'" Fine doubted that the evidence presented by Ingalhalikar's team proved that hardwired differences exist in the male versus the female brain, despite popular misconceptions of cutting-edge neurological research.

For Fine, Ingalhalikar's data were noteworthy, but they failed to demonstrate concrete, measurable associations between male versus female brain connectivity and behavior.[35] For starters, Fine reminded us, gender is a social phenomenon that may or may not correspond to biological sex. In virtually every human society, males and females are socially conditioned at a young age to act, think, and perceive the world in keeping with their assigned gender roles of being either a boy or girl. Do we have here another chicken-and-egg conundrum that is all too familiar in probes of the impact of nature versus nurture on brain, biology, and behavior in humans? If nurture advocates of sex differentiation in the brain are right, then the environment counts as much as—or perhaps more than—nature in shaping male and female behavior. As champions of nurture point out, most young people, commencing in early childhood, are exposed to physical, emotional, mental, and social experiences considered appropriate for their biological sex. From their proenvironmental vantage, the "sex-related" behaviors they develop as they mature into adulthood are not necessarily genetically engrained in their brains. Instead, stereotypic male and female behavior could just as easily be determined largely by social roles—behavioral expectations based on assigned gendered roles.[36]

As Fine explained, some of the purported differences in male and female brains have more to do with brain size than brain sex. Male brains are usually larger than female brains, but a bigger brain is not simply a scaled-up version of a smaller brain. Just as there are engineering considerations when designing a structure in a smaller or larger space, larger brains and smaller brains each have different requirements for meeting energy demands, wiring efficiency, and communication speed. Based on the available evidence, Fine concluded that the structural network differences in the male and female human brain represent alternative wiring solutions to brains of different sizes rather than biological sex. She challenged the accuracy of sex-and-brain stereotypes—notions that female brains are hardwired for enhanced social skills and remembering chatter, whereas male brains are superior for map reading.[37]

In one study, scientists found only minor and statistically insignificant sex differences in such psychological skills as executive control, memory, reasoning, spatial processing, sensory-motor abilities, and social cognition.[38] For Fine, many of the academic and popular narratives on sex differences in brain wiring as fundamentally innate, and therefore as the source of gender differences in behavior, are unproven conjecture that merely reinforces gender stereotypes. Few, if any, studies on sex differences in brain developmental traits, particularly wiring, adequately control for the effects of environmental factors that influence the plastic or, more precisely, "elastic," brain. From Fine's perspective, social phenomena alone, especially socialization experiences, are major contributors to the development of the brain and the formation of individual behaviors.[39]

Today, one of the most pressing demands for researchers is to understand how neuroplasticity—the pliability of the brain and CNS—helps shape the formation of distinct male and female brains. But there is no consensus about the extent to which variation in male and female executive function—including decision-making, emotions, and social roles—reflects genetically determined brain patterns versus environmental and experiential influences.[40] Some researchers claim there are sex-based differences in brain biology, possibly with functional and therefore social ramifications. Other scholars embrace biology, without assuming that genetics is the dominant, let alone final, arbitrator in the nature-versus-nurture dynamics of brain development. This revamped perspective may seem wishy-washy to skeptics whose allegiance lies with hard-core genetic determinism. Nonetheless, this evolving view of the interface between neurogenetics and social behavior is peppered with admonishments about human sociality—and ways to study it. As Fine put it, like other life outcomes that vary between individuals and between populations, gender-linked behaviors stem from a person's developmental path. That developmental trajectory is forged from a continual and dynamic interplay between brain, genetics, and environment.[41]

The developmental path that Fine mentions is also determined by the interactive influence of hormones and both the biotic and social milieu. The environmental components that affect biological development, particularly regarding the brain, range from daily nutrition and personal hygiene to ecological considerations, for example, air and water quality. They span a gamut of influences, ranging from immediate personal experiences to social interactions with family, friends, coworkers, and strangers. To ignore the indisputably powerful effects of the multifaceted environment on the development of the human brain is to dismiss one of the most remarkable aspects of the human condition. No one doubts that socialization influences gender-based behaviors; however, biological factors determine not only the sex of the unborn fetus in the womb, but also possibly gender-linked behavior in young children. Genetic differences in the budding sex organs of the growing fetus result in sex hormones that, in turn, trigger anatomical traits leading to either a male or female body, including distinct differences in sex organs and perhaps the brain.[42]

The amount of prenatal testosterone circulating in the womb may impact sexual orientation and gender identity, as well as certain sex-related cognitive, motor, and personality traits.[43] Although prenatal testosterone is not responsible for all sex-linked personality differences, this steroid hormone is strongly associated with certain psychological and social traits. For instance, females tend to have greater empathy, whereas males display higher levels of physical aggression, a trait correlating strongly with testosterone activity. Testosterone affects specific receptors in various regions of the brain. This male sex hormone may help guide the structure and some of the wiring of the brain as an individual passes through life stages at different ages.[44]

In one study that examined whether prenatal exposure to testosterone and other male hormones influences sex preferences for toys in children's play, the results were both expected and surprising. The findings suggested that the stereotype of boys preferring blue and girls adoring pink might be largely a myth. But the long-held belief, at least in the United States and many other Western industrialized societies, that boys are inclined to play with objects like trucks, whereas girls choose dolls, did not appear inaccurate. When the investigators evaluated children's preferences in such play behaviors as rough-and-tumble games, they found sex differences in psychological traits, for instance, core gender identity, sexual orientation, and childhood play in cognition or personality. Male–female variation in play-related behaviors was even larger than sex difference in height.[45]

These results lent credibility to the claim that gendered behavior stems largely from built-in biology. Brain wiring patterns established before adulthood may contribute to boys' and girls' selection of sex-typed toys.[46] The sex gap in childhood play widens from the ages of two and a half to five years, a stage of early childhood when kids are socialized through learning. This is also a time when cognitive developmental processes are molded to fit gender roles. No serious expert denies the role of environment, especially social and cultural conditioning, in shaping play attitudes and behavior. But multiple lines of research—on genetic diseases, pregnant or breastfeeding women treated with hormones, and hormonal variation in normal individuals—indicate that prenatal testosterone levels also influence childhood preferences of sex-typed toys, playmates, and play activities. Girls exposed to excessively high levels of androgens—male sex hormones—in the womb tend to exhibit play behavior that increases in the male-type and decreases in the female-type. Similar findings have been reported for children whose mothers were prescribed androgenic progestins while pregnant. Progestins are a type of birth control pills that have both estrogen and male hormonal effects. Reversed effects occurred in children whose mothers received antiandrogenic progestins, a class of drugs that can dull the effects of testosterone.[47]

Despite this evidence, scientists are still not sure why boys have sex-typed preferences for certain toys typically linked with maleness. Sex differences have been detected in toy preferences based on the shape or color of the play objects, but boys are seemingly more attracted to toys designed to be mobile in space. Prenatal exposure to testosterone appears to affect the visual system. If so, then higher levels of testosterone might be linked with a pleasurable experience that comes from watching objects move in space. Research on the connections between brain structures and specific mental operations or specific social behaviors is evolving, but notions about inborn sex differences in personality and behavior persist in popular culture. Amid the righteous call for gender equality in society, some of the cultural stereotypes surrounding the mental styles and behavioral tendencies of males and females are, ironically, not unfounded. Indeed, several cause-and-effect relationships between brain anatomy, neural connections, activation patterns, and behavior seem to be reasonably well established. Nonetheless, the biological underpinnings of sex differences in brain structures are not necessarily, and certainly not entirely, due to genetics.

External environmental factors like birth control pills can alter the balance between sex steroid hormones in a female's body, potentially influencing sexual orientation and some behaviors. And changing levels of sex steroids, for example, estrogen or testosterone, can affect a person's performance on cognitive tests for verbal fluency and mental rotation. As multiple studies have shown, females tend to earn higher scores on verbal fluency tasks but lower scores on mental rotation tasks, with the outcomes reversed in males. In some studies, women perform better on verbal fluency during the phases of their menstrual cycle, when levels of the female hormones estrogen and progesterone are highest. Also, their visual-spatial skills, required for mental rotation tasks, may be improved when levels of female sex hormones are lower.[48]

Still, there are contradictions in the scientific studies in this field generated in the last 30 years. From one vantage point, the evidence overall suggests that some human sex differences in behavior, or at least the potential for certain male–female anatomical differences linked with social traits, are hardwired. Consider the hippocampus and amygdala, two structures located deep within the limbic system involved in learning and memory.[49] In some studies, females were found to have a smaller amygdala and hippocampus compared to males, partly contradicting previous results showing no sex differences in the volume of the hippocampus.[50] The limbic system comprises a group of brain structures that govern such basic drives as motivation, as well as emotions like fear, anger, and pleasure. The amygdala and hippocampus help regulate feelings and behavior, and both structures have a large number of sex hormone receptors.[51] This may partly explain why males and females exhibit different patterns of electrical brain activation in the hippocampus when they perform tasks that rely on memory, one of the key functions of the hippocampus.

Males and females tend to use different brain structures when they engage in various cognitive functions, especially memory tasks. In some studies, women, when compared to men, have a thicker cortex due to a proportionately larger amount of gray matter, as well as more complex arrangements of white matter.[52] Or perhaps the reported sex-based functional differences in brain traits might simply be a reflection of females using their white matter—fiber tracts connecting different parts of the brain—more efficiently compared with men.[53] Additionally, the frontal and parietal regions of the cerebral cortex in women contain more extensive folding than in the male brain, increasing the surface area of an otherwise smaller brain.[54] Even before the fetal stage begins in the womb, the cortex of the embryo's still-forming brain starts developing folds with characteristic peaks, or gyri, and troughs, or sulci, which give it a wrinkled exterior. This process of cortical folding, or gyrification, has multiple benefits beyond simply expanding the total surface area of the cerebral cortex. Cortical folding also enhances the brain's functional organization, optimizes its wiring, and allows an oversized cortex to fit in the limited space of the cranium, the part of the skull encasing the brain.[55] Sex-based differences in some, although not all, key features of brain anatomy suggest that multiple pathways within the human brain can be accessed for memory, as well as spatial intelligence and various other cognitive tasks.

Some of the sex-linked anatomical variations in the brain could be related to functional differences that theoretically could translate into behavioral differences between males and females. For example, the hippocampus, known for its involvement in memory, also plays a paramount role in spatial perception. Early studies overwhelmingly suggested that compared to females, males had a larger hippocampus, which might give them a boost in spatial awareness. But other investigations have reported either a larger hippocampus in women or no sex differences in the size of this structure.[56] And compared to men, women have a slightly thicker cerebral cortex, a trait associated with higher intelligence, or at least better performance on cognitive tests in some studies.[57] Yet, links between cortical thickness and IQ can change with age in both sexes, a finding that may partially explain the inconsistent findings reported in this field.

Female brains are proportionately smaller than male brains, but in some studies they exhibit more gyri—ridges surrounded by fissures—on the surface of the cerebral cortex, especially in the frontal and parietal lobes. The corpus callosum is the thick band of nerves that divides the cerebrum into the right and left hemispheres. It is significantly larger in women, particularly in young adult females, leading some researchers to associate it with verbal fluency and (at the risk of stereotyping) "female-type" cognition.[58] To some extent, the rationale for equating "female-type intelligence" with talkativeness is analogous to the reasoning behind linking "superior spatial perception" of males with a relatively larger hippocampus.

Male and female brains indeed differ in several key regions involved in spatial reasoning and motion sensing. Certain visual areas that are activated when a person processes visual input or detects motion are larger in males than in females. And in men's brains, both the volume and surface area of the occipital lobe, located in the back of the head, are more asymmetric, with the right lobe larger than the left.[59] These features of the occipital cortex may be part of the biological underpinnings of movement and spatial perception in humans. These characteristics of the occipital lobe may influence the ability of males to generally outperform females on visual-spatial tasks requiring mental rotation.[60]

In some studies, a thicker cerebral cortex, a trait more common in women, predicts a better IQ score, especially in the left cerebral hemisphere, between about 21 and 29 years of age. But in other research, just the opposite—cortical thinning—has been linked to higher intelligence, particularly in the left hemisphere during childhood.[61] This paradox is partly explained by different rates of growth and atrophy in both the volume and surface area of the cortex as a person ages. Changes in cortical thickness—the depth of the cerebral cortex—are correlated with intelligence throughout a person's life-span, but the association between cortical thickness and intelligence may vary at different ages.[62] In one study, 10-year-old children with higher IQ scores had a slightly thinner cortex than children with a lower IQ. The correlation became stronger as the study participants grew older, with faster cortical thinning occurring in higher-IQ people throughout time. This trend was eventually reversed in highly intelligent young adults, such that by age 42, the thicker the cortex, the greater the intelligence in the higher-IQ adults.[63] This may mean that intelligence is influenced less by brain structure per se and more by the magnitude and timing of developmental changes in brain structure. Developmental changes are never fully finalized but instead are ongoing and closely intertwined with neuroplasticity.[64] In the final analysis, correlations between the male or female brain and a trait like cortical thickness may depend on a person's age and, perhaps more importantly, personality.[65]

In young people, eight to 25 years of age, higher testosterone levels correlated with a smaller volume of gray matter in the anterior cingulate cortex (ACC) of the males but not the females. Higher levels of testosterone in males have also been associated with cortical thinning in the ACC.[66]

Researchers found that six-year-old to eight-year-old American boys with aggressive and defiant behavior had an overall smaller ACC volume and more extensive cortical thinness than boys without these unpalatable personality traits. The scientists speculated that in males, a smaller ACC, combined with increased levels of testosterone, contributes to competitive and dominant behaviors that often surface in males during adolescence.[67]

NAVIGATING THE GENDER NEURONAL DIVIDE

Given the intricate intersectionality between biology and social phenomena in shaping male and female roles, can we truly tell if sex-based differences in brain anatomy, including wiring patterns, are determined more by genes or environment? Are we back where we started? Not necessarily. The plasticity of the brain offers insight into the enigmatic process of nature plus nurture. In biology, plasticity refers to the quality of being sufficiently pliable, flexible, or transformable into a different form, possibly with different functions. Neuroplasticity is the brain's ability to continually reorganize its structure and function by forming new connections between brain cells. Neuroplasticity may take the form of novel and transformative brain wiring patterns or new neural connections between nerve cells either within the same region or between different regions of the brain. Because the structure of the human brain is dynamic rather than static, neuroplasticity becomes a likely and even predictable response of the brain to the ever-changing internal and external experiences throughout a person's lifespan.[68]

Today, a wealth of evidence for the power of experience in modifying the brain's hardwired templates to produce adaptive, neuroplastic softwiring is enabling scientists to fathom the influence of nature plus nurture of sex-linked neural differences. In the process, scientists are honing in on the impact of these differences on gender roles, at least on a case-by-case basis. No single formula fits all brain activity for all social behaviors in all people. As the findings reveal, rough-and-tumble play, aggression, empathy, verbal communication, and spatial perception do not inevitably have equivalent neural underpinnings, let alone the same or even similar environmental influences, in different people, nor in the two sexes.

It remains to be seen which social behaviors are linked with comparable brain processes and which are not, in both males and females. Hardwiring—inherited genetic tendencies—may be dominant in some psychological and social traits, whereas softwiring—the effects of environmental triggers combined with experience and perhaps epigenetics—may show the strongest influence on personality traits. The relative impact of genes versus environment plus experience on attitudes and behaviors, for example, a child's response to toys, may differ not only for each sex. Depending on their strength, influences that are neither purely genetic nor purely environmental—including hormones, epigenetics, and neuroplasticity—seem to exert varying effects on individuals from different cultures or at different life stages, or both.

New maps of the neural territory of the gendered brain illustrate the rich complexity of humankind's three-pound inner galaxy. The links between the brain's structures and its multifold functions reveal an organ incomparably unique in its chameleon quality. With the brain's ability to potentially adapt itself to its surroundings and experiences, it defies descriptions based on either sex or gender alone. Whether an analogous case can be made for ethnic population differences in brain traits—including total volume, quantity of gray matter and white matter, and cortical folding, along with functional differences presumably linked with brain traits—is another issue. This issue takes us into the core of the nature–nurture debate concerning racial, or ethnic population, differences in cognitive performance, social behaviors, and life outcomes. The next four chapters address this topic from multiple perspectives.

The Racialized Brain

INTELLIGENCE: HUMANITY'S HALLMARK

INTELLIGENCE HAS MULTIPLE DEFINITIONS, BUT ITS MOST OBVIOUS ATTRI-butes include cognitive skills for problem solving, future planning, and inductive reasoning. Intelligence is often equated with such mental processes as memory, spatial awareness, and the ability to manipulate symbols for communicating and making mathematical calculations. IQ, or the intelligence quotient, refers to the score a person earns on various standardized tests designed to measure a variety of mental traits[1]; however, an IQ score is not necessarily a precise indicator of a person's innate cognitive ability, even though it is a reasonable assessment of his "intelligence performance"—his demonstrated achievement on a test to evaluate his mental skills. An IQ score is, to some extent, influenced by an individual's inborn mental capacity, but it is also determined by his existing knowledge and level of schooling, and, equally importantly, his level of motivation and attitude while taking the test. In academic settings, apathetic and uncooperative students tend to earn lower IQ scores than their intellectually engaged and proactive, age-matched counterparts.[2]

Conventional Western IQ tests are intended to quantify two aspects of intelligence—g, a mental trait called "general intelligence" and s, or specific intelligence—the knowledge required to perform particular tasks requiring various types of mental skills. But the main component of standardized IQ tests is g—a set of intellectual and cognitive abilities intertwined with abstract thinking, analytical faculties, and, in some cases, accelerated intellectual operations. According to some experts, g correlates with the type of thinking that famed twentieth-century Swiss psychologist Jean Piaget, known for his pioneering work in child development, called formal operations.[3] To some extent, g is a reflection of social, cultural, and mental cues needed to function efficiently in the industrialized world, particularly technologically advanced urban societies. For psychometricians—specialists in developing and administering intelligence tests—an IQ score is a meaningful measure of an individual's ability to solve a wide array of cognitive challenges in the real world. A high g score earned on an IQ test means that a person has a keen potential ability to formulate mental abstractions—ideas created and held in the mind—using one or more sets of written symbols.[4]

In the Western world, traditional IQ tests are structured to assess verbal comprehension (including reading, writing, and communication skills), working memory, perceptual reasoning, and processing speed. Many psychometricians are hereditarians, or genetic determinists, who claim that a person's aptitude for earning a certain number of points for g is innate, or coded in their DNA and hardwired into their brains.[5] But for pronurture critics skeptical of a predominantly pronature view of intelligence, learning and environmental conditioning play a key role, perhaps a more important role, in determining a person's performance on each component of an IQ test.[6]

For almost 100 years since the introduction of IQ testing throughout Western society, the overall IQ reported in numerous nonwhite groups has been lower than the norm for European whites, including white Americans who historically have had an overall IQ of 100.[7] Since the early twentieth century, the IQ scores recorded for African Americans have averaged about 85 points in many reports, with some studies pointing to a score hovering around 70 for native-born African blacks in some African nations.[8] IQ scores can vary substantially across ethnic populations, sometimes differing even in genetically close groups, and scores can change throughout time within the same population. Reported average IQ scores are 85 to 95 for Southeast Asians and Native Americans, 85 for Australian Aborigines, and 85 to 88 for Pacific Islanders. East Asian populations (especially people of Japanese and Chinese descent) tend to have mean IQs well above 100 and as high as 110 in some cases.[9] The overall IQ score of Ashkenazi Jews—the Jewish diaspora that spread throughout various parts of Europe—is about 10 points higher than that of other white ethnic groups.[10] (The ancestors of modern-day Ashkenazi Jews migrated from the Middle East to Eastern Europe between the first and seventh century CE, initially settling in Rome and Southern and

Eastern Europe. Today, Ashkenazi Jews and their descendants comprise a diverse population living primarily in the United States, Israel, Russia, Canada, and several European nations.)[11]

The above-average IQs linked with both East Asians and Ashkenazi Jews have been the source of intense debate in some scientific circles. East Asians, in general, seem to score higher on the nonverbal, including mathematical, components of IQ tests. Ashkenazi Jews tend to excel on both the verbal and mathematical sections of intelligence tests, with statistically significantly higher scores compared to an American mean IQ score of 100 points. Some studies on Ashkenazi Jewish intelligence have reported mean IQs of 107.5 and 110, whereas other research has suggested ranges of 107 to 115, 108 to 115, and 112 to 115. Average IQ scores vary for Ashkenazi Jews in Britain and the United States as opposed to Israeli Jews, with lower scores recorded in Israel.[12] Fueling further controversy, some reports suggest that Ashkenazi "visual-spatial" IQ scores are not superior but instead tend to be lower than the average IQ for Europeans.[13]

Interest in the Ashkenazi IQ in the United States and Britain has burgeoned in the last decade, but much of the academic and public curiosity about ethnic differences in overall IQ revolves around the widely cited (but not universally accepted) 15- to 18-point gap between African Americans and white Americans.[14] For hereditarians, this ethnic gap in IQ is tethered to inborn race-based aptitudes.[15] The brouhaha surrounding this disparity has been tackled from numerous angles and is intimately interwoven with one of the classic nature-versus-nurture debates: Are black/white (and black versus other ethnic) differences in life outcomes due to predominantly, or at least substantially, population or race-based genetics, or the environment, notably experience?

A 15-point divide in IQ scores is statistically significant and corresponds to approximately one standard deviation of the median IQ score of Americans; however, experts are still at odds about the actual size of the IQ gulf between African Americans and white Americans today. Some hereditarians have reported an 18-point discrepancy, with 1.1 standard deviation between these two ethnic populations.[16] And the more than 30-point divide between white Americans and Sub-Saharan black Africans mentioned in some publications is astoundingly large.[17]

However, the evidence for a gap of this magnitude is questionable. In some cases, assertions about the overall IQ of black Africans may be based on racially biased and faulty research methods.[18] If recent counterfindings are any indication, the black African IQ varies widely, possibly averaging about 82 points.[19] More importantly, cognitive performance has been increasing in several countries on the African continent, especially nations that have experienced recent improvements in economic development and health. This trend is consistent with the rise in the average IQs of ethnic populations of the developing world, where educational opportunities and the standard of living have gradually improved in some regions.[20] A similar narrowing of the ethnic IQ gap in the United States has been seen in some studies, albeit not the majority. Still, the fact that ethnic discrepancies in IQ have diminished at all at any point in time is noteworthy and, as leading proenvironmental researchers argue, shows that the black/white IQ divide cannot be significantly explained by population genetics.

In 2006, famed intelligence researchers William Dickens and James Flynn found that the black/white IQ gap in the United States had shrunk to 9.5 points. Although the gap remained wide, it was notably less than the previously reported divide of 15 points.[21] Other world-renowned intelligence scholars, including Eric Turkheimer, Kathryn Paige Harden, and Richard E. Nisbett, have echoed similar views, asserting that nongenetic factors can account for much (or, as Nisbett suggests, all) of the IQ gap between blacks and whites.[22]

The positions of both experts and the public on this issue remain polarized, invariably highlighted by contrasts in black and white, figuratively and literally. Even academics that remain agnostic on this issue—those who consider environment and genetics almost equally accountable for ethnic disparities in IQ—are usually relegated to the hereditarian camp, although not necessarily to a hard and fast genetic determinism. In the early 2000s, researchers discovered that there were practically no IQ gaps between American children from different ethnic groups who came from similar demographic backgrounds through their first three years of schooling.[23] But after early childhood, the ethnic divergence in IQ predicted by some intelligence specialists started to manifest throughout time. Thereafter, the ethnic IQ gulfs remained steadfast throughout various age groups, even among ethnically diverse, higher-income groups.[24]

The predominant causes for the ethnic IQ gap are still not entirely clear, but some evidence suggests that the gap in overall cognitive test performance of whites and blacks, as well as whites and other nonwhite groups, can be explained in entirety by environmental differences between these groups.[25] If intelligence is a complex phenomenon, then measuring and comparing it across different ethnic populations is surely a tortuous task. But it is an issue that deserves our utmost attention if only because it underlies one of the deepest sources of racial bias, both explicit and implicit, in the United States and elsewhere. For more than a century, the statistically significant differences reported in average IQ scores between whites and blacks, and Northeastern Asians, whites, Latinos, and blacks, have fueled a ferocious quarrel concerning alleged racial differences in mental aptitude, particularly higher cognitive abilities. For the past two decades, the continually revamped revival of the race and IQ debate has served as a springboard for an even more contentious controversy: the quarrel over whether population genetics lies at the root of ethnic and racial differences in social behavior, including aggression, violence, and heinous crimes.[26]

RACE AND IQ: THE HEREDITARIAN PLATFORM

In the last three decades, studies on race and IQ have opened the door to a related but distinct set of scientific investigations aimed at finding possible links between race-based genetics and antisocial behavior, especially with regard to people of African black descent. In the early 1970s, Arthur Jensen, a University of California professor and an expert in intelligence testing, became one of the principal, hereditarian architects of research on race and intelligence in the twentieth century. With an unyielding pronature stance, Jensen was revered by some and scorned by others. Although most Americans of this era eschewed ideas promoting race-based genetic determinism, some academics and members of the general public embraced Jensen for boldly vocalizing his view that blacks, including African Americans, are innately less intelligent than whites.

A huge amount of Jensen's body of scholarly work reflected his preoccupation with race, specifically black/white differences in cognitive performance, notably IQ test scores. As a prolific researcher in psychology, however, Jensen's single most important contribution to the field of intelligence testing may have been broadening the concept of *g*—general intelligence—which he claimed was a measurable attribute on IQ tests. As a psychometrician, he argued that *g* provides a practical and scientific definition of intelligence. For

Jensen, IQ tests were meaningful only if they had high levels of *g*-loading—a greater number of tasks that can only be solved by using complex mental transformations and sharper analytical abilities.[27]

Research on intelligence and IQ—on who has it and who lacks it—has rapidly proliferated during the last half-century. In some studies, a person's *g* score is a good predictor of his performance in school, higher educational institutions, the military, and employment settings.[28] Yet, some investigations have failed to show a strong correlation between *g* and adeptness in job performance, suggesting, instead, that emotional competence also influences a person's success in the workforce.[29] But the counterevidence supporting the merit of nonintellectual life outcomes mattered little to Jensen and his supporters. Along with some of his protégés, including racialist scholar J. Philippe Rushton, Jensen was convinced that blacks earn significantly lower scores than whites on "high-*g*-type IQ tests" because, he said, black Africans and their descendants lack the mental aptitude for complex problem solving. This idea had long preceded Jensen's meteoric rise in the academic world. But he sought to make this notion more scientifically credible when he attempted to explain the black/white gap on the *g* components of IQ tests in terms of Spearman's hypothesis. Formulated by hereditarian psychologist Charles Spearman in 1927, Spearman's hypothesis predicts that the black/white IQ divide will be largest on tests with more *g*-loading.[30] Jensen's endeavors to study *g*, particularly in relation to ethnic differences in intelligence tests, helped set the stage for present-day probes into how nature versus nurture shapes cognitive ability and even various social behaviors.

Like much of contemporary hereditarian research, Spearman's hypothesis revolves around the concept of heritability, a method for measuring the relative importance of hereditary and environmental influences on the development of a specific, quantifiable trait. As an aside, heritability is often used in animal breeding to help breeders select parents with the preferred genetic makeup for the next generation of domesticated animals, particularly livestock. If a breeder has information about the genetic relationships, or pedigree, of a ranch or farm animal, for instance, dairy cattle, he can compare this information with the animals' performance in, for example, milk yield and estimate the heritability for the milk production in his herd or population of cattle.[31] A heritability estimate provides a partial description of a single trait in one group of animals at a particular point in time. It indicates that the sum of the genetic differences related to a specific trait account for the variation in milk yield among the cows. A heritability estimate is expressed numerically as a percentage or decimal that lies between 0 and 1, with an estimate of at least 0.30 considered medium to high. A higher heritability estimate means that an increased amount of genetic variance contributes to a trait, whereas a lower heritability estimate corresponds to a decreased level of genetic differences linked with the trait.[32]

A key point here is that heritability is not synonymous with the terms *genetic inheritance*, *inherited*, or *heritable*. Recall that heritability estimates are not a direct measure of the quantity of genetic influence on a trait, nor do they represent fixed values. As biological anthropologist C. Loring Brace explained, heritability is a ratio between the genetic variance and the environmental variance within a particular population, with both types of variance contributing to an outward trait, whether physical or psychological. The variance occurs between members of the same population, such that heritability estimates can vary substantially. The heritability for one population might not hold up for another. And heritability can change under shifting circumstances, for instance during war or famine, when environmental constraints may take a heavier toll on a population than during normal times.[33]

Until recently, heritability estimates for "high-g-type IQ tests" were based almost exclusively on studies in whites, which, as some pronurture advocates note, may mean these estimates do not hold up for nonwhite populations, including blacks. As growing evidence reveals, variation in heritability estimates across different social groups can be partly explained by group differences in socioeconomic status, or SES. Regardless of a child's ethnicity, the demographic background of his family can impact the estimated IQ heritability of his social group. This suggests that genetics has less of an influence on the variance of IQ scores in a group of young children from lower SES groups than on youth from higher SES families.[34]

In Arthur Jensen's dual-level model of cognitive learning, level I IQ tests consist of rote, associative learning. Level II ability tests are designed to assess good problem-solving skills and enhanced academic performance because these tests presumably have more g-loading.[35] Using this model, Jensen observed a larger black/white gap on level II IQ tests than on level I performance tests, a finding that, he asserted, confirms Spearman's hypothesis.[36] But as Lee Cronbach, one of Jensen's rivals in the intelligence testing field, pointed out decades ago, Jensen's level I/level II system failed to take into account multiple learning abilities or environmental constraints. In the 1970s—well before experts realized that adverse environmental exposures could harm the fetus—a handful of researchers like Cronbach suspected that the prenatal environment can influence a person's IQ scores later in life.[37] Multiple interacting factors shape IQ, but there is still no consensus among experts on what constitutes robust abstract thinking. Nonetheless, hereditarians insist that g is a "sound and meaningful" index of cognitive ability. In reality, g may transcend a simplistic binary interpretation. As psychologist Alan Reifman put it, various types of "practical intelligence" that have "real-world outcomes" may coexist with g-type cognition. Ultimately, the issue may not be about the existence of g, but whether a person can function well in modern society without a "reasonably high g score."[38]

RACE, BRAIN, AND INTELLIGENCE: THE BIGGER PICTURE

Humankind is a species of big-brained creatures that have genuine bragging rights for unprecedented, dare we say, superior intelligence compared with (as far as we can presently tell) other living species on the planet. Influenced by a delicate balance between environmental constraints and natural selection, the gradual enlargement of the hominin brain, which eventually transformed into the human brain, was an epoch-making phenomenon. (Hominin is the taxonomic name for a group of primates encompassing humans and their direct ape-like ancestors.) The expansion and refinement of this brain was inarguably one of the most salient watershed events in the emergence of increasingly smarter species within the human lineage, if not *the* paramount watershed event. During the evolutionary trajectory that led to *Homo sapiens*, our current living species, the 3.2-million-year-old australopithecine, popularized as "Lucy," or at least her hominin brain, small that it was, symbolized a critical turning point in our species' biological history.[39]

This was the juncture where a human ancestor paved the way to becoming *Homo sapiens*, literally meaning "the sapient," or "wise human," otherwise known as modern humans. During this time span, the size of the brain in the hominin lineage increased from about 400 milliliters (mL) to an average of 1,350 mL.[40] (One mL is equal to about one cubic centimeter [cm^3]). A larger brain was an obvious asset for Darwinian fitness, namely success in survival, reproduction, and caring for the young. Still, there are limits to the adaptive advantages of a larger brain because the human brain is a metabolically expensive organ. It comprises only 2 percent of the body's mass but requires a considerable amount of oxygen to thrive, using almost 20 percent of the body's fuel.[41]

For more than 150 years, naturalists have been fascinated with not only the brains of different animal species, but also the idea, once accepted and now contested, that different human races are endowed with brains of varying sizes, shapes, and capacities for cognition. Philosophers and biologists alike have explored possible connections between race, intelligence, and the capacity for morality and cultural achievement. As noted in chapter 1, even before Charles Darwin's unprecedented works *Origin of the Species* and *Descent of Man* were published, nineteenth-century American physician and naturalist Samuel Morton studied the crania, or braincases, of human populations by comparing the size of representative skulls of various racial groups. Morton estimated the volume needed to fill the braincase of each specimen of his sample and then, based on this information, calculated an approximate size of the brain. He concluded that whites and Northeastern Asians had the largest cranial capacity, followed by Native Americans, Malays, and ancient Caucasians.[42]

Of all the populations represented in Morton's survey, African blacks displayed the smallest brain volume, leading him to consign African blacks to the bottom of his hierarchy of humanity. Morton's race-based gradient of human intelligence was used to support and reinforce the prevailing racial sentiment of the Victorian Era: the notion—one still entrenched in some sectors of Western society today—that black Africans were intellectually and socially inferior, even subhuman, when compared to virtually every other major living human race. Morton's findings were conveniently leveraged as an intellectual defense of American enslavement of African blacks. It provided white plantation owners and those who endorsed or condoned enslavement with a rationale for why Europeans were justified in maintaining colonial dominance over the world's darker skinned populations—the peoples of Africa, India, Latin America, and the South Pacific.[43]

In the decades that followed, Morton's research drew disapproval, even condemnation, for its alleged fraudulence. Some of Morton's critics suspected that his studies were motivated by an unscrupulous need to prove the superiority of the "Caucasian race." For his supporters, however, he was simply conducting an experiment to objectively measure and compare the brain size of diverse human populations worldwide. Brain scans and other advanced medical technology for examining brain tissue would not appear until more than a century after the peak of Morton's career in the mid-1800s. The racial gradient that Morton proposed—a hierarchy eagerly embraced by proslavery Southerners of the antebellum United States—was debated repeatedly for decades and well into the twentieth century.[44]

In 1981, iconic Harvard paleobiologist Stephen Jay Gould reexamined Morton's data and presented his analysis in *The Mismeasure of Man*. Still considered one of the premier critiques of racial science, *The Mismeasure of Man* was one of the first proenvironmental, antiracist books written for both scholars and lay readers to refute racial superiority on scientific grounds. To the delight of pronurture champions who opposed genetic determinism, Gould claimed he had found evidence that Morton had doctored his findings in an attempt to prove that whites have bigger brains than other groups, notably African blacks. If Gould were correct, then Morton's racialist ranking of brains by size was the result of a sampling error—a biased selection of the skulls evaluated in his study; however, Gould did not actually measure the physical skulls, but instead reanalyzed the data that Morton had originally reported more than 130 years earlier.[45]

Many proenvironmental, pronurture, and antiracist scholars were delighted with Gould's evaluation of Morton's results, but their cerebral glee was short-lived. In 1988, John Michael, at that time an undergraduate (who incidentally never pursued a formal career in academia), published his own reassessment of Morton's data. Michael uncovered fewer errors in Morton's appraisal than Gould had identified, a result that at once seemed to buttress claims of inborn race-based differences in brain size. Michael reportedly had no a priori agenda but was simply presenting the calculations from his own independent analysis.[46]

Many scholars on both sides of the nature–nurture tussle regarding race and brain size were annoyed, if not disturbed, that Gould—at that time one of the world's most brilliant evolutionary biologists—refused to acknowledge his own mistake. Gould's detractors in mainstream academia charged him with sheer political mischief. But this did not preclude hereditarians, race-oriented sociobiologists, and their racialist supporters from scolding Gould for his own mistakes. To the chagrin of proenvironmental writers, who cringed at even faint whispers of race-based differences in brain size, Michael's discovery was a revelation, although an unwelcomed one. It demonstrated that racialist researchers are not the only ones who can make scientific blunders—and serious ones at that. The controversy erupted throughout the scholarly community.

This was not the end of the Morton 2.0 saga. Fast forward 25 years, and the case for Morton's race-based comparisons of skulls was reopened once again—this time by a team of scientists led by Stanford University anthropologist Jason Lewis.[47] These researchers reanalyzed the physical skulls that had sparked the quarrel, meticulously examining Morton's original specimens still housed in the University of Pennsylvania's Museum of Archaeology and Anthropology in Philadelphia. Taking care to avoid risks of bias, perceived or real, Lewis's team was blinded regarding the ethnic populations, or racial groups, to which the skulls belonged. Although these researchers did not measure each individual specimen in the full collection of skulls, the findings they collected were, alas, consistent with Morton's claims. The reevaluation conducted by Lewis and his associates suggested that Morton did not falsify his findings. Instead, from their vantage, Gould appeared to be the one who had erred, whether wittingly or unwittingly, in his calculations of the average skull size of each population.[48]

As an aside, in the late 1980s, John Michael, not surprisingly, was initially celebrated on "hate-filled white supremacist" websites, while simultaneously being discredited by antihereditarian and antiracist scholars. Michael himself, who years earlier had exited academia for a profession in environmental land planning, offered a more tempered and rational outlook on the matter. In response to claims that Gould was motivated by an unconscious bias against any endeavor with overtones of racial science, Michael attributed Gould's errors to "poor scholarship." As Michael opined in a post published in 2012, the evidence he uncovered did not support assertions that Morton's research was tainted by unconscious racial or ideological prejudice. According to Michael, both Morton and Gould were handicapped by a similar confirmation bias.[49]

As a pivotal issue in scientific research today, confirmation bias refers to a person's tendency to interpret new information in a manner that is consistent with his existing beliefs, perspectives, or hypotheses. It is not an objective and unprejudiced approach to evaluating data free from one's personal outlook.[50] Reassessments of Gould's critique of Morton, first published in 1998 and then in 2011, seem to have indicated Morton. With the tables turned and the weight of these scientific mishaps landing firmly in the lap of the pronurture camp, the updated appraisals of Morton's analysis have activated a reset button in the entire field of race, brain size, and intelligence. In some sectors of society, the post-Gould reanalysis of Morton's original findings has invigorated hereditarians, biological determinists, racial scientists, and eugenicists. It has armed folk in these allied pronature camps and, to some extent, soft race-realists who shun hard and fast labels with yet another set of data to buttress their claims that whites and Northeast Asians possess more massive brains than the brains of other continental populations.

Some of the statistics on the brain volumes of specific ethnic populations may have changed, but the perspectives of race-oriented hereditarians and their ideological kin have remained essentially unaltered for almost 150 years. Racialist, genetic determinists continue to subscribe to a narrative that basically says the world's larger-brained populations—whites and Northeast Asians—are more cognitively evolved and, therefore, innately smarter than everyone else.[51] For most racialist hereditarians, the world's fair-skinned populations—people whose ancestors originally migrated to and settled in the world's temperate zones starting about 140,000 years ago—are distinct. By virtue of their evolutionary history, hereditarians proclaim, European whites are born with intrinsically superior mental faculties compared to the descendants of early humans who never ventured far beyond the tropics and subtropics. In this scenario, often drafted in the language of sociobiology, Nature endowed populations in temperate climates with larger braincases for increased intelligence during the evolution of human races. Heightened cognition, according to this narrative, allowed the descendants of Europeans and Northeastern Asians to surmount the severe ecological pressures they faced from harsh climatic environments, for example, the Nordic terrain.[52]

As the author of numerous scientific articles on racial differences, Canadian psychologist J. Philippe Rushton was instrumental in ushering in the late twentieth-century upsurge in scientific racism. His publications helped lay the groundwork for many of the beliefs about race and intelligence, and race and social behavior, that persist today among racialists, race-realists, white nationalists, and white supremacists. For Rushton, brain size signaled a key element of an evolutionary package of life history traits, which, he claimed, varies along racial lines. In his now-largely-debunked model of the evolution of human races, Rushton proposed that fairer-skinned Northeast Asians, along with European whites, evolved bigger brains because they lived in more stable environments. He said that for ethnic populations inhabiting cold, temperate regions, unborn fetuses could spend a longer period of time in the womb, allowing their brains to enlarge before birth, compared to brains of people inhabiting warm climates.[53]

In Rushton's model, increased brain growth was tied to slower maturation, or physical and sexual development, in human babies. Yet, the evidence he cited did not take into account ethnic disparities in nutritional and health status. Environmental factors, including divergent living conditions, can partially explain ethnic population variation in the speed of physical maturation. The potential role of genetics cannot be discounted, but environmental factors are major contributors to the stark population gaps, including black/white/Asian differences, in the rates of infant survival and individual longevity.[54] Rushton maintained that the ancestors of modern whites in the temperate zones faced enhanced cognitive challenges for survival during extended harsh, cold winters. Like the racial scholars of the nineteenth and early twentieth centuries who preceded him, Rushton assumed that human populations that evolved in temperate regions had enhanced mental acumen. These factors, Rushton asserted, would have afforded the ancestors of modern-day whites in Europe an edge for efficiently collecting and storing food, building shelters, manufacturing clothes, and raising children who inherited improved odds for surviving to reproductive age.[55]

During the last few decades, pronature scholars have disagreed among themselves about the effects of the temperate environments of Asia and Europe on the evolution of human populations. Rushton asserted that the severe winters of Europe and Northern Asia would have created more difficult environments for human habitation than those of Africa. Like other commentators whose uninformed stereotypes at times seem to skew their scientific objectivity, Rushton viewed the entire terrain of the vast continent of Africa as if it were homogenous. But as C. Loring Brace pointed out, Rushton's conclusion that the North African climate was more predictable than that of Europe lacked solid support. For Brace, predictable food supplies could have made the temperate habitats of Europe more hospitable to occupy compared to some areas of Sub-Saharan Africa.[56]

RACE AND HEAD SIZE

Scientific findings clearly show that different human populations vary in their overall brain size, but the causes of that variation—nature or nurture—is in dispute. Furthermore, scientific perspectives on brain volume across racial groups may be influenced by changing social climates. As biological anthropologist Leonard Lieberman argued, prior to the mid-twentieth century, most racial scientists believed that whites had the largest average brain size of the human populations. And with the economic expansion of Japan following World War II, some began to counter the long-held assumption that European whites dominated the world's nonwhite peoples, notably Northeast Asians. Faced with the reality of postwar events, academics and intellectuals were reluctant to fully embrace the traditional racial hierarchy of the Western world at this time.[57] Lieberman noted that by the mid-1970s, some scientists reported higher IQs in Asians than in whites. And in the early 1980s, the brain of one Mongolian group—the Buriats of Siberia—was found to weigh more than the average European brain. The notion of larger and heavier brains in Asian populations was not an entirely novel concept. In the early twentieth century, pioneer anthropologists like Franz Boas and Theodore Kroeber conducted research revealing that the brains of some Asian populations weighed more than the brains of other populations. For Boas, this was not a matter of racial superiority or inferiority. But it was for Rushton.[58]

As Lieberman explained in a meticulous deconstruction of Rushton's claims about race and brain size, in the reinvented racialist scholarship of the 1980s, Caucasoid brains shrunk, whereas Mongoloid brains expanded.[59] With the rise of Rushton's racialist revival during this period, whites and Asians traded places on the acme of a new racial hierarchy. But black Africans remained confined to their nineteenth-century position near the bottom rung of a new race-based ladder of intellect, morality, and, indeed, social worth.[60] With this as his core message, Rushton emerged in the mid-1990s as one of the world's most vocal proponents of race-based genetic determinism. One of his trademark claims was that black African populations overall have smaller brain sizes compared with the brains of European whites and Northeast Asians.[61]

Today, MRI studies are commonly used for estimating brain size in individuals and ethnic populations. But MRI research on brain volume was a young science in the mid-1990s when Rushton and his colleague, C. Davison Ankney, published an article based partly on the U.S. National Collaborative Perinatal Project, a study on maternal and child health.[62] This landmark prospective study originally collected data on 48,000 pregnant women between 1959 and 1965, and documented factors influencing pregnancy and birth outcomes of almost 55,000 children born during the study period.[63] Rushton and Ankney's analysis showed that the African American children in the National Collaborative Perinatal Project overall had the smallest head size, lowest average IQ, and largest overall body size. At seven years, East Asian children had the largest average head size and highest average IQ score. The average head size and overall IQ score for seven-year-old white youngsters fell in the middle between those of the blacks and whites.[64] For Ruston, these discrepancies stemmed partly from innate, race-based biological differences. Yet, head size, when used as a proxy for brain volume, may be misleading because head size is influenced by multiple health factors that affect physical development.

Environmental influences are considered sufficient to account for ethnic population disparities in health without resorting to genetic explanations. The U.S. National Collaborative Perinatal Project was published in 1987, but it was based on data collected between 1959 and 1974. The ethnic divide in health in the United States was far greater then than it is today, because most black and white Americans lived in distinctly different environments 40 to 50 years ago.[65] In a follow-up to the U.S. National Collaborative Perinatal Project, researchers reviewed the records of almost 30,000 African American and white children who participated in the original study. As expected, black babies showed the overall lowest birth weight. Although African American mothers were less likely to smoke—a known risk factor for low birth weight—they were more likely to have anemia than white mothers. There were also demographic differences between the African American and white mothers. Low birth weight carried an increased risk for high systolic blood pressure in the seven-year-old African American children, but white children at this age were not affected.[66]

The results of the follow-up study should not be misconstrued to mean that population genetics was irrelevant. Brain volume is indeed influenced by genetics, but it is also shaped by a wide array of environmental factors. These include nutrition in the womb, as well as after birth; mother–infant bonding;

and the infant's general physiological health. A wealth of scientific evidence confirms the obvious: Early childhood conditioning, particularly mental and emotional stimulation, has a formidable impact on the growth and development of the human brain.[67] Yet, Rushton and Ankney avoided in-depth analysis of potential environmental contributions to the population divide in brain volume. Focusing on the biological correlates, they reported an overall correlation of 0.44, or 44 percent, in eight published studies involving MRI brain scans; however, head size is only a rough estimate, at best, of brain size. In some studies, the correlation between head size and IQ is only 20 percent, a finding immediately suggesting that IQ performance is affected by factors other than brain volume, particularly environmental factors. Rushton conceded that brain size and IQ have only a modest correlation, but he rationalized that a relatively small variation in average brain size can translate into significant differences in cognitive performance.[68]

When Rushton estimated the average brain size of enlisted males in the U.S. military, white and Asian men had a similar cranial volume, contrary to his predicted trend in which the largest brain size should have been seen in Asians.[69] His estimates have been questioned because he failed to correct for body weight when he weighed these brains. Rushton later measured the skulls of 6,325 male and female U.S. Army personnel and then adjusted the values for stature and body weight, controlling for race, rank, or sex. His revised values for average skull volumes were 1,359 cm³ for blacks, 1,380 cm³ for whites, and 1,416 cm³ for Asians, consistent with his predicted gradient. (Recall that one cubic centimeter, or 1 cm³, occupies about the same volume as a milliliter, or 1 mL.) The exact brain sizes recorded for the three ethnic populations shifted during the second round of experiments, in keeping with Rushton's predicted trend. Based on his calculations, the cranial capacity of Asians in this study was about 4.2 percent larger than that of blacks.

Today, studies on brain size are hampered by a lack of agreement on what constitutes valid scientific techniques for comparing brain size in different populations within the same species. As neuropsychologist Michael Peters put it, there is no simple formula for correlating body size and brain volume across human populations. According to Peters, the evidence linking the overall higher average IQs of people of Northeast Asian descent and larger brain size is inconclusive.[70] Peters's words echo the position of Harry Jerison, world-renowned UCLA expert on the evolution of the brain. For decades,

Jerison admonished neuroscientists not to draw premature conclusions about population differences in the volume of the human brain. Comparing measurements of brain size across different species is a valid approach because the statistical measurements are very accurate, with a rate of uncertainty of about only 1 percent. But when scientists analyze the brain volume of individuals or populations within the same species, there is a large risk for error.[71]

The difference in brain size between populations within the same species can climb as high as 10 percent. And the correlation for variation in brain size across populations is 16 percent. This means that 84 percent of the variation in cranial volume across ethnic populations *cannot* be explained by genetics, regardless of whether statistics on individual genetics or population genetics are used. The variance in brain size within the same species, including the human species, *Homo sapiens*, is a statistical value, not an absolute value. The variation in brain volume that scientists observe in the laboratory points to a probability statistic. When comparing creatures of different species, the larger-brained animals are predictably smarter than the smaller-brained ones.

As Jerison explained, although brain volume is associated with intelligence when comparing different species, population differences in brain size are not fixed indicators of the level of intelligence in an individual or population, let alone in a wider continental population, or racial group. A larger-brained individual—or even a bigger-brained population—is not necessarily more intelligent on average than other individuals or populations within the same species.[72] Even when brain-to-body ratio is controlled, much of the observed difference between populations arises from developmental factors that occur before a person is born and during early childhood. The in utero health of the unborn infant, combined with such influences as childhood nutrition, exposure to infectious agents, exposure to lead or other environmental toxins, and psychosocial conditions, can have profound effects on a child's physical development. The rate of growth and maturation are crucial parts of brain development.[73]

Returning to Rushton and Ankney's findings, these authors admitted that some studies had reported larger cranial capacities in blacks compared with whites. In fact, of the data that Rushton borrowed for adult male and female brain volumes, the largest cranial capacities were found in the Mokapu, a Pacific Islander group indigenous to Hawaii, and the Xhosa, a

Bantu-speaking black ethnic group of Southern Africa. It is not clear why Rushton ignored evidence showing that some of the largest skulls in his sample came from a black African population. He called the Mokapu and Xhosa outliers, which they were. But as black Africans, the average skull size of the Xhosa should have been factored into his calculations.[74] Despite global trends on cranial capacity showing the smallest skulls in black Africans and the largest in East Asians, there are multiple inconsistencies, if not notable exceptions, to this pattern. Even Rushton reported wide variation in the estimated brain volume among whites, which ranged from 1,356 cm^3 for Iranian soldiers, to 1,455 cm^3 for German air force personnel, to 1,470 cm^3 for white U.S. Army soldiers, to 1,539 cm^3 for U.S. Air Force service personnel.[75]

More recently, anatomy professor I. E. Odokuma and his team in Nigeria measured the cranial volume of multiple indigenous Nigerian ethnic groups and found, as expected, that the males had significantly larger skull sizes than the women.[76] The sex-averaged cranial sizes—the mean for both women and men—were 1,255.89 cm^3 for the Urhobo, 1,273.39 cm^3 for the Ibo, and 1,310.08 cm^3 for the Edo. These findings are consistent with the fairly wide range of cranial sizes cited in other studies of black African populations, but the broad range of brain volume within a single continental population should not be surprising. For example, large brain volumes have frequently been seen in Northeast Asian populations, but some studies have reported an average brain size of 1,299 cm^3 in South Vietnamese soldiers and 1,340 cm^3 in Thai soldiers. These values are considerably lower than the mean cranial volume of 1,406 cm^3 that Rushton approximated for East Asians. The brain case volume of the Edo was slightly larger than that of the South Vietnamese soldiers, even though the difference was not statistically significant.[77]

(In his earlier publications, Rushton often referred to the Asian samples he studied as the "Mongoloid" race, a quaint if not derogatory and scientifically misguided term resurrected from nineteenth-century anthropology. For Rushton, "Mongoloids" were often East Asians of Chinese and Japanese, and occasionally Indonesian, descent. Each of these three East Asian populations is genetically diverse within their individual groups, and there is further diversity between them. Chinese, Japanese, and Korean ethnic populations overall are genetically fairly close. These groups are genetically more distant from other Asian ethnic groups, for example, Indonesians and Vietnamese, than they are from one another.)[78]

Inaccurate calculations of brain size may occur when researchers attempt to quantify a population's brain size by averaging the brain volume of its male and female members combined.[79] Most estimates of brain volume in a population are based on the mean, or statistical average, of the individual specimens in the sample being analyzed. But the participants in research samples do not always have an equal number of males and females. If a population has more than twice as many males as females, the population mean will be closer to the male than the female mean.[80] Vice versa holds true as well. And problems can arise if brain size values are based on disparate sampling of the sexes and unweighted data, or values that do not control for differences in age (and age-related height) of individuals in the sample. Errors can easily become compounded if the cranial capacity of genetically diverse ethnic populations within the same continental population—for example, such diverse European white populations as English, Italians, Swedes, and Hungarians—are further averaged to produce an aggregated mean.[81]

WEIGHING IN ON WEIGHING BRAINS

When neuroscientists estimate the cranial capacity in different populations, their findings may be mismatched if they rely on inadequate methods of measuring skull volume. Some techniques used to calculate skull size or brain weight may be based on flawed assumptions, inadvertently producing misleading, if not patently false, results. This is evident when researchers attempt to weigh or measure the volume of brain tissue from deceased individuals. More than 40 years ago, esteemed South African physical anthropologist Phillip Tobias advised researchers to document their methods with the utmost precision when they correlated race with cranial size, based on autopsied brains.[82] A host of factors, including not only sex and body size, but also the individual's age at the time of death, as well as the temperature of the corpse, can alter the postmortem brain volume. The size of an autopsied brain is also influenced by the deceased person's nutritional status (which starts during or before early childhood), his occupations throughout his lifespan, and the exact cause of his death. The brain may shrink only slightly or, in some cases, substantially, depending on the length of time between death and autopsy. Tobias warned that population studies on brain size might not be credible unless researchers consider all of these variables when estimating brain volume, whether in genetically similar or diverse groups of people.[83]

Based on Tobias's assessment, much of the older research on the average brain weight of different populations, particularly nonwhite groups, conducted in the twentieth century might need to be reevaluated. As he noted, the majority of comparative studies on brain volume or brain weight across ethnic populations have failed to provide adequate background information on the social, educational, and nutritional status of the men and women whose brains have been examined. Most scientific studies historically have provided only minimal demographic information about the deceased participants, particularly when the studies have focused on African, African American, or other mixed-race populations of African descent whose brains were inspected at autopsy.[84]

Like Tobias, other scientists have since called for caution in interpreting correlations between IQ and brain size in blacks. They argue that data on the brains of different populations are not fully viable unless researchers also document the historical circumstances, for instance, the health and nutritional status, of the populations being investigated.[85] Highlighting evidence from a 1970 study of Tobias, psychiatry professor Zack Cernovsky discovered that blacks residing in countries with superior nutritional standards had cranial sizes comparable to those of whites and larger than the cranial capacities of "Caucasians" in Cairo.[86] Although this evidence was gleaned from older research, it suggests there are exceptions to the race-based trend in brain size predicted by Rushton. It reveals that in societies in which blacks and whites have a similar nutritional status and standard of living, blacks do not necessarily have statistically smaller brains than whites.[87]

In addition, skull volume provides only a rough and imperfect approximation of true brain size. Inadequate techniques for measuring skull volume can inadvertently distort the results of a study, even when researchers endeavor to conduct unbiased research. In the early twentieth century, brain size was sometimes estimated using the "German formula," originally designed to calculate the volume of a brachycephalic, or short-headed, shaped skull. This technique was well suited for measuring the size of skulls in Western Europeans, including ethnic Germans, since people of German ancestry often have brachycephalic heads. But the formula proved less precise, in some cases visibly flawed, when used to quantify the skull size of non-Europeans who are not short-headed. As Michael Peters explained, the German formula can produce incorrect measurements of most black African skulls, which tend to be dolichocephalic, or long-headed. The tendency of some researchers to indiscriminately rely on a single formula to measure differently shaped skulls can lead to systematic errors in their calculations of brain volume.[88]

To remedy this issue, Peters proposed that researchers use multiple cranial formulas rather than a single, idealized formula intended to measure only one set of racial characteristics. For instance, when researchers employed the German formula to measure the skull volume of African blacks, they calculated a value of 1,359 cm³. But when they relied on the Ainu formula—a method that places more weight on the length of the skull—the skull size of blacks in the same study averaged 1,418 cm³.[89] Nonetheless, selecting the optimal formula for estimating skull volume in a specific ethnic population can be challenging. The skulls of descendants of black African populations vary widely, reflecting the extensive genetic diversity of black ethnic populations in Sub-Saharan Africa and their diasporas throughout the world. The vast genetic variation found among blacks is not limited to blacks in Africa and African Americans. It also extends to diverse African-derived populations throughout the world.[90]

In one study, Rushton estimated the average number of neurons in the cerebral cortex, as a proxy for cranial capacity, in different continental populations. After aggregating values calculated by other researchers, he presented the unweighted, sex-combined averages of overall brain size as 1,406 cm³ for Asians, 1,385 cm³ for whites, and 1,331 cm³ for blacks.[91] He then translated these estimates of average brain volume into a cortical neuron or nerve cell count that averaged 13,767 billion in Asians, 13,665 billion in whites, and 13,185 in blacks.[92] In a separate analysis, Rushton calculated the "excess number of neurons" in millions for each of these three continental populations as 8,900 for Asians, 8,650 for Europeans, and 8,550 for Africans.[93] (Excess neurons refer to nerve cells that can help a living creature, human or nonhuman, adjust to its environment, above and beyond the number of neurons needed for transmitting nerve signals in the brain.[94]) According to Rushton, as few as 100 to 200 million neurons were sufficient to afford Asians and whites an advantage over blacks in cognitive skills needed to reach important cultural milestones. He claimed that the reported ethnic population differences in neuron count could account for different levels of intellectual and cultural achievement between these populations. The divide between blacks and whites was wide. But the gap between whites and Asians was small, leading Rushton to surmise that differences in social accomplishment between these two populations would be minimal.[95]

Still, Rushton's calculations were problematic from the start. As Lieberman retorted, Rushton's value of 8,550 million "excess neurons" for Africans represented an aggregated figure.[96] To calculate the African mean for "excess neurons," Rushton averaged values of 8,400 and 8,700 "excess million neurons" in Kenyans and African Americans, respectively. (Incidentally, the value of 8,700 million excess neurons reported in African Americans was more than the 8,500 and 8,600 reported for white Americans and white Frenchmen, respectively, in this same study.[97])

Rushton frequently used meta-analysis, sometimes termed aggregation, a research method for calculating the mathematical mean for measurements of brain size and other physiological traits across different populations. This approach can be an invaluable statistical tool, especially in medical research. But its validity and utility are sometimes limited by publication biases, as well as different study designs that can confound, and possibly even distort, the results of the overall study.[98]

Rushton insisted that his research was impartial, but on multiple occasions he omitted crucial pieces of evidence needed to accurately interpret the data he analyzed. He never mentioned that the dura mater—the thicker outer membrane covering the brain—might be responsible for at least part of the reported larger cranial volume in some Asian populations.[99] Although input from nerves reaches the dura mater from other sites in the brain, there are no neurons in the dura mater. According to Lieberman, Rushton's projected neuron count for the average brain volume of East Asian populations may have been artificially inflated.[100] Rushton's overall values for average brain size and average neuron count across the three continental populations were consistent with the race-based gradient he predicted. But his failure to explain anomalies that contradicted this trend, for example, the higher excess neuron count in African Americans, is worrisome, at least for champions of unbiased scientific research based on integrity, objectivity, and parsimony.

Rushton speculated that five cubic inches, or approximately 80 cm³ of brain volume—about the size of two golf balls—was responsible for the black/white difference in IQ scores. Even molecular anthropologist Vincent Sarich, who shared many of Rushton's racialist perspectives, disputed the accuracy of his calculations. As Sarich explained, if the correlation between brain size and IQ is 40 percent, then the difference in brain volume between blacks and whites would have to be 130 cm³ to account for the IQ divide between these two ethnic populations. Instead, the smaller cranial size of blacks reported in some studies would correspond to only about one-third— five out of 15 points—of the usually cited black/white gap in IQ score.[101] Psychologist Earl Hunt is sometimes credited with staking out a middle ground in the arena of intelligence testing—a territory niched between hereditarianism, epitomized by the research of Arthur Jensen and J. Philippe Rushton, and antirace, proenvironmental models, exemplified in the work of Steven Jay Gould, Leonard Lieberman, and C. Loring Brace.[102]

Hunt seemed to champion the hereditarian more than the environmental position, but in a stance not unlike that of Sarich, he rejected Rushton's calculations for brain size and IQ across ethnic populations. Hunt estimated that the population differences in head size Rushton used could account for only 0.09, or 9 percent, of the 1.0 to 1.1 standard deviation of the IQ gap between blacks and whites in the United States. Hunt based his estimate on a 0.35 correlation between brain size and IQ. As he stated, if the African American/white IQ divide is 15 to 18 points, then the gap would correlate with only 1.35 to 1.62 points. Even if a higher brain size–IQ correlation of 0.44 were applied, the proportion of the gap attributable to genetic differences would still be too small to be statistically significant. It could not account for the IQ gulf between blacks and whites in the United States.[103]

THE MULTIFACETED HUMAN BRAIN: ADAPTATION RECONSIDERED

Inconsistent population differences in brain size are not unexpected in light of the problems that arise when the world's numerous diverse populations are lumped into antiquated and simplistic categories of white, Asian, and black. (Adding Australian Aborigines, a fourth major category of the traditional racial classifications of the nineteenth and early twentieth centuries, expands the number of groups but does not increase the validity or accuracy of racial categorizations.) Lieberman, like many scientists who have contested hereditarian claims, called Rushton's findings meaningless because they are based on what renowned anthropologist Ashley Montagu called an "an omelette of populations representing nothing that exists in nature."[104] Cranial capacity is likely to vary among different smaller populations within a major race—a continental population—if only because of the diverse climatic regions within each continent.[105]

The evolution of human cranial features, including overall brain volume and the brain's surface-to-volume ratio, has been heavily influenced by climate. Even the shape of the human brain may vary slightly between populations that originally lived in temperate rather than tropical regions. Peoples inhabiting temperate zones tend to have more spherical braincases because this shape is preferable for conserving heat. It is an advantageous adaptation in cold regions, where as much as 80 percent of body heat can dissipate through the head, primarily its blood vessels, on cold days. But in extremely hot climates, a modified engineering design—essentially a smaller human brain—may be optimal because it is more effective than a larger brain in regulating excess heat. A slight reduction in cranial size, then, may be a better metabolic and physiological fit for populations exposed to intense solar radiation in habitats near the equator. Depending upon the climate of a population's original environment, a smaller brain overall is not necessarily less efficient than a larger brain. The most appropriate brain volume for a species, humans included, is one that achieves a balance between energetic demands and ecological conditions.[106]

Still, the potential advantages of a slightly smaller human brain must be weighed against the disadvantages. According to the University of Missouri's David Geary, the size of the skull, and therefore the size of the brain, has decreased in the human lineage in the last 30,000 years. As human societies grew more complex during the last few thousand years, eventually leading to increased population density, people found novel ways to solve problems. With a larger overall brain size no longer a necessity, the human brain shrunk almost 10 percent during this period of time, from a volume of 1,500 cm^3 to 1,359 cm^3, or roughly the size of a tennis ball. Ironically, with the advent of such watershed events as sedentary agriculture, the industrial revolution, and the information age, humans have not had to depend on their intelligence to the same degree that they did in their prehistoric past.[107]

Claims of race-based differences in the organization and function of the brain, as well as its size, are highly explosive. Yet, some brain-related features, especially disease-related traits, are not evenly distributed among the planet's continental populations. Ethnic populations vary in their rates of certain neurological disorders, for instance, susceptibility to Alzheimer's disease and stroke—two conditions known to have genetic underpinnings. Even the average loss of brain weight associated with aging can vary substantially across ethnic populations.[108] Understanding how genes contribute to brain abnormalities in psychiatric disorders may help to improve the quality of life of people from every ethnic background.

Equipped with an ever-expanding armamentarium of technologically advanced genomic tools, some scientists are currently exploring ethnic population differences in the frequency of several brain-related conditions and their associated genes. Evolutionary geneticists are examining whether any of these differences are linked to natural selection—physical, disease-related, cognitive, or behavioral adaptations that human populations have experienced within the past 100,000 to 150,000 years. But even when researchers can convincingly demonstrate genetic variation across populations, rarely is there a consensus on precisely what that variation means. Experts still do not agree on whether differences in the average brain volume of diverse ethnic populations stem primarily from genetic or environmental influences, or equally from both. Until recently, conventional perspectives on this debate rarely considered the potentially momentous impact of epigenetics and plasticity on brain size, as well as the amount of white matter and gray matter, and the connectivity of various brain regions, on cognition and psychiatric disorders.

During the evolutionary trajectory of humans and their immediate primate ancestors, the size of the human brain increased from about 400 mL to an average of 1,350 mL in the present human species. The primary evolutionary driving force behind this extraordinary gain in brain mass, epitomized by increased quantities of gray matter, or densely packed neurons, was inarguably the advantage of possessing enhanced intelligence. In Darwinian parlance, smarter (and presumably bigger-brained) species were more fit than smaller-brained hominids (accounting, of course, for a similar brain-to-body size ratio). In the words of evolutionary theorists, compared to smaller-brained individuals, those with keen wit, presumably due to a more massive cerebral cortex, were more adept at surviving, reproducing, and caring for their young.[109]

The notion that a larger brain has selective advantages is logical, but a bigger brain was, and is, not an inevitably, favorable, adaptive trait in every situation or every environment inhabited by *Homo sapiens*. During the evolution of the human lineage, the brain—especially the cerebral cortex—enlarged, but there were limits to its expanding volume. To compensate for this limitation, the development of the cortex in primates evolved novel anatomical features: cortical folding coupled with connectivity, or wiring, that allowed nerve signals to be rapidly sent back and forth between disparate parts of the brain. (Cortical convolution refers to the grooves and bumps that form on the surface of the cortex when it is folded. Cortical convolution substantially increases the brain's surface area.)[110]

There are obvious limits to the maximum size of the human brain, particularly during birth, since a newborn's head must be small enough to safely pass through its mother's birth canal. Given these constraints, from an evolutionary perspective, the best operational human brain might not be the largest brain, but, instead, the most fuel-efficient brain that can squeeze into the smallest size possible. The utility of a smaller brain in some environments weakens and may even overturn the idea that a larger human brain is automatically a smarter human brain.[111] Today, the mean size of the brain in healthy human adults averages 1,345 to 1,400 cm^3, but with individual and population differences the volume of the human brain ranges from 1,200 to 1,500 cm^3.[112] Larger brains gain their additional volume through enlarged areas of the cortex but at the expense of slower growing areas of the "lower brain," which govern emotional, sensory, and motor functions.[113] Determining which human brain design is ideal—or, in the evolutionary past, was optimal—must ultimately be based on the brain's ability to allow a population to successfully adapt to its own environment.

The overarching question for the human species as a whole may be less about the utility of a smaller versus larger brain. Instead, it may be more about the interrelationships between various components that make up the living brain and its remarkable capacity for plasticity, or rewiring the connections within it. With increased access to sophisticated brain imaging technology, some researchers investigating intelligence today are turning their attention from gross brain size to various components of the human brain. They are examining specific brain structures and anatomical regions—for example, neuron-rich gray matter, the fiber tracts of white matter, and network circuits—between different areas in the brain in relation to both genetics and social factors, including SES and school attainment. In the next chapter we will explore some of their discoveries.

Brain Traits

The View from MRIs

THE RISE OF MRIs: FROM MEDICINE TO BRAIN SIZE

BEFORE THE WIDESPREAD USE OF MRI IN MEDICAL DIAGNOSTICS STARTING in the 1980s, research comparing the brain size of healthy adults across different ethnic populations was virtually nonexistent.[1] Today, innovative imaging tools are sometimes used for this purpose, but studies on brain volume in various racial groups are limited. Current state-of-the-art techniques to quantify brain volume include magnetic resonance imaging, or MRI, scans, which create detailed three-dimensional (3-D) pictures of the outer surface and internal structures of the living brain. Sophisticated functional MRI (fMRI) scans reveal activity in various parts of the brain by measuring changes in blood flow.[2] With modern imaging tools, neuroscientists can assess if there are links between brain volume and cognitive performance, including IQ test scores.[3] MRI scans provide far more accurate and much more precise calculations of brain volume than either external measurements of the head's perimeter or the quantity of mustard seeds needed to fill up the brain case. But even neuroimaging technologies are not totally free of potential error. An MRI scan of a patient may yield a false reading if he involuntarily moves his body while undergoing the scan. Such minor actions might explain why estimates of brain size in different ethnic populations varied in some of the early MRI studies.[4]

In some cases, MRI hardware may be unable to precisely divide brain tissue into gray matter and white matter. The ratio of these two types of brain tissue detected in the same individual brain may vary, depending on the laboratory techniques used. Additionally, the contrast between gray matter, or groups of nerve cells, and white matter, or myelin-covered fiber tracts, is not always clearly defined on conventional MRI scans. Brain regions composed of mixtures of cerebrospinal fluid (CSF) and gray matter, or gray matter plus white matter, may be difficult to tease apart and accurately read on MRI scans. To identify specific areas of the brain, MRI segmentation allows an image to be partitioned into meaningful and homogeneous anatomical regions that share similar attributes like intensity, depth, color, or texture. But even with this technological enhancement, scans of the brain or other organs can result in misleading and possibly false findings if radiologists are unable to correctly interpret the clinical significance of each image.[5] This problem can become magnified if researchers do not acknowledge the limitations of their interpretations. Despite innovative advances in this field, the art and science of delineating brain tissue into discrete regions, including distinct areas of white matter and gray matter, are still evolving. If the data generated from neuroimaging studies are imprecise or misinterpreted, neuroscientists might unknowingly draw misleading conclusions about variation in the average brain size across different ethnic populations.[6]

Still, MRI studies published in the past 15 years have greatly enhanced the accuracy of brain size estimates. But even with expanding data on brain volume, experts do not agree on whether correlations typically reported between brain size and intelligence test scores are statistically significant.[7] One of the earliest studies to calculate the correlation between brain size in living humans and IQ scores was published in 1991, and reported a relatively high value of 0.51.[8] Subsequent studies have generally found a lower correlation between brain volume, estimated from MRI scans, and intelligence (usually based on IQ test scores). Although most correlations are positive, they generally yield only modest values ranging between 0.3 to 0.4 (30 percent to 40 percent).[9] In a 2015 meta-analysis of 148 samples with more than 8,000 participants, University of Vienna researcher Jakob Pietschnig and his colleagues found a positive but weak association of only 0.24, or 24 percent, between total brain volume (based on MRI brain scans) and IQ.[10] This is considerably lower than the estimates from previous studies that measured brain size either with or without MRI. For Pietschnig's team, brain volume accounted for only a minor portion of IQ test scores, or intelligence performance, in humans. However, when the findings were reanalyzed by other cognitive science researchers—the University of Western Australia's Gilles Gignac and the University of Edinburgh's Timothy Bates—the overall correlation between brain volume and IQ was notably higher, at 0.31. And when the quality of the IQ tests, including the standard deviations, used by Pietschnig's team was considered, the correlation between brain volume and IQ increased even further, to 0.4. According to Gignac and Bates, the association between whole brain volume and intelligence was underestimated in Pietschnig's meta-analysis.[11]

Still, for some scientists, the increased value of the recalculated correlation between brain size and IQ was unimpressive and not strong enough to demonstrate a definitive cause-and-effect relationship. These critics do not accept brain size per se as the ultimate neurobiological foundation of intelligence. In and of itself, the overall size of the human brain across global ethnic populations does not appear to have a measurable impact on higher executive functions, for example, the intellectual acuity to correctly answer IQ test questions. Instead, the brain-related traits most closely associated with higher cognitive abilities encompass features other than overall size, for instance, the speed of development of specific regions within the brain, localized areas or structures in the brain, and the thickness of the cerebral cortex.[12] Without a strong correlation, say some experts, gross brain size per se does not adequately predict IQ.[13]

However, different aspects of intelligence have been linked with other traits, including specific regions and structures within the brain. In a recent neuroimaging study using MRI, researchers found that people with larger brains usually earned higher scores on general intelligence—g.[14] Overall brain size was only one of several brain-related factors that contributed to the variance in IQ scores, including scores for both g and fluid intelligence—logical reasoning, and problem-solving ability. Cortical thickness was responsible for an estimated 5 percent of the difference in IQ scores, whereas lesions in the white matter explained about 2 percent of it. Brain volume was the strongest single predictor of cognitive ability, accounting for about 12 percent of the variance in IQ. These structural brain features combined, based on quantifying the details in MRI images, explained approximately 20 percent of the variance in general intelligence scores among the participants in the study.

Nevertheless, as the authors admitted, these factors applied to a particular population, in this case a group of older individuals. Therefore, the same brain-related structural features might not be significant in other populations largely comprising younger people.[15]

More than a decade ago, neuroscientists at the National Institutes of Health (NIH) discovered that the brain size of adolescents with extraordinarily high IQs was not larger than that of their age-matched counterparts with average IQs. But the brains of the high-IQ individuals exhibited a different pattern of brain development than those with average IQs. The cerebral cortex of youth with exceptionally elevated IQs appears to mature more quickly than in adolescents with average or lower IQs.[16] Genes influence brain development and presumably contribute to accelerated brain growth in high-IQ individuals. At the same time, environmental factors can influence gene expression, which in turn can affect the speed of brain growth.[17]

Intellectual skill is associated with two general areas of the brain: gray matter—concentrations of neurons—and white matter, or axons—the long, slender fibers projecting from nerve cell bodies and encased in fatty, or myelin, sheaths that transmit electrical impulses between nerve cells. Men have an estimated 6.5 times as much gray matter as women, notably in the frontal and parietal lobes. (As the main center of higher cognitive function, the frontal lobe plays a key role in personality, learning, memory, and attention.[18]) Women have roughly 10 times more intelligence-related white matter than men, with most of their IQ-related gray matter located in the frontal lobe, including the area associated with speech and language processing.[19] The adult male brain is larger than the adult female brain, but there are usually no statistically significant sex differences in the correlation between brain volume and IQ between men and women of the same ethnic population.[20]

In chapter 6, we discussed evidence for possible sex differences in the brain's functional attributes, with men supposedly exhibiting overall better ability in spatial problem-solving and women on average excelling in verbal and memory tasks.[21] The paradoxes evident in the gendered human brain underscore the fact that the correlations found to date between specific brain traits and cognitive functions, including IQ, are sometimes equivocal or inconclusive. It remains to be seen if similar paradoxes will emerge in future research on interrelationships between genetic ancestry, brain traits, and cognitive performance or educational achievement.

BRAIN SIZE ACROSS POPULATIONS: THROUGH THE LENS OF MRI

A small number of investigations published in the last decade have leveraged MRI to examine ethnic population differences in the size of various brain structures, as well as the overall brain volume. Most of the research in this area conducted thus far has involved a small number of participants. Only a few MRI studies have compared specific brain features other than overall brain volume across ethnic populations. Despite these limitations, some research has demonstrated ethnic population differences in brain structure, for instance, slight differences in the shape of key areas of the brain.[22] In some investigations, the average brain of Korean or Chinese adults has a shorter front–back length and a rounder shape compared with Caucasian adult brains.[23] The mean volume of specific regions of the brain may also vary among ethnic populations. For example, Chinese and Caucasians show differences in widespread areas of the frontal and temporal lobes (located in the front and above the ears), as well as a region called the midbrain basal ganglia, which is defective in Parkinson's disease.[24]

Similarly, African Americans and Caucasians may differ in the size of specific regions of the brain—the cerebellum (an area involved in motor or movement control), the amygdala (a structure in the limbic system that helps regulate emotions), and a part of the cerebral cortex called the orbital frontal cortex, or OFC, which plays a role in social behavior.[25] In an innovative MRI brain study, Nneka Isamah and her research team examined population differences in the volume of various brain structures in 44 white and 25 African American adults.[26] The two groups of participants did not differ significantly in sex or education level, although the average amount of schooling was 15.2 years for the whites versus 14.8 years for the African Americans.

The blacks were significantly younger, with an average age of 35.6 years, compared with 46.4 years for the whites. After controlling for age, sex, and education level, the researchers found no statistically significant differences between the two groups in the total amount of either gray matter or white matter, or the quantity of CSF housed in the brain's cavities, or ventricles. But compared to the white participants, the African Americans in the study had a significantly smaller total cerebral volume—or size of the two cerebral hemispheres—but a larger left OFC.[27] The average total cerebral volume was 1,076.0 mL (milliliters) in the African Americans versus 1,178.3 mL in the whites.[28] (A volume of 1,000 mL is almost one quart of liquid, based on U.S. standard measurements. One mL is approximately one-fifth of a teaspoon.)

Located in the brain's limbic system, the amygdala are two small almond shaped structures activated during emotional experiences. In this study, the amygdala in the blacks were smaller, but they accounted for a larger proportion of the total cerebral volume compared to the whites. The OFC lies within the prefrontal cortex, where it sits just above the orbits, or eye sockets. The authors did not speculate on what the significantly larger OFC in the African American group might mean. But they noted that the functions of the OFC and amygdala are linked, with the two structures involved in emotions and facial recognition. Both the OFC and amygdala accounted for a significantly larger proportion of the total cerebral volume in the African American participants compared to the whites.[29]

The prefrontal cortex, the seat of higher thought, is pivotal for decision-making. The OFC is implicated in a variety of functions, including anxiety, mood, addictive disorders, and impulse control. This region also seems to be activated when a person has to reevaluate new information to determine if he should alter his decision, should a situation change. The OFC, when stimulated, reportedly sends a person a bodily signal that triggers an emotional response to help him become aware of the likely consequences of his actions if he is about to make a poor choice.[30] In another study, researchers who were curious about how emotions influence daily decision-making recorded the electrical activity in the brains of test subjects while they were making a financial decision. The participants' heart rates and electrical activation patterns in the OFC varied, depending upon whether they chose a riskier or less risky action.[31] Electrical activation patterns in different regions of the brain, along with the size of brain structures, represent two distinct features that can influence brain function.

Isamah's team demonstrated that anatomical differences occur in the size of the OFC—an emotion-linked brain structure—between ethnic populations. Based on these preliminary findings, we could speculate about population differences in brain function, but our interpretation could easily be misleading, if not patently false, because it would be based on limited data. Isamah's team never denied that population genetics might contribute to variation in either the size of certain brain structures or the activation of specific areas of the brain. But as they admonished, correlations between specific regions of the brain and the function of those regions have been inconsistent in multiple studies throughout the years.[32]

Nonetheless, this type of research can be fruitful because it can potentially shed light on risks for neuropsychiatric diseases and symptoms of brain-related disorders in different ethnic groups. The findings of Isamah's research team neither proved nor disproved a cause-and-effect link between population genetics and brain structures that differed between the black and white participants. The sample size was small, even though it was similar to that of most other MRI studies on ethnic population variation in brain structures. The participants were grouped by ancestry but only superficially. Their racial identity was based on their gross physical features, as perceived by the researchers, and their own self-reported categories.[33]

The overall smaller brain volume observed in the blacks accords with other findings on populations of African black descent. Still, the scientists could not determine the degree to which environment and plasticity might have contributed to variance in the sizes of the cerebrum and OFC, and even the amygdala. Socioeconomic status, or SES, has been associated with various features of the brain, including the size of the amygdala and the hippocampus (a structure involved in memory), and the volume of white matter; although the results have been inconsistent.[34] Isamah's team used education as a proxy for SES, recognizing that it is an imperfect substitute. Additional environmental factors like childhood exposure to various adversities, trauma, or health habits also might influence correlations between specific features, including the size and shape, of various brain structures, and cognitive or behavioral functions.[35] And while no one can deny that ethnic population variation in some brain-related traits exists, the causes of that variation and, more importantly, the meaning of any differences await further clarification.

Diffusion tensor imaging, or DTI, is a specialized form of MRI used to reveal complex and intricate details of the brain, including the orientation of tracts of white matter and minute structural changes in blood flow while a person performs tasks. Brain specialists sometimes use DTI to identify the underlying causes of brain abnormalities and select appropriate treatments for various clinical conditions. DTI provides an exquisite method for estimating the volume of the total brain and specific regions within it. Using MRI and DTI, researchers in Singapore compared global brain volume, brain structure, and white matter in healthy Chinese, Malay, and Indian newborns.[36] The size of both the total brain and most anatomical regions of the brain were similar in all three groups. But compared to the Malay newborns, the Indian neonatal brains were more elongated in the front–back axis than in the top–bottom axis. And the brains of the Malay newborns overall were significantly more rounded than those of the Indian newborns. This difference held even after controlling for household income and prenatal exposure to tobacco. Variance was also seen between the Indian neonates, on the one hand, and the Chinese and Malay newborns, on the other hand, in neural circuits and anatomical areas linked with sensory and motor, or movement, functions.[37]

The comparison of the brains in these newborns suggested that biological differences in ethnic populations may first appear at an early age, but more research is needed to determine if these differences persist throughout the life span. Other factors, including a host of environmental factors, possibly influence when, where, how, and perhaps why population variance in brain-related traits develop, even at a later age. Individual differences in brain structure, including its size and shape, are strongly impacted by environmental forces and biological development, for example, the plasticity of the brain. Amid mounting scientific evidence, the idea that some ethnic populations may vary in certain brain features, including the volume and shape of various brain structures, has become increasingly more credible.

The evidence is still limited, but it appears to support some of the claims advanced during the last 150 years, especially the provocative notion that the average size of the human brain follows a "race-based gradient." Some of the findings based on advanced imaging techniques that have been reported to date seem to confirm, at least at first glance, the trend predicted by Rushton and like-minded racialist hereditarians. The largest brain volume has generally been observed primarily in Northeastern Asians, with an intermediate size in European whites and a relatively smaller volume in Sub-Saharan black Africans.[38] And yet, this trend says nothing about the complex interactions of myriad environmental, epigenetic, and plasticity influences that affect more than brain size. Perhaps more importantly, it reveals nothing about patterns of brain cell firing and the brain's wiring, or circuitry, which ultimately influence an individual's cognition, emotions, and social behaviors.

From Size to Synapses

More than a decade ago, Michael McDaniel published a comprehensive meta-analysis of the association between brain size and intelligence.[39] As one of the first investigations to correlate the brain volume of a sample of living humans, based on MRI scans, with full-scale IQ test scores, his study immediately drew criticism and intensified the nature-versus-nurture debate concerning cognitive ability. Drawing from 37 previously published studies with 1,530 people, McDaniel examined a potential connection between IQ scores and brain volume in relation to both sex and age. The ethnic composition of the samples included in the meta-analysis was mostly white, but a few studies had adults or children who were either black, racially mixed, or of unknown racial ancestry. McDaniel reported an overall correlation of 0.33, or 33 percent, for the studies as a whole. He found a positive association, which he concluded demonstrates that brain size influences IQ.

And yet, when the individual studies in this meta-analysis were examined separately, correlations between the overall size of living brains and average IQ varied widely, sometimes revealing inconsistencies even for the same sex or same ethnic group.[40] For example, the correlation was weakly negative in a sample of white male and female adults (–0.08) and a sample of male white children (–0.13). The correlation was somewhat positive, although relatively low, in a study of male white children (0.27) but slightly higher (0.39) in another sample of white adult males. Some of the sample sizes assessed in the meta-analysis were quite small, including those with black participants, which had highly variable correlations between brain volume and IQ—values of –0.86, 0.17, and 0.46.[41] The broad spread evidenced in the correlations of separate studies in this meta-analysis may make the results of these studies questionable. But as McDaniel explained, the variance in the association between brain size and IQ reported across studies may reflect differences in brain growth at various ages of childhood. After all, the participants in the studies spanned a wide age range, from childhood to adulthood.[42]

The brain achieves 50 percent of its adult weight by six months of age, 70 percent at 18 months, 80 percent at 3 years, and 90 percent in middle childhood at 5 to 8 years.[43] Research on neuroplasticity—the plasticity or pliable nature of the brain—reveals that much of the human brain is not completely hardwired with fixed nerve cell connections at birth. Rather, brain wiring in humans is, to some extent, mutable and based largely on a person's life experiences. Brain growth and development can be modified, substantially at times, depending on numerous environmental influences, both positive nurturing and adverse experiences, especially when they occur during early childhood.

Neuroplasticity allows a growing person to adapt to his surroundings, a process that accounts for much of the brain's remaining growth after birth. New neurons and synapses, or junctions between neurons, develop rapidly in the human brain starting in the embryo before birth. This continues at an accelerated pace until a person is about two years of age. As the child gains more experiences, the synapses form tree-like branches that increase the connectivity of different areas of the brain. This extensive branching contributes to an expanding neural network and the increased volume of a person's brain. During this early formative period, as many as 40,000 new synapses can form per second, eventually leaving a young child with more neurons and synapses than his brain needs to function effectively.[44] When a young child is maturing into an adult, some of the excess connectivity that he no longer uses or needs is eliminated, or pruned, permitting nerve messages to be transmitted more efficiently throughout the brain's neural network. About half of the synapses formed at two years of age have been destroyed by the time a child is 10 years old. The process of connections between synapses first increasing and then decreasing is called synaptic arborization and pruning.[45]

The number of neurons and synapses can affect the efficiency of information processing, but so does the molecular complexity within these synapses. Learning and ultimately the plasticity of the brain unfold when molecular connections in the nerve cell junctions are altered starting at a young age.[46] The synapses proliferate remarkably during the first one to two years of life as an infant learns about the world around him. Throughout time, as he grows older, the number of synapses may continue to expand in certain areas of the brain. But in other regions, the quantity of synapses may diminish, leading to an economic conservation of nerve cell junctions. This "downsizing" of synapses is one reason why the volume of the human brain changes throughout life.[47]

No single region of the cerebral cortex correlates strongly with intellectual ability, but areas packed with gray matter have initially seemed to be a prime candidate for the biological underpinnings of cognition in the brain. Once again, the amount of gray matter, or density of nerve cells, in various regions of the brain predicts cognitive skills better than the total brain volume. Recall that the human male adult brain houses, on average, 86 billion neurons,[48] but they are not evenly distributed throughout this organ. The frontal lobe—the seat of problem solving, memory, language, and impulse control—is rich in gray matter and seems to correlate more than overall brain size with IQ. White matter is not irrelevant here because various areas of the brain containing gray matter and white matter are linked together. And yet, in some studies the correlation between these regions and intelligence is weak, ranging between 0.14 to 0.27. The correlation tends to be smaller in children than in adults because, as we just learned, brain volume changes considerably during childhood and adolescence.[49]

THE BRAIN'S TERRAIN: PEAKS AND VALLEYS

The human female brain, when compared to the male brain, is characterized by additional folds in the cerebral cortex that give it not only a more wrinkled external appearance, but also added volume. The increased folding might compensate for the relatively smaller volume of the female brain by providing extra geographical space in the cortex for more gray matter, or neurons. Cortical folds may partly explain why adult females earn IQ scores comparable to those of adult males in the same population, even though overall men have larger brains. Do ethnic populations differ in the degree of cortical folding such that, theoretically, the average size of their total brain, as well as specific regions and structures within it, might be larger or perhaps smaller than estimated by spatial volume alone? In other words, do ethnic populations with an overall smaller brain size have relatively more cortical folding than "larger-brained" human populations? The second question, and perhaps the more important one, is: Does it really matter, one way or another?

With only limited research currently available on ethnic population variance in cortical folding, there are no definitive answers to these questions at this time. However, a team of University of California, San Diego, School of Medicine scientists led by Chun Chieh Fan found that genetic ancestry—racial ancestry—was associated with the 3-D shape of the surface of the cerebral cortex, or cortical surface. The researchers analyzed data on this brain feature in an ethnically diverse sample of 562 children and adolescents, aged 12 years and older, from various locations in the United

States. The brain scans, along with the participants' DNA data, were part of the Pediatric Imaging, Neurocognition, and Genetics (PING) study, conducted in collaboration with the National Institute on Drug Abuse and the National Institute of Child Health and Human Development. The youth were grouped by genetic ancestry, traceable to one or more of four continental populations: West African, Northern and Western European, East Asian, and Native American.[50]

Although some of the study participants had mixed genetic ancestries, the complex geometric pattern of cortical folding of each participant correlated closely with his genetic ancestry. Research designed to map unique patterns in the grooves and peaks within the folds of the cortical surface may be beneficial for diagnosing neurological and psychiatric disorders in patients, regardless of their ancestry. For physicians and healthcare professionals, the findings gleaned from this research can be used to determine if a brain-related disorder exists based on cortical surface patterns linked with different genetic ancestries. As this study demonstrated, the geography of cortical folding that may appear to be abnormal in one patient may be normal for a patient from another population and vice versa. Patterns of cortical folds, then, can help experts tease apart differences in brain shape linked with a potential disorder from differences associated with his genetic ancestry.[51]

Some biological reductionists, including pronature advocates, may wonder if patterns of cortical folds, either on the entire surface of the cerebral cortex or the outer layer of specific lobes, correlate with IQ. In the study conducted by Fan's team, the pattern of cortical folds across populations was not associated with functional or cognitive abilities.[52] Race-based comparisons of connections between specific brain traits, other than total brain volume, and social outcomes, such as IQ score or mental rotation task performance, are sparse. This type of research carries a risk for misuse, but it also offers possible dividends for improved healthcare. If proposed links between brain traits, genetic ancestry, and social outcomes are to be accurate and meaningful, experts must look beyond simplistic correlations.

Thus far, the evidence strongly suggests that population variance in key brain traits, including gross brain volume and cortical folds, is not consistently or unequivocally related to cognitive abilities. But even if, in the future, statistically significant correlations are found between genetic ancestry, brain traits, and IQ, researchers must ensure that they control for possibly confounding factors that could distort how they interpret their data. To prove that population differences in certain traits contribute to ethnic population differences in IQ and educational achievement, researchers must account for the influence of a full spectrum of brain features, notably the architecture of the brain, that could affect the target traits they are investigating. If scientists do not consider continental population differences in the width versus length of the braincase—the overall geometrical proportions of the skull—when estimating the average brain size in different ethnic populations, they might unintentionally make a miscalculation. Without adequate knowledge of ancestral variation in brain shapes, researchers may overlook subtle influences of the brain's architecture on such features as white matter, or fiber tracts.

FIRING, WIRING, AND THE PLASTICITY PLAYERS

Neuroplasticity refers to the brain's exceptional ability to alter its own structure and function in response to changes and new experiences arising from either inside the body or the external environment. The driving force behind plasticity in the human brain is captured in the popular adage, "Nerve cells that fire together wire together."[53] Coined in 1948, by Canadian neuropsychologist Donald Hebb, this widely accepted axiom explains the core mechanism underlying the brain's ability to learn—to transform by adapting to a variety of experiences. The brain does this by rewiring itself, which includes modifying the strength of synapses by changing the connections between neurons situated within gray matter. But plasticity also involves altering the structure of white matter.[54]

Prolific in multiple neuroscientific fields, Donald Hebb was one of the principal architects of several theories on the brain and learning—theories of neural networks, developmental psychology, and human behavior. He is even credited with influencing Raymond Cattell's theory of fluid and crystallized intelligence, which today remains a core principle of the cognitive sciences.[55] (Despite this area of overlap, however, these two giants of mid-twentieth-century intelligence research were polarized on other issues. Ironically, while some of Catell's work on intelligence was used to promote eugenics, Hebb's findings revealed the benefits of early environmental stimulation for cognitive function, paving the way for Operation Headstart to help uplift children from disadvantaged families.[56])

Hebb's description of nerve cell activity explained plasticity at the level of the synapse, where nerve messages are transmitted across the synaptic junction. Plasticity in the human brain can occur in various regions and at various levels, from molecular actions underlying the synaptic firing of neurons to white matter—the fiber tracts located throughout the brain that govern longer-distance messaging between different regions of the brain. As new insights into plasticity emerge, experts are discovering previously unknown attributes of its effects on the brain's anatomy and abilities. In a recent laboratory study, scientists examined higher vision centers of the cerebral cortex and found that certain "neurons that fired together were, in fact, not wired together."[57] This study was preliminary, but it may mean that some neighboring groups of neurons, even when they tend to respond in concert to the same sensory stimuli, do not always interact with one another.[58] Another finding challenging the traditional view is that new skills or tasks can be learned without involving plasticity. Apparently, some changes in brain form and function that occur after a learning experience are due to innate predispositions, even though the specific gene variants may not be identified as yet. This new evidence does not overturn the fundamentals of brain plasticity in learning, but it illustrates the dynamic quality of research in this field.

White matter is not the only source of plasticity, although it is an important element, given that it accounts for half of the brain's volume. White matter is crucial for the brain's capacity to learn—an ability that manifests as new neural circuits form in the brain.[59] Scientists do not understand exactly how this process happens, but alterations in white matter are sometimes associated with improved performance on various tasks. We have noted that the white color of the brain's fiber tracts comes from myelin, the fatty substance covering some axons, although not all axons are fully encased in myelin. This is noteworthy because some brain changes related to learning involve myelin-coated axons, whereas others affect axons with partial myelin coats or axons that lack myelin altogether. Altered connections in white matter occur as a person learns a new task or practices a task throughout time to improve upon an existing skill, whether school-related, work-related, or recreational.

Still, the structural modifications that appear in white matter during learning are not always consistent, and they can vary across studies. In some situations, when a person is acquiring a novel skill, the amount of white matter increases, while in other cases it decreases. Changes can also occur in gray matter, which can either expand or shrink. In pianists, the white matter fibers connecting the motor areas on both sides of the cortex correlated with the number of hours a person had practiced during childhood.[60] The more a piano student practiced, the more fibers he had in his corpus callosum—the broad band of nerve fibers connecting the two cerebral hemispheres of the brain. In another study, white matter was positively linked with the number of hours a professional musician practiced, but the largest changes were observed in brain regions not yet fully encased in myelin.[61] Similarly, enhancing such motor skills as dancing or typing has been linked with an increased volume of white matter.[62] And yet, in one study, professional ballet dancers showed smaller amounts of white matter and gray matter, as well as lower brain activation, compared to nondancers.[63]

Researchers are unable to explain discrepancies observed in correlations between the quantity of either white matter or gray matter, on the one hand, and improvement in performing a similar task, on the other. Some of the contradictory findings could be due to the amount of training, the stage of learning, or the learning strategy.[64] In academic tasks involving memory, reading ability, learning grammar, or mental rotation, experts cannot always determine if the structural features of white matter detected on brain imaging scans are due to plasticity, based on experience, or caused by preexisting features.[65]

When scientists studied a group of adults learning to read, they found multiple changes in the white matter—increases in the volume, anatomical organization, and functional connections between the white fiber tracts and areas of the cortex activated during reading. This confirms the findings of a growing number of studies: When an adult learns a novel skill, the structure of the brain's white matter is altered. The changes in brain function required to learn a new skill may correlate with shifts in the structure of either gray matter or white matter, or both. These transformations in brain features and behavior collectively represent functional plasticity.[66]

Brain Traits and Educational Attainment

For years, scientists have known that parental education is strongly linked with IQ in children four to eight years of age, but they have not been able to clarify why this link exists. Could it be a reflection of genetic or environmental influences, or perhaps both? By the close of the twentieth century, there were indications that the educational level of a child's mother affects the child's IQ score, primarily, some experts surmised, through genetic mechanisms. Experts soon realized that environmental influences shared between a mother and her child also impacted the child's brain development, although only slightly so. One study suggested that parental education has a modest shared environmental effect, accounting for no more than 3 to 4 percent of the variation in verbal IQ scores in a large representative sample of developing children and adolescents in the United States.[67] In a study published almost 10 years later, researchers examined links between the volume of the total brain or specific brain regions, age, sex, family income, and parental education in American adolescents. The correlation between parental education and the child's IQ was positive but lower than that reported in adults. Brain volume—both the total brain size and regional brain volumes—as calculated from MRI scans did not mediate the correlation between parental education and IQ. The level of education of a child's parents seemed to influence certain aspects of his IQ, particularly verbal IQ, but it did not appear to impact full-scale IQ. Brain traits played a role, but that role tended to be indirect and difficult to precisely define.[68]

In this scenario, genetics undoubtedly contributed to the correlation between IQ and parental education, but so did such nongenetic influences as maternal education, prenatal care, and infant and child nutrition. The nongenetic, or environmental, pathways comprise well-known sources of enrichment, or nurturing, that can at times immensely alter the development of a young person's brain, for better or worse. Key brain traits like the structure of white matter, brain activation patterns, brain wiring patterns, or neural network connectivity are conditioned by delicate and dynamic interactions between biology and environment.[69] Currently, some experts are focusing on not only correlations between brain traits and intelligence, but also possible links between brain traits and behavior. They are investigating where, exactly, such social factors as family income and education fit into a multicomponent equation involving the brain, IQ, and psychosocial development. The correlation between the size of the cerebral cortex and full-scale IQ in children is small, typically only around 0.20 to 0.27, and lower than in adults.[70] This association is strongest between certain areas of the brain—the frontal and temporal lobes—which play a pivotal role in executive function, language, and other cognitive skills evaluated on IQ tests.[71]

Unlike other regions of the brain, the frontal and temporal lobes develop as a child ages. Even within the frontal and temporal areas, the correlation between brain volume and IQ may be either stronger or weaker, depending on whether the area contains primarily white matter or gray matter. IQ may be related to enhanced brain size—increases in the amount of both gray matter and white matter in the temporal and frontal lobes. But verbal IQ may be tied more to white matter—tracts of nerve fibers connecting different brain regions—and nerve cell transmission than brain size per se.[72] Nerve pathway connections with the frontal and temporal areas seem to explain some of the IQ variation in children and adolescents.[73] During both prenatal and postnatal development, environmental exposures to the fetus and newborn appear to affect verbal IQ more than full-scale IQ. Early nutrition is a major environmental factor that may have a greater impact on brain formation—a trait related to verbal IQ—than brain size—a trait linked with performance IQ.[74]

Neuroscientists Kimberly Noble of Columbia University and Elizabeth Sowell of Children's Hospital Los Angeles are among the innovative researchers exploring the intersection between biology, environment, and social outcomes, for instance, IQ and educational achievement. In a seminal study, their interdisciplinary team examined the relationship between a key brain feature and family income of 1,099 children and young adults, aged between three and 20 years.[75] The scientists found a positive association between surface area of the brain's cerebral cortex and parental educational attainment, with the strongest link occurring in children from the lowest-income families.

The brain surface areas of children whose families earned less than $25,000 a year was 6 percent smaller than those of families earning $150,000 or more. In the children from low-income backgrounds, even small differences in income correlated with relatively large differences in the surface area of various brain regions related to academic skills. These regions are crucial for language, memory, spatial skills, and reasoning.[76] The children from low-income backgrounds had overall lower scores on cognitive tests compared to the children from higher-income families. In children from higher-income households, incremental increases in income level were associated with much smaller differences in surface area. The correlation between income and brain structure was strongest in the most disadvantaged youth.[77]

The thickness of the cerebral cortex of the brain, at least in certain regions, is also associated with improved performance on intelligence tests. In another study, neuroscientists John Gabrieli, Allyson Mackey, and their associates from the Massachusetts Institute of Technology and Harvard University searched for correlates between brain anatomy—specifically the thickness of the cerebral cortex—and the income achievement gap in American students. (Cortical thickness is one of several metrics that experts use to measure the form and shape of the layers comprising the brain. In humans, the cerebral cortex is essentially a highly folded sheet of neurons. Cortical thickness, combined with the surface area of the cerebral cortex, determines the volume of gray matter in the brain. But cortical thickness can vary in different regions and different lobes within the brain.)[78]

To identify potential three-way links between brain anatomy, family income, and school achievement, Gabrieli and Mackey's team performed MRI brain scans on a group of 23 low-income and 35 higher-income teenagers in Massachusetts who had taken annual math and reading tests. The scans revealed that cortical thickness varied between the two groups, with students from higher-income backgrounds exhibiting greater cortical thickness in every brain lobe compared with those from lower-income backgrounds. A thicker cortex, especially in the temporal and occipital lobes, predicted better test performance. In keeping with this trend, 91 percent of the higher-income students versus 57 percent of the low-income students obtained proficient scores on math and reading tests.[79]

During adolescence, a thicker cortex, particularly in the temporal and occipital lobes, is associated with better test scholastic performance.[80] As previously noted, the correlation between cortical thickness and IQ appears to be either weaker or stronger at different ages.[81] But the study of Gabrieli's team was one of the first to demonstrate that cortical thickness in different regions of the brain varies between higher- and lower-income students, and that cortical thickness correlates with academic achievement tests scores. The focus on scholastic performance is refining research on the achievement gap, at least the gap between children of the "haves" and those of the "have-nots." In a published interview, Gabrieli noted that emerging findings underscore the powerful impact of economic influences on a physical trait as basic as brain structure.[82] According to experts at the forefront of research on SES and brain size, family income is associated with a broad array of other social factors—nutrition, health care, schools, play areas, and, sometimes, air quality.[83] And the critical thinking parts of a child's brain, but not necessarily his total brain size, along with cortical thickness, are influenced by his parents' educational attainment. These brain features combined seem to impact a child's performance on academic tests.[84]

In a similar study, cognitive neuroscientists used fMRI to evaluate the brain activity of adolescents from higher-income and lower-income families while they performed a working memory exercise—the *n*-back task.[85] Compared to the youth from lower-income families, those from higher-income backgrounds showed more activation in the brain's executive network located in the frontal and parietal regions. This neural network is activated to meet the need for working memory—short-term recollection required for such mental tasks as reasoning, learning, and comprehension. Higher family income also correlated with a greater capacity for working memory, as well as higher scores on a statewide test of math achievement. The youth who earned higher scores on the math achievement test were more likely to exhibit greater parietal activation while they performed the working memory tasks. For these authors, future studies can explore whether interventions to enhance working memory and its underlying neural circuitry improve the cognitive performance of lower-income students and narrow the achievement gap.[86]

Today, as neuroscientists explore how nature and nurture interact to determine brain function in humans, some are focusing less on total brain volume and instead turning their attention to the interplay between specific regions or structures within the brain and environment, particularly SES.[87] Some investigators are evaluating environmental triggers that may harm brain growth, as well as environmental interventions that may slow down or even reverse deleterious effects and promote healthy brain development.[88] Many of these interventions are intricately intertwined with the achievement gap. According to some experts, the physical differences in the brains of students from higher- and lower-income backgrounds correlate with monetary resources needed for prenatal care, nutritious food, healthcare, quality education, and social support, including quality education and residence in a safe and secure neighborhood. These factors are essential for promoting healthy brain development, as well as balanced physical and psychological growth. And all of them, or a lack thereof, can impact neuroplasticity, especially early in life. In some areas of the brain, neuroplasticity appears to last well beyond its peak of early childhood and into the formative years of adolescence and possibly even later adulthood. And while not all brain plasticity is beneficial, some adaptive changes of the brain can lead to appreciable improvements in life outcomes.[89]

Still other studies that compare youth, and sometimes adults, from different SES levels aim to find correlations between genetic variation, the size of the brain or specific brain structures, such as white matter, and educational attainment or cognitive performance, including memory, and IQ.[90] While both types of research examine the impact of biological and environmental influences on social outcomes, especially scholastic skills, they take somewhat different approaches. The two orientations overlap to some extent, but studies tracking genetics tend to reflect more of a pronature perspective than those that assess other biological influences such as neuroplasticity. In the next chapter we will dig deeper into recently discovered links between brain features; demographic or environmental factors like SES; gene variants, in some cases; and social outcomes—especially IQ and academic achievement.

The Search for IQ-Linked Genes

INTELLIGENCE: DEMYSTIFYING THE "H FACTOR"

MOST EXPERTS INVESTIGATING THE IMPACT OF NURTURE VERSUS NATURE on human social traits agree that intelligence, whether an IQ score or some other indicator of cognitive ability, is shaped by genes, environment, and the interplay between them. Exactly how much of a role genetics or environment plays may vary from one scientist to another. Hereditarians are quick to point out that intelligence has a high heritability, with estimates ranging from 45 percent in childhood to 80 percent in adulthood. The current overall range of IQ heritability is generally accepted as 20 to 50 percent[1]; however, the relatively high heritability of IQ test scores is sometimes misinterpreted to mean that IQ is determined mainly by genetics. But heritability—the illusive "H factor" of intelligence research—is not a measure of the direct influence of genetics on a given trait. It says nothing about the exact degree to which genetics determines a particular trait, whether a physical, disease-related, personality, or behavioral trait.[2] Heritability estimates simply tell us if genes are decisive in shaping a specific trait. A key point here is that heritability pertains to individuals in a population and not a single individual.[3]

As we learned in chapter 7, heritability is an estimate of the degree to which genes account for variation in a trait within the same population. The technical definition of heritability is the proportion of the total variation—the total amount of differentness—for a given observable trait between individuals in a specific population due to genetic variation (as opposed to environmental influences or chance) between the individuals in that population. A heritability estimate also indirectly allows experts to determine the extent to which environmental factors explain the variation in a trait.

By comparing genes and environment in identical twins and unrelated individuals, scientists have been able to estimate the heritability of numerous traits that are influenced by both genetics and environment. Height is a classic example. Heritability estimates for humans overall range from 60 to 80 percent, such that environmental factors, especially nutrition, account for 20 to 40 percent of the differences in stature of people throughout the world.[4] And yet, height heritability varies across ethnic populations, just as it does between men and women. Height heritability tends to be lower in African black populations, partly because such environmental factors as diet and nutrition, access to healthcare, and climate can vary considerably in developing nations.[5]

The heritability of traits can differ so dramatically from one population to the next that many experts, including some hereditarians, now admit that it is a meaningless construct for comparing genetic variance in two or more different populations.[6] Also, heritability estimates for traits such as height can be difficult to grasp because the estimates may vary for different ages. The environmental contribution to variance in height is strongest during childhood, with the genetic contribution increasing as a child grows, eventually peaking in adolescence. This pattern may be modified in different ethnic populations living in different social circumstances and different ecological conditions.[7]

The concept of heritability is marred by so much confusion that some experts suggest abandoning its use in scientific research, particularly with regard to studies on IQ and cognitive performance.[8] But the quest to identify and explain variance in cognitive abilities and personality traits is so pivotal to behavioral genetics, perhaps to genetics, in general, that notions of heritability cannot be avoided. Observable human traits include almost any imaginable characteristic, from biochemical to anatomical to social characteristics that are part of a person's makeup. Height, dimples, preference of either the right or left hand, the ability to taste the chemical PTC (phenylthiourea), red-green colorblindness, earlobe attachment, and curly hair are classic textbook examples of heritable traits in humans.

According to geneticist Tinca Polderman and her colleagues, essentially every human trait is heritable, meaning that none of them has a heritability estimate of zero.[9] In some cases, the amount of genetic influence on a trait may be very small. As a proportion, heritability estimates can range from a value of 0 (where no genetic influence is involved) to a value of 1 (where the different manifestations of a trait are attributed to genetic variation). Heritability values can also be represented as percentages, with a heritability estimate of 0.35 equal to 35 percent, an estimate of 0.5 equivalent to 50 percent, and so forth.[10]

Polderman's team conducted a detailed meta-analysis of almost every twin-based heritability study on both identical and fraternal twins during a 54-year span. (In a meta-analysis, the combined findings of multiple studies on the same topic are analyzed to produce a more statistically accurate analysis of the data, collectively.) The highest overall heritability occurred in traits related to brain function, eyes and vision, and ear, nose, and throat, while the lowest were found for skin, cardiovascular function, and blood pressure.[11] The lowest average heritabilities were observed in traits linked with, for example, height, social interactions, mental and behavioral disorders due to alcohol consumption, hyperkinetic disorders, and various anxiety disorders. The researchers found that heritability was close to 50 percent for traits related to skeletal development and function, weight maintenance functions, and mental and behavioral disorders due to tobacco use. This last set of traits appeared to be equally conditioned, at least on first glance, by both genetic and nongenetic factors.[12]

Polderman and colleagues also calculated high heritability values for specific personality disorders, as well as temperament and personality functions. But genetic contributions to complex psychological and social traits are not fully known. These traits vary within a given population, but the reasons for the variance may involve more than just an additive interaction between genes and environment. Other factors, like epigenetics—genetic inheritance without DNA—are thought to influence whether a genetic tendency in a particular individual manifests. In other words, multiple factors, which include, but also go beyond, genetic and environmental differences, can account for a population's variance in a trait, be it height or conduct disorder or IQ. And heritability estimates for the same exact trait sometimes differ from one ethnic population to the next and, in some circumstances, can even fluctuate within the same population at different times.[13]

Much of our knowledge about the heritability of various traits comes from studies of identical twins that share virtually 100 percent of their genes. And yet, the heritability of a particular trait for identical twins in one ethnic population may not be the same as the heritability for that same trait in a set of identical twins from another ethnic group. As in the case of height heritability, IQ heritability varies for different socioeconomic groups, ethnic populations, and age groups. IQ heritability is an estimate of the amount of variation in IQ scores in a single population explained by differences in IQ-linked genes between individuals in that population.

Almost 50 years ago, Sandra Scarr, the prominent researcher who helped propel behavioral genetics into the national spotlight, found that IQ heritability is decreased in relatively low-income environments. Youth (regardless of their ethnicity) raised in families with lower socioeconomic status (SES), as well as African Americans, had a lower IQ heritability relative to middle-class whites.[14] These results have since been confirmed by other experts, including University of Virginia psychology professor, Eric Turkheimer, who reported striking differences in the IQ heritability of 7-year old twins from higher and lower SES families. Published in 2003, the study of Turkheimer's team revealed almost inverted proportions of the genetic versus environmental contributions to IQ variance in the two income groups.[15] In impoverished families, shared environment—an indication of their common lifestyle—accounted for 58 percent of the IQ variance, with the estimated heritability close to zero, at only 10 percent. The proportions were reversed in the children from affluent families, where the IQ heritability was 7 percent and shared environment accounted for only 15 percent of the variance.[16]

The results of this study underscore the momentous role of environmental factors, even subtle ones, in shaping intelligence, especially in young people raised in marginalized or poverty-stricken environments. The lower IQ heritability reported in low-income social groups suggests that environmental factors have a relatively larger impact than genetics on the range of IQ scores in poorer communities—including many African American neighborhoods—compared to middle-class and affluent white and Asian American communities.[17] This is a widely accepted finding among experts today, but it has not altered the views of some researchers who claim that the genetic underpinnings of intelligence are so dominant that genes actually influence how individuals respond to environmental factors. However, the problem with this perspective is that it ignores the potentially important influences of epigenetic markers and brain plasticity.[18] We will delve more deeply into these topics later in this chapter.

Even though heritability cannot be equated with genetic inheritance, it offers a useful starting point for appreciating the fact that cognitive and behavioral traits, like numerous physical traits and certain diseases, are ultimately determined by both genes and environment. In and of itself, however, heritability does not shed light on the intricate activity of specific genes, or the interaction between genes and environment, or the interaction between genes and other nongenetic forces (e.g., epigenetics) that influence how well a person performs on an intelligence test.

Tracking Illusive IQ Genes

Almost 20 years after Robert Plomin, one of the world's leading behavioral geneticists, discovered a gene variant linked with above-average IQs, scientists still have not been able to single out any gene variant that is inarguably responsible for a substantial portion of IQ scores.[19] In the last two decades, several genetic factors have been considered top contenders for genes and gene variants potentially linked with various measures of intelligence, including IQ test scores and, according to some studies, overall brain size. In some cases, experts have earmarked a handful of isolated genes associated with both intelligence and brain traits presumed to play a role in cognition.

According to one recent estimate, there are almost 8,000 potential IQ-related genes, 158 of them core intelligence genes.[20] But to date, scientists have detected very few specific IQ-linked genes known to unequivocally either raise or lower IQ. In the early 2000s, premature assertions, sometimes bordering on outright fake science news about intelligence genes, became so rampant that several leading scientists in the field barely stopped short of calling for a global moratorium on the quest to find single IQ-related genes. At times, individual genes that once excited experts (and occasionally incited intense debate well outside the scientific community) have been touted as major breakthroughs only to be later dismissed as tantalizing but meaningless findings or, worse yet, relegated to the obsolete. To this day, there is still no scientific consensus about solid cause-and-effect relationships between individual gene variants and overall IQ scores.[21]

Recall that single nucleotide polymorphisms (SNPs) are one of the main types of genetic markers scientists use to help estimate a person's risk of developing a particular disease. Genes are composed of strips of DNA. No matter how long or short, each strand of DNA is composed of four bases, or

nucleotides, often designated by four letters—A, C, T, and G—which constitute the DNA alphabet. A SNP by itself represents only a minuscule difference within one source of DNA when compared to other sources of genetic
material. A SNP is a variation in a single nucleotide—a DNA building
block—that occurs between the members of a population or individuals from
different populations. A SNP can also refer to variation in a single nucleotide
between the DNA on paired chromosomes in the same individual.[22]

Often arising through mutations, SNPs are the most common form of
variation found in corresponding genes in people. When two or more people from at least two different populations have different SNPs within the
same gene—on one or more locations in the two analogous stretches of the
DNA double helix—there is genetic variation between populations. SNPs
can reveal important information about the genetic relatedness or genetic
distance between two or more people, or two or more populations, regarding
a particular trait, for instance, the risk for breast cancer or prostate cancer.[23]

With sophisticated technical tools now readily available, scientists
are applying the same methods used to study the genetics of disease to
investigations of intelligence-linked genes. Initial reports published in the
mid-2000s suggested that certain populations might have different frequencies of SNPs located in gene variants that seem to affect intelligence.[24]
But efforts to find single or isolated SNPs as a potential source of novel
genes that influence intelligence, personality, and social behavior are time-
consuming. They are akin to looking for a needle, not just in a haystack, but
also in all the haystacks stored inside a barn, and perhaps all the barns in
an entire community. With an estimated 3 billion base pairs, or chemical
letters of DNA, in the human genome—the complete set of DNA in our
species—how can scientists possibly hone in on key gene variants involved
in intelligence and, by implication, genes that contribute to brain traits,
including the size of various brain structures?

Both intelligence and social traits involve emotion, thought, and behavior, which are intertwined to mold an individual's personality. Cognitive
skills revolve around mental processes, including attention, memory, and
reasoning. Social traits linked with personality are sometimes called the "Big
Five": openness, conscientiousness, extraversion, agreeableness, and neurot-
icism. Some psychologists rank individuals based on the degree to which
they display each of these social characteristics and then share their findings

with geneticists and brain scientists. Ongoing collaborative research among specialists from these different fields is under way to identify multidirectional correlations between genes, brain features, and personality traits.[25]

The search for candidate genes—specific genes that are causally linked with a specific trait, for example, IQ—is viable when screening preselected gene variants in a small population. But these types of studies are inadequate for detecting a broad array of SNPs. Studies on links between brain volume and individual genes represent one way but not the only way of demonstrating correlations between the genetic underpinnings of social traits, notably cognitive ability, or IQ. Genome-wide association studies (GWAS), on the other hand, offer a plausible alternative to the traditional route of candidate gene association studies (CGAS) of disease, intelligence, and, more recently, social traits. With expanding databases of genetic data on massive numbers of individuals, including identical and nonidentical twins, scientists are increasingly conducting large-scale GWAS on the prevalence of genetic markers, particularly SNPs, in large numbers of individuals. These investigations equip researchers working at the interface between genetics, neuroscience, and psychology with a tool for tracking and mapping the genes contributing to the variance of both disease-related traits and cognitive traits.[26]

Project ENIGMA (Enhancing Neuroimaging Genetics through Meta-Analysis) is a GWAS, launched in 2009, to explore brain-related gene variants, brain connectivity, and brain function. On a practical level, the information generated from Project ENIGMA may help scientists identify genes associated with mental illnesses and Alzheimer's disease, and, more importantly, lead to effective new treatments for these disorders.[27] GWAS reportedly offer a more robust method for finding SNPs linked with cognition and behavior. Despite some critics who question their validity, these investigations seemingly have fewer built-in biases than CGAS.[28]

For almost a decade, neurogeneticists have been collecting evidence that demonstrates that intelligence, typically in the form of IQ scores, is heritable, or genetically inherited, and shaped by multiple, even a plethora of, individual gene variants. In 2011, a largely British team of researchers led by University of Edinburgh statistical geneticist Gail Davies conducted a GWAS of almost 550,000 SNPs in more than 3,500 unrelated middle to older adults.[29] At the risk of oversimplifying, their goal was to analyze the connection between common gene variants, specifically closely linked SNPs usually inherited together, and both crystallized-type intelligence and fluid-type intelligence. These are the two principal types of cognition assessed in IQ tests. Crystallized-type intelligence refers to acquired culturally specific knowledge gleaned through experience, whereas fluid-type intelligence is associated with innate problem-solving ability, including memory.[30] As in other studies in this field, no individual genes or gene variants, including SNPs, were identified that were strongly associated with intelligence. However, the linked SNPs, combined with unknown influences, accounted for 40 percent of the variation in crystallized-type intelligence and 51 percent of the variation in fluid-type intelligence between the individuals in this study.

In 2012, Project ENIGMA scientists, including UCLA neuroscientist Paul Thompson and University of North Carolina at Chapel Hill neurogeneticist Jason Stein, published their results of a massive GWAS study on the genetics of brain disease in healthy people of European ancestry.[31] With access to brain scans and genetic data for more than 19,000 individuals, researchers tracked close associations between specific genes and brain volume. The scientists collected DNA from the study participants and examined the DNA for individual SNPs. Each participant underwent an MRI scan, and the resulting data on the size and features of each individual's brain were cataloged and subsequently correlated with his DNA. The findings yielded novel information—previously undiscovered links between gene variants and brain size, including total brain volume, as well as the size of the hippocampus, one of the brain's main memory storage centers. With the promise that the data possibly could be used in the battle against aging memory loss, some of the gene variants identified in this study may become potential targets for therapies to combat Alzheimer's disease and other forms of dementia.[32]

Thompson's team also found that certain gene variants were associated with both brain size and intelligence, specifically IQ as a measure of intelligence performance. One of these variants, HMGA2, is a gene with multiple functions, including increased risk for cancer, diet-related obesity, and height. By 2012, HMGA2 provided the best evidence up until that time of a single gene influencing IQ.[33] The HMGA2 gene has two alternate variants that differ by only one base pair, or chemical letter of DNA. One version has a C instead of a T. Scientists already knew that this gene affected a person's height, but its link with mental aptitude was a new revelation. The C-variant of HMGA2 was associated with both height and brain volume, although not with the size of the memory center or the size of the entire brain. Early research had shown that people with the C-version of HMGA2 had larger brains and higher IQ scores.[34] According to the Hapmap database—a now largely defunct dataset of millions of SNPs—the frequency of the C-version HMGA2 SNP is highest in some African black populations, reaching 60 percent in the Yoruba of Nigeria and 54 percent in African Americans. This variant is reportedly lowest in Asian populations, occurring in 11 percent of the Han Chinese in Beijing and 12 percent of Japanese in Tokyo. Populations with frequencies of the C-variant intermediate between the rates of Nigerians and East Asians include the Luhya and Maasai of Kenya, the Tuscan of Italy, and residents of Utah with Northern and Western European ancestry.[35]

Regardless of its distribution, the SNP in the C-version of the HMGA2 gene had only a small impact on intelligence and brain size. The variant's link with intelligence was weak, at best, with each C-variant accounting for only about 1.29 IQ points. (The average association produced a gain of about 2.58 IQ points in individuals who inherited two C-variants of the HMGA2 gene, or one variant from each parent.) But skeptics surmised that the correlation between this variant and brain size could simply be due to a link between a gene and general body size. They claimed the association might not have anything to do with enhancing cognition by influencing brain volume.[36] At one time, HMGA2 signaled the most promising intelligence gene discovered to date. Diminished in status and seemingly swept under the rug, today HMGA2 is no longer considered a viable candidate for an IQ-linked gene.

Enthusiasm about this gene variant has waned throughout the years and with good reason. Individual intelligence-linked genes, particularly those that are unique because of only a single SNP, have a minuscule effect on IQ. It soon became apparent that the fate of a small-time genetic loner, an isolated SNP, is virtually meaningless in the realm of genetics and cognition. Researchers eventually began their search to track multiple genes that in concert might yield combined or perhaps synergistic effects on brain traits tied to intelligence. Today, efforts to clarify how IQ-linked genes are expressed, and if and how different variants of these genes might interact or possibly counteract one another to impact IQ, are more intense than ever. In the past, methods used to analyze large data sets have generated many false positive associations between specific genes and cognitive traits, but laboratory techniques and computerized analyses have improved during the last decade.[37]

Only a few specific gene variants underlying IQ have been confirmed, but another research strategy—genome-wide association studies—is increasingly being used today to identify genetic influences on educational achievement. Researchers can now examine DNA data on unrelated individuals in an attempt to find correlations between SNPs and signs of educational success.[38] One research team found a gene variant that correlated significantly with years of education in a group of more than 20,000 healthy individuals.[39] The gene variant was a SNP, which in turn was significantly associated with general cognitive ability ("*g*") in overlapping groups of participants in the study. With the use of large-scale datasets, scientists were able to find a key association in two different cohorts of more than 43,000 and 68,000 individuals, respectively. According to the authors of this study, the association between a SNP and cognitive function, and that particular SNP and schooling, demonstrated that educational attainment is a valid proxy for cognitive ability when using GWAS to assess intelligence levels.[40] Not surprisingly, previous research had revealed that this same SNP was linked with years of schooling.

Other investigators, often from diverse academic disciplines, are currently leveraging similar approaches to map genetic correlates of schooling.[41] The Social Sciences Genetic Association Consortium, an international collaboration of researchers, identified 69 genes linked with higher educational attainment.[42] With Cornell University's Daniel Benjamin and the Netherlands' Cornelius Rietveld as lead coinvestigators, this study included about 106,000 people, all of them Caucasian. The results were compared with the findings from another study on a sample of 24,000 people. After cross-checking, researchers found that three of the 69 genes seemed to have a direct connection to slightly elevated cognitive abilities. But each of the three variants could account for an average of only about 0.3 points on an IQ exam.[43] Each gene variant had about the same amount of influence that a SNP has on educational attainment—that is, approximately one month of schooling per variant. The impact of these three education-linked SNPs was far less that the effects of SNPs on such complex physical traits as height and body mass.[44]

The score from the schooling-related SNPs was responsible for about only 2 percent of the variance in both educational attainment and cognitive function.[45] As behavioral scientist Daniel Benjamin explained, these variants have a mild influence on IQ, about one-twentieth of the effect of genes have on traits like height. Putting this information into perspective, this means that to identify all the common gene variants responsible for the average 15-point gap in IQ between African Americans and white Americans, researchers would have to analyze the IQ scores, educational attainment, and possibly related SNPs of more than 1 million people from diverse genetic ancestries.[46] The findings from this study were preliminary, but they support previous claims that the role of genetics in intelligence, regardless of what that role may be, involves a sizeable number of genes. Sorting out precisely what these genes mean and do not mean is still unclear, however. We cannot ignore the fact that multiple genes may interact with each other as well as with nongenetic influences, including environmental forces and epigenetics in shaping cognitive ability, for instance, performance on an IQ test.[47]

Daniel Benjamin's team may seem to have completed their study empty-handed, but their endeavor was hardly futile. Their work has provided tangible evidence that individual IQ-linked genes can indeed each have a small, though minute, effect on intelligence performance. The knowledge gleaned from their investigation and others like it is helping to chart a revamped course for future scientific explorations, while also tempering expectations of what experts are likely to unveil. GWAS have merit, but they have not turned out to be a panacea for uncovering the inner workings of myriad genes involved in the biology of intellect. More importantly, these studies have yet to explain just how a massive number of genes is influenced by epigenetics, or how gene expression is shaped, even transformed, by intersecting, even crisscrossed environmental forces, or ultimately how gene variants collectively limit or augment the plasticity of the brain.

Unfortunately, studies designed to detect such small genetic effects have been difficult to replicate. Another serious problem is that the genetic links with intelligence reported in these investigations tend to be false positives. Even though IQ score and perhaps educational achievement are influenced in varying degrees by genetics, it is not entirely clear if the same genes contribute to both educational attainment and cognitive function in the same person, let alone different people or different populations. And it is even less clear whether these genes play a role in additional aspects of intelligence related to social behaviors.[48]

What, then, can we conclude about the variation in adult IQ scores that is not influenced by DNA? Can nongenetic forces modify genetic tendencies toward higher or lower cognitive performance? To answer this question, some researchers are now honing in on genes that appear to influence not only intelligence, but also a wide range of personality traits.[49] Are behavioral geneticists co-opting the pronurture arsenal of environmental influences on intelligence and social traits? Are they justified in subjugating a wide array of environmental forces to what they call the power, even supremacy, of the gene? Perhaps. But other experts, some of them geneticists themselves, admonish that it is premature to infer cause-and-effect relationships between specific genes and intelligence levels, or specific genes and social traits. According to psychological researcher Christopher Chabris and colleagues, most of the genes associated with intelligence probably do not affect a person's IQ. These experts do not deny a crucial role of genetics in intelligence. And they acknowledge that the advanced technology used today in GWAS has overcome many of the flaws that in the past led to false positive associations between specific genes and cognitive traits. But they caution that even these types of studies have their limitations.

When Chabris's team examined large sets of data on both intelligence testing and genetics, they, like other researchers, could not pinpoint particular genes associated with mental ability. They analyzed links between g—general intelligence—and 12 specific genetic variants in more than 9,700 people. Based on their calculations, they expected to find 10 to 15 significant associations. And yet, only one gene in their entire study was tied to cognitive ability. Even then, the effect of this gene on intelligence was noticeably small. As Chabris explained, intelligence, like height, is most likely influenced by thousands of genes, as well as multiple variants for some of these genes. And the effects may extend well beyond the direct influence of single genes. Chabris's conclusions confirmed what other research findings suggest: Genes interact with one another and the environment, as well as with non-DNA mechanisms of inheritance—epigenetics.[50]

Chabris's perspective resonates with the view that Robert Plomin expressed in the 1980s, long before the advent of GWAS, that numerous genes acting in concert influence cognitive performance. Intelligence, like height and various diseases, is polygenic, or influenced by multiple genes. Today, GWAS are stringent enough to allow researchers to perceive connections between specific SNPs and complex diseases. Some scientists are now using these same methods to identify even more detailed associations between SNPs and either increases or decreases in intelligence performance, especially at different ages in the life span.[51] According to recent estimates, SNPs account for 40 to 50 percent of the individual variation in IQ scores in a single population.[52] Estimates of the influence of SNPs on intelligence performance in children range between 22 and 46 percent,[53] but once again, single SNPs have an almost negligible effect on IQ scores. This may be indirect evidence that aggregate or combined effects—perhaps synergistic effects—ultimately may explain the genetic contributions to individual differences in IQ scores.[54] Still, even additive or synergistic genetic influences can account for only a small fraction of the variance in IQ scores among individuals in the same population.

By the early 2010s, some researchers proclaimed that many of the published findings of the previous decade on behavioral genetics were outright "wrong or misleading" and failed to further "advances in knowledge." Serious scientists were reluctant to abandon their laboratories, but they stood ready and waiting for the next great transformation in the quest to discover the genetics of intelligence. In 2017, the tide suddenly turned when an international team of 30 scientists, notably Amsterdam's Vrije Universiteit neuroscientists, Danielle Posthuma and Suzanne Sniekers, published their findings on the DNA of more than 78,000 participants—about 60,000 adults and 20,000 children—of European descent.[55] The study identified 52 genes for intelligence, 40 of them newly uncovered variants that are mainly involved in activating brain tissue.

This fresh bucket of "smart genes" reportedly helps to regulate multiple functions of brain biology—neuron growth, the formation of brain cell gateways or pathways, and development of synapses. (Recall that synapses are the junctions between brain cells where messages are transmitted from one neuron to the next.) These genes collectively showed more of an impact on the biological underpinnings of intelligence than any study had previously revealed, but still they accounted for only as much as 5 percent of the variation, or heritability, of the participants' IQ scores.[56] The genes detected in this study were also linked with several other factors, for example, increased educational attainment, a larger head circumference in infancy, a longer life, taller height, overcoming tobacco addiction, and, quite unexpectedly, autism spectrum disorder. Also, people lacking certain high-IQ genes were found to have a higher risk for schizophrenia or obesity.[57]

GENES AND IQ: A RACE-BASED REINTERPRETATION

The studies discussed so far in this chapter reflect a new direction in behavioral genetics—research designed to integrate neuroscience, genetics, and social science in explaining variable levels of intelligence performance, especially IQ scores, and, to some extent, educational attainment, in different groups. As noted in chapter 8, some experts have started examining the genetic and nongenetic correlates of socioeconomic divides in IQ and educational achievement. In contrast to much of the behavioral genetics research of the 1980s and 1990s, the recent published scientific literature in this field pays minimal, if any, attention to ethnicity or population ancestry, or racial background, in determining IQ and academic success.

Emerging evidence suggests that genetics can explain some of the differences in cognitive performance between social groups as correlates of SES. This evidence comes largely from studies conducted on demographically diverse white Europeans. The goal of most of this research is not to explore links between genetic ancestry, or ethnicity, and IQ. Could SES, at least in some studies, serve as a proxy for race and ethnicity? Perhaps, to some extent, but SES and race are not synonymous, despite their overlapping attributes. The implications that the new wave of genetics and SES studies may have for the ethnic achievement gap, specifically social outcomes of African Americans and blacks in other societies, remains to be seen. Some authors have interpreted emerging research findings to mean that lower SES individuals seem to be drawn to harsh and impoverished life circumstances because, despite adverse environmental influences on their life circumstances, they have a genetic propensity to gravitate toward these types of environments.[58]

Today, most conventional behavioral geneticists and their closely aligned peers in the neurosciences are not actively, or at least conspicuously in the public domain, exploring possible links between population genetics—race-based genetics—and such social outcomes as IQ and educational attainment. The same cannot be said for some of those outside the mainstream. Davide Piffer, an Italian evolutionary theorist, has authored multiple articles in which he has reanalyzed other researchers' data in an attempt to prove that innate race-based differences in intelligence exist.[59] As an active supporter of the alt-right, Piffer has been criticized for not only his views on race and IQ, but also his unorthodox research interests that fall outside the canons of established scientific inquiry. Nonetheless, Piffer has gained a small but dedicated following of race-realists and racialists who are anxiously awaiting empirical research findings that once and for all will settle the fierce debate at hand. Some of Piffer's supporters apparently believe that his analyses can, or will, validate the intellectual superiority of whites, coupled with proof of the mental inferiority of people of African black descent.

In one article, Piffer ranked the average frequency of seven supposedly IQ-raising gene variants in different ethnic populations, using gene variants, specifically SNPs, culled from previously published mainstream data.[60] Piffer based the average IQ of each ethnic population in his analysis on the country-level average IQs compiled by racialist scholars Richard Lynn and Tatu Vanhanen in their 2012 book *Intelligence: A Unifying Construct for the Social Sciences*.[61] According to Piffer's calculations, the frequencies of gene variants linked with elevated IQ scores were highest in East Asians, followed in descending order by Europeans, Native Americans, Melanesians, and Papuan New Guineans, as well as Sub-Saharan black Africans. The lowest prevalence occurred in the Khoisan (sometimes derogatorily called the San Bushmen) and Mbuti Pygmies, the indigenous hunter-gatherers of the tropical rainforest in Africa's Congo region. Piffer claimed that these gene variants are tied to natural selection of intelligence, implying that populations with a low prevalence of advantageous, "IQ-increasing" genes did not evolve the same level of increased cognitive abilities as those with a higher prevalence.[62]

One of the immediate flaws with Piffer's analysis was that he relied only on the IQ-enhancing SNPs from European populations. They were, after all, the only SNP data recorded in the published GWAS he reviewed. But with literally hundreds and perhaps thousands of IQ-linked SNPs in the species-wide human genome, no one knows if other SNPs contribute to cognitive skills in non-European populations. Piffer's paper was initially rejected by the journal *Intelligence* because, according to the reviewers, the assumptions and methods of his study were fraught with multiple errors. Incidentally, *Intelligence*, although a peer reviewed scholarly publication in psychology, is closely linked with Richard Lynn, one of the world's most widely known racialist scholars and eugenicists. Piffer's paper was later accepted and published in this journal in 2015, under the title, "A Review of

Intelligence GWAS Hits: Their Relationship to Country IQ and the Issue of Spatial Autocorrelation."[63]

Piffer's publications on genes and intelligence have been recycled, to some degree, through social media outlets for the alt-right, human biodiversity forums, and various white supremacy organizations. But publicly, his work has not received any acknowledgment, nor has it garnered professional praise from leading neuroscientists, geneticists, and other experts exploring human cognition. It bears repeating that researchers can be pronature and engaged in genetic research on intelligence and social behavior without being racialist, race-realist, or racist, at least overtly so.

However, Piffer did have a brief, missive exchange with Danielle Posthuma, the aforementioned specialist in statistical neurogenetics. For more than a decade, Posthuma has been involved in groundbreaking GWAS (including one of those leveraged by Piffer) that explore the association between genes, brain features, and cognitive ability. In an effort to clarify what GWAS can and cannot reveal, Posthuma wrote a commentary published in the *Unz Review*, an alternative media blogging platform. The audience of the *Unz Review* includes largely, but not exclusively, anti-Semites and race-realists. In her commentary, Posthuma pointed out the obvious—a fact already stated in the article she coauthored. As she put it, the results of this study, quite simply, do not provide information that can explain genetic population difference. There was no evidence that the particular gene variants identified in this article could also explain genetic variance, or heritability, in non-European populations. In addition, she emphasized, linkage patterns—nonrandom links of certain gene variants—found in one population do not necessarily hold up for other populations.[64]

As Posthuma explained, gene variants linked with educational achievement or IQ in a European population have different frequencies in other populations. She noted that this holds true for a lot of gene variants, regardless of whether they are tied to educational attainment/IQ. "If our discovery sample were to be non-European," she wrote,

> we would probably also predict the European population to have a lower intelligence, simply because the alleles [gene variants] that explain differences within the non-European population may have

a lower frequency in the European population, due to historical reasons. Thus, that prediction is meaningless for explaining between group differences.[65]

Posthuma's clarification may have seemed to settle the issue, at least for rational thinkers, whether pronature, pronurture, or somewhere in between. She was not denying that human cognition has genetic underpinnings. Recall that genes linked with intelligence in general are reportedly associated with about 20 percent (and some sources say as much as 50 percent) of the IQ variation in a given population.[66] Although these genes appear to be the main sources of DNA linked to IQ, additional gene variants may well be revealed in the future. On a cautionary note, Posthuma warned, "What's written in our genes is not what determines our lives." As she clarified, although scientists usually examine genetic effects in isolation, a certain pattern of genetic variants rather than the sheer number of variants—along, of course, with a respectable quotient of gray matter—may be what makes a person more intelligent.[67]

Despite Posthuma's eloquent elucidation, in 2018 Piffer uploaded another article online in which he used a new set of similar data to reiterate the same assertions.[68] Once again, few, if any, conventional, let alone prestigious, researchers have commented publicly on Piffer's reanalysis. But some fringe onlookers, seemingly impatiently seeking scientific proof to abet racial divisiveness, are buying into notions, perhaps unknowingly, that are not backed by rigorous scientific research or, for that matter, good, old-fashioned, sound reasoning.

SES AND SNPs: A DEEPER LOOK AT BIOLOGY, IQ, AND SCHOOL ACHIEVEMENT

Long before the advent of genetic studies on personality and behavior, educators recognized that a family's socioeconomic status—SES—is one of the best predictors of a child's ability to master knowledge and skills for literacy, numeracy, and science.[69] For most social scientists, the interplay between family SES and a child's educational achievement signals an interaction between two environmental conditions, without any mandatory role for genetics. But that perspective is changing. Some may say it is evolving, but for others it is devolving. Today, some experts no longer view SES as an isolated environmental phenomenon shaped solely by a host of experiential and ecological influences. Instead, by mapping associations between genes identified through genome-wide SNPs and various social outcomes, psychologists like King's College London's Robert Plomin and Eva Krapohl have shown that links between children's educational achievement and family SES correlate with genetics. Krapohl and Plomin first analyzed large amounts of genetic data and then used their findings to calculate a score for the impact of multiple SNPs on the variation of social traits, for example, SES and children's educational achievement, within a population. Their goal has been to determine the degree to which multiple SNPs can explain differences in educational achievement or SES in a large number of individuals. GWAS provide a convenient technique for collecting this type of data.[70] We will discuss some of their research momentarily.

Findings from other studies come from emerging endeavors to map the genetic underpinnings of social and behavioral traits. Scientists in England studied more than 6,000 pairs of twins and found that genes affect academic test scores and grades.[71] About nine out of 83 various traits related to scholastic performance, including health and happiness, were highly heritable, which technically means they had a high heritability estimate. By contrasting traits in identical twins with those of nonidentical twins, researchers were able to tell which social traits correlated with test scores on the General Certificate of Secondary Education (GCSE), a language and math exam given to secondary students in England, Wales, and Northern Ireland. Just fewer than two-thirds—about 62 percent—of the individual differences in the GCSE scores could be attributed to genetic factors. These differences pertained to heritability, but they were not a direct measure of heredity. Furthermore, the heritability of social traits can vary, depending on the society, culture, or country being studied. As in the case of height heritability, genetics may have a larger effect on differences in IQ or scholastic test performance in developed countries compared with underdeveloped societies, where educational opportunities are limited.[72]

As noted earlier, multiple studies have reported a lower heritability for IQ in low-SES families, suggesting that environmental effects account more than genetics for the variance in intelligence performance in people from low-income families. Vice versa has also occurred, with higher heritability estimates calculated for high-SES families compared to low-SES families.

The implication here is that genetics contributes more than environment to the differences in the IQs of individuals from middle-class and upper-class backgrounds. But the overall results from research in this field have been inconsistent. Some studies have produced findings that are not just equivocal but sometimes have even trended in the opposite direction predicted by experts. In other cases, lower SES has been associated with either higher IQ heritability in children or no impact on the genetic effect on intelligence.[73]

In a study led by Krapohl that assessed children's educational attainment and family SES, researchers could not pinpoint particular genes that might contribute to academic performance. But they suggested that, as in research on the genetics of IQ, literally hundreds of genes acting in concert might influence specific traits that in turn could impact test sores. These scientists did not ignore nongenetic factors that potentially contribute to intelligence performance. They were well aware that a child's success in school is also shaped by self-efficacy—the ability to believe in himself—and such traits as personality, well-being, behavior problems acknowledged by the child's parents, and the child's own awareness of his behavioral issues.[74] Self-efficacy affects multiple dimensions of a person's life. A child's health, as well as his attitudes toward home and school, can have direct bearing on his educational attainment in life.[75]

Intelligence, at least as manifested in IQ test scores, is influenced by not only genetics but also environment and experience. And a family's SES in the local and wider community is a strong predictor of a child's educational achievement. These findings combined suggest that cognitive performance is strongly conditioned by interlocking environmental and other nongenetic forces. Today, the once-simplistic dichotomy of genes versus environment in shaping cognitive and even social traits is disintegrating. This knowledge is widely known. But ironically, amid rapid advances in the intersection between genetics, neuroscience, and behavioral research, some of the influences on intellect that were once considered purely environmental have been redefined. For behavioral geneticists like Robert Plomin and Ian Deary, the correlation between education, social class, and intelligence may be due to shared environmental causes or shared genetic influences, or both.[76] This is a novel way of thinking about education and social class—distinct social conditions that have traditionally been conceptualized as only environmental. As a result, some researchers are now examining how genes might mediate the association between two environmental influences—family SES and educational attainment.

In a landmark study, Krapohl's team identified the SNPs of 3,000 unrelated 16-year-old children who had taken the national examination of educational achievement in the United Kingdom.[77] When the DNA results of the youth were analyzed in relation to their family SES and educational achievement, there was a 0.50 correlation between educational achievement and family SES. The correlation with intelligence was 0.55 for educational achievement and 0.38 for SES. The children's cognitive ability accounted for one-third of the SNP link between family SES and the youngsters' educational achievement. This study was one of the first of its kind to use molecular data to show that genes can explain some of the variation in the educational attainment of secondary school children. The correlations between the gene variants and the two social outcomes were hardly perfect, as most fell well below 1.00. Still, the authors concluded that genetics exerts a substantial influence on differences in children's educational achievement and its association with family SES. Then they went one step further. They suggested that their empirical evidence for links between family SES and genetics data could help inform policy decisions to benefit the education of all children.[78]

In a related study, scientists tested DNA from 3,000 unrelated children in Britain to determine how genes are linked with family SES and its association with the children's IQ at ages seven and 12.[79] The goal of this study was to find the degree to which the same genes affect family SES and children's IQ. Researchers reported a genetic correlation of 0.66, which they claimed indicates that the same genes are largely responsible for genetic effects on both family SES and children's IQ. As one investigator put it, the results suggest that once genes linked with a child's IQ are identified, the same genes will probably also be associated with family SES.[80]

In this study, genetics accounted for 94 percent of the correlation between family SES and children's IQ at age seven but only 56 percent of that correlation at age 12. As the scientists admitted, however, further studies are needed to explain why such a difference, if indeed the difference is real, exists. To reconcile this discrepancy, they reasoned that the genetic contribution to the SES–IQ correlation might be lower at 12 years of age because environmental influences outside the family are increased at this time. Examples of environmental effects outside of home life include peers in the classroom, as well as teachers. For some researchers, SNP data alone, at least in aggregate, are sufficient to predict intelligence performance in individuals, particularly during middle to older adulthood.[81] But other experts are not so sure.

In many behavioral genetics studies on the heritability of IQ, SES is based on the educational level and occupation of the children's parents, and on the household income of the child's family. But this is a severely narrow concept of SES that may unintentionally miss nuanced, but nonetheless critical, social and economic forces that influence intellect, emotions, and behavior.[82] To their credit, some pronature researchers recognize that children from economically disadvantaged families may be socially and culturally less prepared for academic life than middle-class youth. Some evidence suggests that children reared in low-income homes are less likely to have caregivers or siblings that read to them, and they are reportedly more inclined to watch more television. They generally attend substandard schools where they are exposed to fewer books and have less access to computers.

In addition, there is some evidence that caregivers in low-income families tend to be more authoritarian and less involved in children's school activities compared to parents of middle-class children. To some degree, mainstream researchers' (along with the public's) perceptions of low-income African American households are based on hyperbolic, if not distorted, social and ethnic stereotypes. On the other hand, low-income households overall, compared to middle-class lifestyles, do tend to have more instability, more crowding, and a noisier atmosphere. They are often located in impoverished neighborhoods that lack safe facilities, safe play areas, and well-stocked retail businesses.[83]

For pronurture scholars, the lower IQ heritability often reported in low-SES groups means that environment is more responsible than genetics for variance in the IQ scores of the group members as a whole. These groups are typically low-income demographic groups or predominantly lower-income ethnic minority populations. From the traditional pronurture perspective, family oriented environmental interventions starting early in life might be more beneficial for increasing cognitive skills in children from lower SES backgrounds.[84] However, as just noted, some behavioral geneticists, embracing a pronature stance, are interpreting IQ heritability trends in a different manner and, in fact, reaching the opposite conclusion. One study reported that shared environmental experiences of parental SES—environment–environment interaction—explained the variance in the IQs of children from low-SES background more than gene–environment interaction. The authors found only a modest difference in the shared environmental influence between lower-versus-higher SES families.[85] They also claimed that the shared environmental effects on IQ decreased from childhood to adulthood.

Based on these findings, some pronature-oriented researchers have concluded that environmental engineering is unlikely to have a favorable impact on increasing IQ throughout time.[86] For many of them, these findings provide sufficient grounds for rejecting, or at least reassessing, policies supporting childhood environmental interventions. This may be why some behavioral geneticists and their like-minded colleagues immersed in research on genetics, SES, and IQ do not believe that full-scale environmental support—intensive nurturing programs—is sufficiently effective or economically sustainable to justify implementing in low-SES families.[87]

Other evidence, however, paints a contrasting picture that underscores the psychosocial benefits of early environmental, antipoverty interventions. In the last decade, several studies have reported significant links between SES and changes in the developing brain, particularly in areas related to memory, executive control, and emotion.[88] In chapter 8, we mentioned the large, multiinstitution study helmed by Kimberly Noble of Columbia University and Elizabeth Sowell of Children's Hospital Los Angeles.[89] Although these researchers did not analyze gene variants of their study participants, previous research has shown that chronic poverty explains a six- to 13-point IQ decline in children by five years of age. And SES accounts for an estimated 20 percent of the variance in children's IQ.[90]

Noble's team found that parental education level and family income correlated directly with brain size in developing children and adolescents. The scientists conducted MRI scans of the brains of 1,099 children and young adults, from three to 20 years old. Correlations between brain volume and IQ are not precise, but this study revealed that children of higher-SES families had either larger or more developed key brain areas, or both, including regions involved in memory or language.[91] In addition, children from higher SES backgrounds performed better than lower SES children on IQ tests, reading and language exams, and tasks requiring focused attention. As these experts pointed out, low SES is associated with negative environmental effects—family stress, increased exposure to environmental toxins, or inadequate nutrition—that could halt normal brain growth. Children from higher status backgrounds are spared these adversities, while also usually benefiting from ample "cognitive stimulation" from their parents.[92]

In this study, the children of parents with only a high school education—12 years of schooling or less—had 3 percent less cortical surface area than children whose parents had attended universities with at least 15 years of education. (The cortical surface area is the outer covering of the entire cerebral cortex that houses key thinking centers of the brain. It is rich in gray matter, meaning it has a dense concentration of neurons, or nerve cells.) Cortical surface area and family income levels were also correlated, although not in a linear fashion. Instead, larger incremental increases in cortical surface area were linked with the lowest income levels, but the amount of increase tended to level off at higher income levels. Still, the difference was striking at the two poles of lower and higher incomes. Children from families earning $25,000 a year or less had cortical surface regions approximately 6 percent smaller than those of families earning more than $150,000. The correlations between SES and the size of various brain regions or structures were not affected by race and ethnicity but remained the same across individuals, regardless of their genetic ancestry.[93] This may well be a paramount finding. Measurable differences in brain-related traits during childhood were associated with environmental factors linked with family finances and social status rather than population genetics.

NEUROPLASTICITY: REORGANIZING THE BRAIN'S SYNAPTIC CONNECTIONS

The research of Kimberly Noble and associates showcases the profound effects of environmental and possibly other nongenetic forces on flexible change in brain form and function, which in turn can dramatically impact cognitive performance. As noted, this capacity for malleability is called neuroplasticity—the brain's ability to reorganize itself by forming new connections between neurons and synapses throughout a person's life. These connections allow the brain to respond to injury or environmental changes.[94] In contrast to older concepts of brain development and its impact on intelligence and social behavior, the emerging model of neuroplasticity revolves around a brain perceived as dynamic and adaptive through much of a person's life. Despite some exaggerated claims and unfounded hyperbole surrounding the narrative of neuroplasticity in popular culture, the concept of the plastic brain is largely accurate. It has already radically changed the landscape of neuroscience research.

Some, although not all, aspects of the human brain directly affect cognition by reprogramming its softwiring, sometimes resulting in improved abilities but, conversely, potentially leading to reduced mental or psychological coping. Well-known examples of neuroplasticity in humans include changes in the visual and spatial regions of the brain used in navigation, bilingual skills, and cognitive recovery after a brain injury. Nutrition, especially severe deficits, can have profound effects on the brain's overall growth, as well as connections that form between nerve cells in the brain. In 2003, South Korean anthropologist Sunyoung Pak documented the shorter stature of North Korean refugee children living in China compared to their counterparts in South Korea. The two Korean populations have historically been considered genetically homogeneous, but decades of substandard living, with poor nutrition, have taken their toll on North Koreans.[95]

As *Los Angeles Times* journalist Barbara Demick reported in 2004, although public comment had been scarce, there were indications that the overall aptitude in intelligence performance—IQ scores—was substantially lower in North Korean refugee youth compared with age-matched youth in South Korea.[96] There were no published reports in academic journals or the popular press on this matter, but apparently international health workers and policy makers acknowledged the decline in the cognitive skills of North Korean school-age children. Amid the continued secrecy pervasive in North Korean society, accurate data on the average IQ of North Koreans by age have not been readily available. It is not clear if the situation today is any different than it presumably was 15 years ago. The divergence in IQ between two populations once considered genetically identical is anecdotal but, from all accounts, tenable. It is a vivid reminder of the haunting effects that poverty and its ramifications—famine, psychosocial stress, polluted air and contaminated water, and numerous other adverse life circumstances—can have on brain and body, especially during childhood. Detrimental stressful life events seemingly have had adverse physical and psychological consequences for some North Korean children, affecting not only their growth and development, but also their cognitive abilities.

Although the human brain can often adapt to external pressures and modify some of its connections, or softwiring, particularly in a young person, there are limits to the remedial effects of plasticity. Irrespective of environmental influences, basic hardware, or hardwiring, is essential for a proper functioning brain. In an anecdotal report in the medical literature, neurologists described how a young boy given the pseudonym "VJ" suffered from selective impairment of memory—episodic memory—throughout his lifetime.[97] A native French speaker from a lower-middle-class family, VJ was diagnosed at eight years of age with an atrophied hippocampus on both sides of his brain. (The plural for "hippocampus" is "hippocampi." The normal human brain has two hippocampi, one on each side of the brain, located deep within the limbic system.) Despite a normal IQ, proficient reading skills, and good performance on recognition tests, VJ lacked sufficient memory recall needed for everyday functions. He had to depend on "clues" or "reminders" associated with past events and experiences, which improved his recall, but only partially. An intact temporal lobe, one of the brain regions involved in long-term memory, may have helped trigger some of his recall ability. But, overall, the brain's higher centers, notably the temporal lobe, in the cerebrum could not compensate for the diminished, or lost, memory-related function of his hippocampi.[98]

VJ's case seems to affirm the complaint of skeptics that the plasticity of the human brain is exaggerated in the popular media because it cannot remedy every problem.[99] They are not wrong. But VJ's disorder—hippocampal atrophy—is extreme, a condition that requires corrective action far beyond the brain's usual capacity to rewire itself. In normal circumstances the hippocampus exhibits considerable plasticity, especially early in life, which can reportedly improve spatial memory.[100]

In 1997, British neuroscientist Eleanor Maguire and colleagues published a seminal research study on the spatial navigation properties of the right hippocampus in London taxi drivers. Researchers used positron emission tomography (PET), an advanced imaging technique that measures changes in brain blood flow, to visualize this structure. This study was one of the first to demonstrate that the right hippocampus is a localized area of the brain that facilitates spatial memory for navigation.[101] Although other brain structures are also involved in memory, apparently only the right hippocampus houses specialized neural networks for recalling routes for traveling, or navigation. Befitting the needs of big-city taxi drivers, this brain structure is used to memorize complex routes leading to specific destinations; however, it is not activated when other types of complex memory recall are required. Scientists soon realized that some of the properties of the hippocampi resulted from experience and neuroplasticity. And while taxi drivers in this study had larger hippocampi than the public, in general, they probably were not born this way.[102]

Three years later, Maguire's research team published an MRI study showing key differences in the size of specific areas of the hippocampus between licensed London taxi drivers and controls that did not drive taxis.[103] (All the participants were right-handed.) Compared with the control, or nontaxi, driver, the taxi drivers had a larger region in the rear, or posterior, hippocampi on both sides of the brain. Conversely, an area in the front, or anterior, part of the hippocampus was greater in the nontaxi drivers but smaller in the taxi drivers.[104]

The size of the hippocampi correlated with the amount of time spent as a taxi driver, with a positive correlation seen in the posterior region and a negative correlation in the anterior area. Based on this evidence, it seemed likely that the hippocampi, and perhaps other localized regions of the brain, are plastic and can structurally transform, becoming either larger or smaller, to meet the demands of an individual's environment. As predicted, the volume of gray matter, or densely packed neurons, was significantly increased in the hippocampi.[105] Compared to controls, no alterations were evident in other brain regions of the taxi drivers. Recall that specific areas within the hippocampi changed in a different direction in each of the study groups. The front area of the hippocampi was larger overall in the controls, especially on the left side, whereas the back region of the hippocampi was larger in the taxi drivers than in the controls. And yet, there was no difference in the total volume of the hippocampi between the taxi drivers and nontaxi drivers; however, the longer a taxi driver had worked in this profession, the larger the size of the right posterior hippocampus.[106] Apparently, the gray matter was redistributed in the hippocampi of the taxi drivers, increasing in the rear but decreasing in the front. This provided further proof that the changes in the hippocampi of taxi drivers were acquired. The structural change detected in the hippocampi did not seem to be due to genetics but instead appeared to reflect a plastic response to an environmental or experiential trigger.[107]

No scientifically informed scientist disputes the existence of brain plasticity, but the magnitude of its impact, even on such well-studied structures as the hippocampi, is still controversial. According to a study published in 2011, taxi drivers exhibited a similar increase in the gray matter volume of the posterior hippocampi, but the brain scans failed to show the previously reported overall decrease in the size of the anterior hippocampi. The researchers who conducted this study acknowledged that the hippocampi exhibit plasticity. But they cautioned their readers to consider that some individuals, including tax driver trainees who passed their navigation tests, may have a genetic predisposition toward plasticity that might be lacking in the applicants who were unsuccessful in navigation training.[108] The authors seemed to imply that individuals may have inborn differences in their "hippocampal capacity" or "hippocampal potential."

In another study not involving taxi drivers, researchers found that the outer surface and front region of the hippocampi are inversely correlated with IQ. The relationship between a smaller hippocampus and higher intelligence performance was not necessarily causal, but a link was found. Apparently, the anterior hippocampus is active in processing spatial and temporal associations between different sensory experiences, for instance, seeing, hearing, and smelling. As such, a smaller anterior hippocampus is thought to increase the efficiency of information transmitted through neural connections in the brain, possibly by pruning, or eliminating, excess synapses and neurons. Regardless of their exact functions, the hippocampi, and not necessarily larger hippocampi, contribute to a person's overall intelligence.

The optimal size of the hippocampus, whether on the right or left side, for spatial learning and cognition may depend partly on the study and partly on the area of the hippocampus being investigated. For example, a recent GWAS examined gene variants linked with hippocampal size and found that multiple gene variants may influence the size and structure of this brain organ.[109] A SNP—a small variable region of DNA within a gene—implicated in hippocampal volume is currently being scrutinized as a biomarker for Alzheimer's disease. The genetic differences that explain the variance in hippocampal size among individuals in a population may help to further pharmacological research on therapies for memory-related disorders. The genetics of hippocampal neuroplasticity, on the other hand, is still in its infancy, but scientists are investigating this structure from other angles that do not directly involve an analysis of gene variants.

The hippocampi are not the only pliable structures in the brain, but their ability to respond to environmental stimuli has made them a prime target for studies related to not only memory, but also learning in school and successful life achievement in the real world. In an MRI study that investigated links between family income and brain traits, children from lower-income families had a lower density of gray matter, or neurons, in the hippocampi.[110] Recall that more tightly packed neurons correspond to an increased volume in the hippocampi. The adverse experience that occurs in a child who is separated from his mother or caregiver can lead to structural changes in the hippocampus, which in turn can affect how it regulates stress and memory later in that person's life. By contrast, warm, emotional nurturing, supportive parental care, and an intellectually stimulating environment that promotes learning—where books are present in a child's home and parents read to a child—predict improved performance on memory tests.[111]

In the brain, both densely packed regions of gray matter, or neurons, and white matter, or neural connections, have the capacity to respond to various environmental cues or experiences. This quality is called softwiring—the acquired, flexible changes that occur in some areas of the brain. Hardwiring, by contrast, refers to the brain's innate, genetically driven anatomy that is not plastic and is therefore fairly fixed.[112] Several studies have tracked plasticity in specific areas of the brain's white matter, or neural circuitry, as a child improves in literacy. In 2018, the University of Washington's Elizabeth Huber, a postdoctoral neuropsychologist, and colleagues used diffusion MRI scans to detect brain changes in grade-school-age, below-average readers.[113] The white matter was rapidly strengthened in two of three wide areas of the children's brains after only eight weeks of an intensive reading intervention. Their reading skills improved an average of one full grade level, correlating with measurable changes in the white matter that connects various regions of the brain. Within a few weeks of starting the tutoring program, the children exhibited new neural features, including an increased density of fiber tracts, or white matter, and better organized wiring. These brain changes were not observed in the participants in the control group, who did not receive the intensive reading instruction.[114]

These findings are encouraging for anyone who thinks that effective and efficient solutions for correcting remedial reading are possible. But as Huber and associates warned, the correlation between brain anatomy and a social behavior like reading reflects just a single point in time.[115] Other patterns of plasticity may also occur, affecting different brain regions and perhaps promoting or possibly disrupting various abilities. Still, their study offers strong support for the power of plasticity in triggering favorable cognitive and behavioral skills advantageous for success in school and potentially in life. As Huber told *ScienceDaily*, "We tend to think of these connections as being fixed, but . . . in reality, different experiences can shape the brain in dramatic ways throughout development."[116]

THE EPIGENETIC LYNCHPIN: LINK BETWEEN GENES, ENVIRONMENT, AND PLASTICITY

Gene–environment interactions are pivotal in stimulating the brain's plasticity as it responds to ever-changing internal and external stimuli, whether maltreatment or emotional nourishment, poverty or plentitude, or toxic air and water or clean air and water. With increasingly sophisticated technology at their disposal today, scientists are accelerating their probes into the genetic architecture underlying cognitive and social traits. Progress in this field has been impressive, but the science of mapping the steps of gene–environment interactions is still young. One avenue of research helping to advance this field is epigenetics, or non-DNA inheritance, which is fostering innovative approaches to studying the interrelationship between gene–environment interaction, gene expression, and plasticity. These processes separately and synergistically seem to influence intelligence performance, notably IQ. In the last decade, the field of epigenetics has burgeoned exponentially, giving rise to such new terms as *epigenetic genes*. The building blocks of epigenetics are not genes in the traditional sense of DNA strips composed of a specialized chemical alphabet. But the ability of epigenetic units to modify the activity of DNA-containing genes, sometimes substantially altering this activity, allows these units to potentially affect gene expression.

Consider the DNMT family, a group of chemical substances that modifies the structure of DNA. These substances are controlled by several DNMT epigenetic genes. Together, the DNMT epigenetic genes and the chemicals they produce help regulate gene expression—the degree to which a gene exerts its effects on a trait. The normal function of certain DNMT genes can be disrupted in some types of cancer growth.[117] At least two DNMT epigenetic genes are associated with intelligence.[118] Versions of one of these DNMT epigenetic signals was measured in Scottish elders and correlated with their childhood intelligence tests scores. One form was linked with higher cognitive performance, whereas the other was associated with lower scores, suggesting that DNMT gene variants may play an important role in normal-range intelligence.[119]

A related DNMT epigenetic gene may help to explain why certain individuals vary in their IQ scores, depending on whether they were raised in an enriched or impoverished environment.[120] Scientists were able to evaluate the activity of DNMT3B by studying identical twins who essentially have the same, or at least very similar, DNA. The researchers correlated the twins' DNA and epigenetic profiles with their IQ tests scores or working memory performance. Depending on which epigenetic gene variant each twin possessed, he or she tended to be either more sensitive or less responsive to environmental surroundings.[121]

Scientists do not yet understand exactly how an epigenetic unit influences gene expression. Epigenetic marks have the potential to tweak gene expression, transforming the manner in which a particular gene responds to its environment. They can do this without changing DNA—without affecting the sequence of the chemical letters in a segment of DNA that corresponds to a gene. Importantly, epigenetic mechanisms may be affected by lifestyle, for example, diet and alcohol. With the capacity to profoundly affect epigenetic activity, a range of environmental factors potentially can shift the effects of standard DNA-containing genes. This process is called epigenetic imprinting. Depending on gene expression, some gene versions may function as risk-raising variants, whereas others may serve as protective variants.[122]

The unprecedented discoveries of epigenetic effects have, to some extent, broken conventional rules about the biology of DNA. When applied to real-world situations, epigenetic activity might mean that the heritable aspects of intelligence can be changed by such environmental influences as diet in newborns and young children. If, as one researcher speculated, the high heritability estimate of IQ is due to epigenetic imprinting rather than DNA, then research on intelligence genes would be completely turned on its head. It is up for debate whether this has actually happened as yet. But one of the DNMT epigenetic genes linked with intelligence affected gene expression in utero.[123] Like DNA, epigenetic traits are inherited by offspring from their parents. But instead of directly affecting a trait the way DNA-containing genes do, most epigenetic genes alter the expression of DNA, in some cases starting with the newborn and even the unborn progeny in the mother's womb. Epigenetic processes can be retained through childhood and in some cases on into adulthood. If certain epigenetic patterns affect poor health, for instance, increasing the risk for obesity or diabetes, they can reduce a person's quality of life from womb to death. Diseases influenced by epigenetics, even in cases where epigenetic markers are passed from parents to children, may not develop until later in life, with the same markers potentially passed down across generations to the grown children's own offspring.[124]

The role of epigenetics is less clear in intelligence than in physical diseases, but the discoveries made to date in this field, although still somewhat nascent, may help unravel some of the interminable mysteries surrounding the interplay of genes and environment. Epigenetics is not likely to single-handedly solve the "missing heritability" problem—the fact that genetic variants identified through GWAS cannot fully explain the heritability of complex traits, including IQ.[125] But epigenetics, as a subset of genetics and more so as a field in its own right, is now etched into popular discourses about the human capacity to overcome the limitations of what hereditarians historically have called the genetic destiny, or at least the inherited proclivity, of certain populations toward lower intelligence and poorer educational achievement. Most importantly, epigenetics is now increasingly recognized, even by some genetic determinists, as a data-driven science that cannot be ignored. In the realm of human intelligence, epigenetics has literally and figuratively made enough of a mark that scientists today are forced to ponder the intricate interconnections—connections that may vary across individuals and ethnic populations—between environment, genes, gene expression, and plasticity.

BACK TO THE FUTURE: SURMOUNTING ETHNIC GAPS IN IQ AND EDUCATIONAL ACHIEVEMENT

Much of the current research on the interplay between nature, nurture, and intelligence performance focuses on links between genetics, SES, and educational attainment or IQ. Other contemporary studies in this field are exploring the association between environment; brain plasticity; and cognitive skills, for example memory and reading. These two approaches, although not mutually exclusive, are distinctly different, reflecting to some degree contrasting ideologies regarding the role of nature versus nurture in cognitive and social traits. Still, future endeavors that strive to integrate both approaches may be constructive. They might yield unprecedented insights to help answer unanswered questions about group differences, specifically ethnic gaps in IQ and educational attainment.

For experts who incorporate SES into research on cognitive and scholastic performance, it may be advantageous to consider SES within a broader context that views elements comprising the environment as dynamic rather than static. The participants in the current wave of genetics, brain, intelligence, and SES studies, no matter their demographic or ethnic background, are living individuals whose real-life experiences mold their changing exposures to, and shifting perceptions of, people, places, and ideas. Examining separate but intersecting environmental components of SES that venture beyond just income, educational level, and neighborhood residence—and in addition consider the impact of constantly moving upward and downward shifts in both family and personal aspiration, motivation, and resilience on IQ and school success—may shed new light on one of the biggest conundrums of them all: the black/white gap in IQ scores and scholastic achievement. This approach need not focus on genetic variants, particularly when a study includes participants of different racial backgrounds or divergent genetic ancestries. After all, the gene variants that account for IQ variance in one population are not likely to be the same as the gene variants that explain IQ differences in other ethnic populations.

SES is emerging as a pivotal nexus in links between brain traits and school performance, and either IQ or educational achievement and multiple IQ-enhancing gene variants. There is no denying that genetics influences these links, but the expression of gene variants affecting intelligence and academic performance seems to be mediated, partly at least, by other processes. This may be why a growing chorus of expert voices today continually reminds us that to fathom the origins of human intelligence, we have to delve deeply into the nongenetic contributions that encompass, but also transcend, the environment. Paradoxically, this brings us back to the more malleable side of Nature—those other "biologies" of epigenetics, gene expression, and plasticity.

CHAPTER TEN

Smart Genes, Big Brains, and Civilization

IQ-Linked Genes: The Saga in a Nutshell

If DNA is the blueprint of life, then mutations, or changes within DNA, are the driving force behind evolution, including adaptation to new niches, as well as the survival of a population or species. In chapter 9, we explored several pivotal genes reportedly linked with IQ. Since the search for the DNA underlying human intelligence began, the last three decades have witnessed the rise and eventual fall of multiple candidate genes purportedly involved in human cognition. Numerous gene variants have been tentatively associated with intelligence, but even when a genetic trait has been shown to contribute to IQ, the strength of that contribution is so small that it is often barely detectable. Variation in DNA comes from multiple sources and includes genetic changes that range from single nucleotide polymorphisms (SNPs), to insertions or deletions of sections of DNA, to movement of one part of DNA to another part, to chromosome mutations.[1] A SNP—a variation in a *single* DNA building block, or chemical letter—is the most common type of mutation and increasingly used to help identify a person's risk of developing certain diseases. SNPs also play a role in cognitive performance, sometimes either raising or lowering a person's IQ score, depending on which gene variants have IQ-linked SNPs. Several candidate genes, including multiple SNPs, are thought to influence brain size or cognitive performance, or both. FADS2, a variant of a gene involved in reproduction, initially appeared to increase IQ in people who were breastfed, but later studies have contradicted this finding.[2]

In the last chapter, we noted that in a recent study on the genetics of cognitive ability, only three out of 69 candidate genes were implicated in increased intelligence. Each of these three variants was responsible for raising an IQ score slightly less than one-third of a point.[3] For the majority of genes associated with cognition or social behavior, gene expression—the way in which a trait like smartness manifests—relies to some extent on the interactions between different genes and between genes and environment. Despite notable advances in clarifying the genetic underpinnings of intelligence, experts have yet to demonstrate that specific individual genes are responsible for a substantial portion of a person's IQ, regardless of whether his score is at, above, or below the norm. Most importantly, to date, no one has pinpointed specific gene variants that are irrefutably responsible for differences in average IQ score across diverse ethnic populations.

The scarcity of SNPs firmly connected to intelligence performance may be comforting to antihereditarian, pronurture supporters who view ethnic differences in IQ through a lens of environment and experience. For some experts, the majority of gene candidates once considered crucial in the race and IQ debate have been debunked and dismissed as unviable prospects for the biology of intelligence. This does not negate the existence of IQ-linked genes, however. As neuropsychologist Christopher Chabris points out, the failure to find genes linked with g, or general intelligence, does not mean that g has no genetic basis. It simply means that intelligence is a highly polygenic trait—a trait influenced by multiple genes—and the common genetic variants that act on this trait individually have only small effects.[4] This makes all the more sense in light of the astounding results of a large genome-wide association study, published in 2018, on more than 240,000 people. Scientists identified 538 gene variants linked with intellectual ability, the largest number of intelligence-related genes discovered to date.[5] In addition, some of these gene variants were also associated with other physiological traits, for instance, the activity of the brain's white matter, as well as such health-related outcomes as longevity.

The search for intelligence genes continues, as it rightfully should. But closely paralleling that search are the efforts of racially oriented hereditarians intent on finding the genetic machinery, which, they claim, divides human populations into the haves and have-nots of intellectual prowess, empathy, morality, and, ultimately, the capacity to build civilization. Some of these researchers no longer confine their search for population-specific, or racially distinguishing, gene variants to the brain. After all, social behavior is also influenced by genetics not directly tied to IQ. The quest to discover the genetic intricacies behind the making of the human brain is not a fringe or frivolous pursuit. It is essential research, pure and simple. And yet, as mainstream science has shown, some brain traits, including brain size and brain activity, are not identical in every situation for every human population. Instead, the overall volume, as well as the size of certain regions of the brain, varies at times, albeit slightly, from one ethnic population to the next.

Findings like these have ushered in a new heyday for genetic determinists, especially those who argue there is ample scientific proof for causal links between race, genes, brain size, and intelligence. Amid emerging research on population variability in genes and brains, genetic determinists and their race-realist supporters are quick to blame the ethnic gap in life outcomes on what they call each racial group's innate capacity for cultural achievement—a capacity that, in their words, is deeply entrenched in ancestry or race-based genetics.

BRAIN-LINKED GENES IN THE HEREDITARIAN ARSENAL

For almost a decade and a half, two brain-linked genes, commonly known by their abbreviations, ASPM and MCPH1, have become (at least for some researchers) prime candidates for the biological underpinnings of population differences in intelligence. ASPM stands for "abnormal spindle-like, microcephaly associated protein," a gene that codes for a protein involved in cell division, especially in the developing brain.[6] In normal circumstances, MCPH1, a key variant of the microcephalin gene, is involved in cell division needed for ordinary growth of the body, including the fetal brain. In the early 2000s, scientists discovered that some mutant variants of both ASPM and MCPH1 were associated with microcephaly, a rare congenital neurological condition characterized by a smaller head circumference than normal. The cerebral cortex—the thinking part of the brain—is underdeveloped in this disorder, preventing the head from reaching its normal size at any given age.[7]

ASPM and MCPH1 are each distinguished by their own unique functions, but both variants of these two genes also share several key features, notably a relatively recent appearance in the human lineage. During the evolution of anatomically modern humans, these two gene variants reportedly contributed to the increasingly expanding volume of the brain, at least in certain human populations. According to some estimates, the time lines for the evolution of ASPM and microcephalin correspond to major watersheds in human cultural evolution. As such, ASPM and microcephalin are thought to have been crucial genetic forces that propelled the emergence of an enlarged *Homo sapiens* brain—a brain that during hundreds of millennia has accelerated the cultural advancement of the human species. Specific versions of these two genes are said to have accelerated the pace of social and technological achievements in some human populations, endowing their members with a genetic makeup that heightened their cognitive skills.[8]

For more than a decade, scientific narratives revolving around the ASPM and microcephalin genes have stirred a raging controversy because the favorable variants of these two genes are not distributed equally across all human populations. Both variants appear to be virtually absent in Sub-Saharan black Africans. Does this mean, as some writers have proposed, that populations without these two gene variants lack the intellectual capacity to create advanced civilizations? This question has framed one of the most volatile debates concerning the causes of ethnic differences in social behavior and cultural achievement to surface in the scientific community in the last 20 years.[9]

In 2005, Bruce Lahn, a genetics professor at the University of Chicago, and his coworkers introduced to the scientific community the then-novel idea that certain gene forms of ASPM and microcephalin played a crucial but previously unrecognized role in the evolution of human cognition and, ultimately, human culture. Earlier research had shown that some versions of ASPM and MCPH1 are associated with microcephaly—a smaller brain approximately one-third the normal size. Lahn reasoned that if these gene variants are associated with brain size, and if brain size is linked to intelligence, then the two variants must have been pivotal in the evolution of human intelligence. Yet, as several studies have repeatedly revealed, the correlation between brain size and IQ overall is modest at best and, according to some experts, rather weak.

The idea that ongoing, accelerated evolution may be occurring in modern living humans is not entirely new, although this notion has been revived. In the 1990s, molecular anthropologist Vincent Sarich proposed that human races evolved rapidly in the past few thousand years.[10] Sarich's time line was considerably shorter than the prevailing scientific wisdom predicted at that time. But Lahn's study provided the first possibly concrete evidence that human microevolution—the evolution of genetically divergent human populations—has continued and even sped up in recent millennia. Using quantitative techniques, Lahn's team identified a "DNA signature" that seemed to show that these two mutated genes emerged through positive selection, or favorable natural selection. In a bold interpretation of the findings, Lahn and colleagues asserted that both new gene variants evolved to enhance the overall intelligence of the populations carrying these particular versions. Although each gene variant evolved independently, Lahn noted, both spread rapidly throughout much of the globe, reaching high frequencies in the populations of some regions but bypassing other populations in different areas of the world. If Lahn were correct, some human races, to their detriment, were losers in the lottery of bigger brains for a better and smarter human. Lahn suggested that these populations missed out on their share of these fast-track, brain-expanding genes.[11]

The rapidly evolved ASPM gene variant is estimated to have emerged in humans about 5,800 years ago, but the time line that some scientists have proposed is vague, spanning a broad period ranging between 500 and 14,000 years ago. Lahn claimed that the recently evolved ASPM and MCPH1 gene variants coincided with major watersheds in human cultural history. ASPM reportedly emerged at about the time when agriculture spread in the Middle East and the earliest cave paintings first appeared. The dispersal of this gene variant into human populations seemingly corresponded with the development of cities and written language.[12]

According to some studies, this version of the ASPM gene exists in about 30 percent of humans worldwide, including an estimated 50 percent of Middle Easterners and Europeans, but it is less prevalent in East Asians and rare in some Sub-Saharan African peoples. If Lahn were correct, a recent gene that evolved as quickly as ASPM would have been more likely to have arisen through natural selection than genetic drift, or neutral selection. It would travel throughout a population if it were sufficiently adaptive, or advantageous, for a significant number of people in a population to inherit it. If so, it would be a prime illustration of strong positive selection. The adaptive value of intelligence would have rendered such a gene variant a desirable selected trait in the humans who possessed it.[13]

The time line for MCPH1 is less certain, with estimates varying between 14,000 and 60,000 years ago. The genetic alteration that led to MCPH1 is usually dated at about 37,000 years ago, a period when symbolic cave art appeared in Europe. Like ASPM, the frequency of MCPH1 varies across continental populations, occurring in at least 70 percent of people in Europe and East Asia, and 100 percent of some Indian populations of South America. As in the case of ASPM, MCPH1 is reportedly uncommon in most Sub-Saharan African populations, at least the populations that to date have been sampled for this genetic mutation.[14]

Lahn speculated that the origin of ASPM and MCPH1 contributed to the cultural advancement in the populations where these gene forms are prevalent today, giving them an edge over other populations, notably African blacks, in their genetic makeup and related cognitive skills. Referring to what some writers have called the "cultural lag" of certain human races and ethnic groups, Lahn warned that society eventually will have to deal with the "very difficult facts" of population differences, or "racial differences," in recent genetic evolution. The publication of Lahn's article was met with numerous rebuttals, with critics skeptical of his conclusion and some forcefully accusing him of racial prejudice. Several studies soon challenged his claim, providing counterevidence suggesting that neither ASPM nor MCPH1 is associated with brain size, let alone mental ability or behavior. Even racialist psychologist J. Philippe Rushton and his colleagues admitted they could not find a statistically meaningful association between either gene form and individual variation in head circumference, mental ability, or altruism.[15]

As the controversy regarding Lahn's provocative claims continued, experts offered diverse opinions about a plausible role, or lack thereof, of ASPM and MCPH1 in the expansion of the human brain, particularly the cerebral cortex. Stanford University archeology professor Richard Klein suggested that the rapid spread of ASPM could have been triggered by another adaptation totally unrelated to intelligence, for example, adaptation to much colder climates. University of Maryland geneticist Sarah Tishkoff, a coauthor with Lahn on a study of these gene variants, recommended caution in drawing implications from the reportedly unequal distribution of these gene variants across ethnic populations. Tishkoff complained that the idea a single gene could determine complex cultural traits is simplistic, emphasizing that there is no evidence linking these two genetic mutations to the rise of civilization.[16]

To date, the samples of people tested for these two mutant genes have included only a very small number of black Africans, of whom very few, if any, are apparently from East Africa or Central Africa. For some critics, it is naive and perhaps ignorant to presume that one, two, or only a handful of genes could ultimately be responsible for the social behaviors and cultural organizations that have spawned civilizations throughout the world.[17]

ASPM and MCPH1 across Populations

The controversy concerning ASPM and MCPH1, two gene variants supposedly associated with increased brain volume, centers around an idea filled with racial connotations. According to Lahn and other proponents of the hypothesis that ASPM and MCPH1 determine brain size in some human populations, the apparent absence of these variants in African black populations may contribute to the average lower IQ of black populations and their diaspora globally. Early research suggested that certain versions of the ASPM and MCPH1 genes were involved in the adaptive evolution of the human brain, including brain volume.[18] But other evidence strongly demonstrated that these two gene variants do not influence brain size, even though these variants may have subtle effects on brain biology or perhaps an impact outside the brain.[19] Some SNPs, for example, in certain dopamine-linked genes can adversely affect the executive function of the brain's prefrontal cortex, which in turn compromises normal cognitive function.[20] Still, scientists have not found a firm connection between specific SNPs or other variants within either ASPM or microcephalin and such higher-level brain functions as attention, memory, abstract thinking, planning, or problem solving.

This does not mean that these genes were, in the evolutionary past or the present, irrelevant to human culture. Since the early 2000s, scientists have known that certain variants of ASPM and microcephalin are closely tethered to the ability to speak tonal languages like the Chinese tongue. Native speakers of nontonal languages tend to have two distinct variants of these genes, whereas those who speak tonal tongues have a different set of variants of these genes.[21]

Various aspects of language learning and language processing are at least partly controlled by multiple genes. ASPM appears to be involved in sound perception, or comprehension of linguistic tone, a trait that can affect a person's aptitude in verbal communication.[22] Using functional magnetic resonance imaging, or fMRI, studies of the brain, neuroscientists have discovered a strong connection between ASPM and activated regions in the right temporal cortex—the seat of the brain's auditory centers. This association was not seen with MCPH1.[23]

In tonal languages, whether in China or Africa, words are differentiated by changing their pitch. It now seems that ASPM may influence voice pitch, or the higher and lower sounds used to make lexical or grammatical distinctions when speaking, which are prevalent features of the highly tonal Chinese oral languages. However, the specific versions linked with such tonal human languages as Chinese are the older ASPM gene variants. They are not the most recently evolved variants described by Lahn's research team.[24]

In another study, researchers found that white Americans who did not speak a tonal language had a higher prevalence of the ASPM adaptive variants and greater tone perception than those with lower frequencies of this gene variant. This may mean that the recently evolved ASPM gene variant is not related to brain size, but instead to higher-level functions in the auditory processing centers of the brain—the temporal cortex, a region related to human communication.[25] Although ASPM and MCPH1 are crucial for the growth and development of nervous tissue, there is no conclusive proof that either of these gene variants directly influenced the adaptive evolution of the size of the entire brain. And there is no indisputable evidence for a role of these two gene forms in the evolution of brain-related traits governing cognition, including the volume of the cerebral cortex and the quantity of gray matter, or nerve cells.[26]

Amid ongoing debate, it remains to be seen whether microcephalin and ASPM are causally related to brain size in humans, and, perhaps more importantly, associated with IQ. An early investigation showed no link between the recent adaptive evolution of either ASPM or microcephalin and shifts in IQ.[27] But in a study published in 2014, researchers found a significant correlation between IQ and microcephalin (0.790 and 0.847) at the population-level, even though no association was detected between microcephalin and individual IQ.[28] Sweden's Umea University professor, Michael Woodley, claimed that by knowing the overall frequency of microcephalin in a population, scientists can predict with significant accuracy whether the members of that population, or a nation, will have a lower or higher IQ. His research team found that populations with higher levels of microcephalin tended to have higher IQ averages.[29]

Woodley and colleagues have published articles that some scholars may relegate to the fringes of mainstream genetics and intelligence research, but the definition of fringe may depend on one's identity politics. These academics are known affiliates of prominent racial scientists, eugenicists, and the alt-right.[30] Still, their main research finding in this study—a significant association between microcephalin and average IQ score across 24 nations—appeared to be based on objective, quantifiable data. Many of the countries in the survey, which spanned the world, are populated by a predominant continental population, such that the category of "nation" in this investigation served as a proxy for race.

If, as Woodley's team proposed, microcephalin affects intelligence, what mechanism might account for that connection? One possibility, the authors hypothesized, is that microcephalin could improve cognitive performance through its role in both DNA repair and antiviral immunity. The researchers speculated that thousands of years ago enhanced immune resistance against disease, for example, viral disorders, elicited by a mutation in the microcephalin gene, could have helped hunter-gatherer populations grow and expand. The ability of these populations to survive in environments heavily ridden with zoonoses—diseases transmitted from animals to humans—would have been a beneficial adaptation.[31]

Early hunter-gatherer societies that successfully transitioned from a nomadic to a sedentary way of life were able to access not only a new physical and biotic environment, but also a novel way of thinking about themselves and their surroundings. Combining their increasingly rational understanding of the interrelationships between plants and animals in their environment with improved skills for social cooperation would have fostered practical strategies for manipulating their environment. This strategy is sometimes called adapting to a new cognitive niche.[32] Microcephalin may well correlate with humans evolving new cognitive niches. But proving that this gene variant is responsible for the increased intelligence needed for humanity's cultural advancement requires more than arm-chair theorizing.

Woodley and his colleagues' hypothesis may be appealing to some readers intrigued by evolutionary psychology, behavioral genetics, and other Darwinian-inspired scenarios for the development of the human mind. The problem is that the correlation between the microcephalin gene and the average IQ in 24 countries does not prove a cause-and-effect relationship predicting that the lower IQ in African populations is due to MCPH1. In Sub-Saharan Africa, the heavy burden of pathogens that cause infectious diseases is an important contributor to the overall lower IQ of black populations. According to some experts, such infectious diseases as malaria, tetanus, and hepatitis may exert as much of, or even more of, an impact than wealth and access to formal schooling in determining IQ.[33]

As of this writing, there is still no scientific consensus on the exact function of the recent ASPM and MCPH1 gene variants in the human species. Assuming that both genes are in some way tied to brain activity, most experts agree that genes rarely operate in a vacuum, particularly genes that influence the development and plasticity of the brain and central nervous system. To determine the impact of a particular gene variant on a physical or psychological trait, scientists would have to be able to map the influence of each environmental effect alone, as well as gene–environmental and gene–gene interactions, which affect that particular gene variant. They would have to account for gene expression, which is ultimately conditioned by both types of interactions. With 538 genes now firmly linked with intellect, the number of genes specifically involved in brain function is enormous, making the number of interactions between genes and other genes, genes and epigenetic signals, and genes and environment potentially vast.[34]

For the sake of argument, let us assume that ASPM and MCPH1 have specific cognitive functions and, as a result, played a role in the human evolutionary past of promoting social intelligence. Let us also assume that both gene variants evolved rapidly through natural selection in certain human populations, but not others. A positively selected behavior-linked genetic trait that produces a favorable effect on human intelligence and cultural innovation would be an advantageous adaptation in populations where the trait is widespread. But this does not necessarily reveal anything about the populations in which either or both gene variants are uncommon, rare, or even totally absent, since the lack of either gene variant in and of itself need not be detrimental.

Other alternative genes and gene variants not yet discovered may perform similar brain functions or even the same exact brain function across different ethnic populations. For decades, scientists have been stymied about why, in some cases, the same brain-linked gene variant produces different effects in different populations. For example, the same version of the serotonin transporter gene has variable effects on patients' response to antidepressant treatment across ethnic populations.[35] When researchers engage in such contentious topics as the unequal dispersal of ASPM and MCPH1 in modern human populations, it is crucial that they clarify the specific gene variants to which they are referring. Considering the limitations in current knowledge about the genetics of neurobiology, rational and prudent scientists typically exercise restraint in rendering value-laden judgments about the presence or absence of certain gene forms. There is still much to learn about alternative genes involved in brain development and brain function, and in their close ally, intelligence.

Bruce Lahn quickly became a household name in the small but quasi-influential community of genetic determinists who study nature, nurture, and race. He drew attention because he declared that black Africans overall had missed out on brain expansion and therefore had gotten shortchanged in enhanced intelligence, since they reportedly lack the most recently evolved versions of ASPM and MCPH1. But the links between both gene versions and increased brain volume or, for that matter, any other specific brain function, are tentative and not definitively established. Gene variants, including those related to favorable brain features, evolve and usually persist in environments if they are an asset to survival and reproduction.

In fairness to Lahn, he noted that even though these two gene variants may augment the fitness of brain function in certain niches, they are not necessarily superior versions. This is not the only important caveat to be heeded in regard to the saga of ASPM and MCPH1. The seemingly uneven spread of these two brain-related genes throughout the world does not automatically translate to inborn population differences in intelligence. Only a few blacks were included in Lahn's original sample, and reports on the worldwide frequencies of these gene variants are limited, generally confined to small sample of individuals in only a few African countries.[36]

The lid has been put back on this Pandora's box, for now at least, but it has hardly been tightly shut. Novel cases of population variation in brain- and personality-linked gene variants will undoubtedly reappear in the future again and again. And when they do, old controversies will be reignited that speak to the core dilemma in nature, nurture, and race: Are ethnic population differences, particularly differences in social behaviors and life outcomes, rooted in genetics, emerging through natural selection, such that they have become fixed and unchangeable, and therefore not likely to be altered through environmental engineering?

Whether ASPM or microcephalin gene variants are undeniably associated with brain size, let alone mental ability, may be a matter of perspective. In the immediate future, the answer may be tied to each researcher's (and each reader's) own implicit bias, or lack thereof. Despite the risk of prematurely drawing conclusions about genes, intelligence, and populations, investigations into population differences in cognition-linked genes are fair game. They are a legitimate scientific enterprise. But if studies are conducted from a pigeon-holed vantage, they run the risk of becoming distorted by investigators who are heavily steeped, perhaps unconsciously, in a priori biased beliefs.

We cannot wish away population genetic differences, including the small number of links between gene variants and personality or social traits that have been empirically confirmed. But we can opt to be professionally objective, ethically transparent, and socially responsible to avoid rash and explosive pronouncements amid preliminary, contradictory, and, at best, partial research results. Some cognitive and social adaptations are clearly related, even indirectly, to such biological traits as brain function, which evolved through natural selection. But to conclude, without exhaustively investigating comparable gene versions across diverse populations, that an ethnic population is mentally superior or inferior, or culturally advanced or culturally lagging, simply because one or two or even more specific gene variants are more prevalent or less prevalent, is preposterous. This type of reasoning belongs to a speculative realm that is more reminiscent of unsettled reasoning than settled science.

Lest we forget, East Asian populations have lower frequencies of the recent ASPM gene variant than several other populations. Yet, they have some of the highest overall IQ scores in the world. And recall that in 2012, a primary candidate gene—the C variant of HMGA2—linked with increased intracranial volume and IQ was reported as more common in African blacks than in whites.[37] The fact that one of the most promising of the recently discovered IQ-enhancing gene versions may be more prevalent in a population stereotyped globally for its lower overall IQ is noteworthy. It illustrates the conflicting evidence sometimes observed in contemporary research on genes, cognition, and ethnic populations.

Two decades ago, Robert Plomin proposed that intellect may be influenced by at least 50 genes.[38] Today the number of gene variants known to shape cognitive ability has risen exponentially and is now more than tenfold higher than estimated in 1998.[39] Simultaneously, recent breakthroughs in the genetics of intelligence have also revealed that the frequency of certain IQ-linked genes varies, sometimes only minimally, across ethnic populations.[40]

Suppose that in the future, specific gene variants or a cluster of genes associated with a substantial rise in intelligence performance, for example, IQ score, are discovered in some populations but not in others. Would this conclusively prove that certain ethnic populations, or geographical racial groups, are intrinsically and significantly smarter than others? Hypothetically, it could. But to arrive at a data-driven, definitive answer, scientists would have to first be able to meticulously map the additive and synergistic effects of a multitude of intelligence-related genes. They would have to examine each IQ-related gene's potential interactions with numerous other genes, epigenetic processes, and myriad environmental forces, as well as the overarching consequences of these interactions on the plasticity of the developing brain.

Although lofty efforts in this direction have begun, no one to date has performed this intricate level of analysis specifically on intelligence-linked genes and compared the results across a broad spectrum of ethnic populations. As we wait for unequivocal answers regarding the role of one or two isolated genes, for instance, ASPM and microcephalin, or even multiple genes in advancing human cultural achievement, let us remember the profound impact of environment on gene expression. As scientists recently reported, some genes linked with intellect may also influence a person's—and by implication, a population's—susceptibility to environmental conditions that directly affect health and longevity.[41] In the ongoing search for the genetics of intelligence, we cannot afford to lose sight of the impact of a wide gamut of environmental constraints on social outcomes, either in the human evolutionary past or humankind today. Environment will inevitably remain a confounding factor in any scientific quest to quantify the genetic contributions to human intellect and the myriad manifestations of cultural achievement.

Mean Genes

The (Un)Holy Grail of Race-Realism

TABOO 101: RACE, GENES, AND VIOLENCE

THE NATURE–NURTURE DEBATE CONCERNING THE SOURCE OF ETHNIC DIF-ferences in IQ and social patterns revolves around a web of life circumstances and life outcomes, ranging from family cohesiveness to educational attainment to financial stability to interactions with the justice system. Nowhere is this debate more controversial than in the realm of violent behavior, an area dominated by contentious stereotypic notions of ethnic and racial differences in proclivities toward aggression. If the blatant contrast in ethnic social outcomes is an unspeakable taboo topic, then surely the alarming statistics for violent crime in some ethnic minority groups, particularly trends reported for African Americans, are the elephant in the room.

The statistics on violent crime, especially homicides, in the United States are alarming. They are striking for multiple ethnic groups. In 2015, the homicide rate for all ethnicities in the United States was 5.7 percent per 100,000 population, with rates per 100,000 population of 20.9 percent for non-Hispanic blacks, 4.9 percent for Hispanics, and 2.6 percent for non-Hispanic whites.[1] According to the FBI's 2016 "Report on Crime in the United States," the rate of violent crime rose by 3.4 percent nationwide in 2016, representing the greatest climb for a single-year increase in 25 years.[2] The nationwide homicide rate rose by 7.9 percent, accounting for an increase of more than 20 percent in the national homicide rate since 2014.[3] Some critics question the accuracy of ethnic differences in rates for violence across ethnic groups, arguing that personal prejudice and skewed statistics paint a racially biased picture of crime, notably violent crime, in the United States. Although the black murder rate remains exceptionally high in cities like Baltimore and Chicago, which have large black populations, the rise in the white homicide rate accounts for much of the overall increase.[4] As criminologist Richard Rosenfeld put it, if there had been no increase in the white murder rate from 2014 to 2016, the nationwide homicide rate would have declined.[5] Still, the fact remains that almost eight times as many African Americans as whites are murdered each year.

Frequently cited statistics, some say, even when factually true on the surface, fail to capture the underlying causes of ethnic discrepancies in crime rates. Pronurture proponents point to the environment as the primary, if not the sole, trigger of the widespread violence that has all but ravished some traditional, predominantly African American communities. Some evidence backs them up. Childhood experiences and environmental conditions shape social behaviors, regardless of ancestry or ethnic heritage. As numerous studies show, maltreatment early in life can lead to aggressive, deviant, and societally unproductive behavior in later years.[6]

Research in human genetics has identified multiple gene variants linked with aggression that may shed light on the biology behind brutality. Like the heritability of IQ, the heritability of human social behaviors is not an exact science. Some estimates suggest that genetic influences account for about half of the variation in aggression among individuals within a given population.[7] As experts unveil more genes associated with mean-spirited or vicious behavior, new and seemingly scientific narratives are emerging on innate tendencies toward violence in certain personality types and certain racial groups. Investigations in this area span broad geographical areas and are not confined to particular regions of the world. But the sound bites of research findings trickling down from the Ivory Tower to the popular media that have most piqued the public's attention usually pertain to black African descended peoples. The blacks who have captured the world's imagination, for good and bad, are not only African Americans, but also Afro-Latinos, black West Indians, and a long list of ethnically distinct populations of African descent, both inside and outside Africa. They also include traditional Sub-Saharan African nomads and pastoralists far removed from the modern world.

During the last decade, gene candidates for personality traits and social behaviors have come and gone from scientists' armamentarium. But a small number of gene variants have retained their cherished status as likely biological drivers of violence and aggression. These are the versions of various genes that reportedly correlate strongly with increased prevalence of antisocial acts, for example, physical brutality, across different ethnic populations. A few of these gene versions occur disproportionately in some populations but are virtually absent in others. Most of the differences manifest as variations in single nucleotide polymorphisms—SNPs—or the number of tandem repeats in a gene. (Tandem repeats are short, identical DNA segments in which DNA letters are repeated multiple times within a gene.[8]) Still, genes are not necessarily the ultimate arbitrator of an individual's personality and repertoire of social behaviors.[9]

Matching specific genes with violent tendencies is an exercise in finding correlations but not necessarily in determining definitive cause-and-effect relationships. When race is brought into the mix, ethnic populations are often compared, wittingly or unwittingly, for their risk of antisocial behavior based on the frequency of aggression- or violence-related gene variants in each group. However, links between the prevalence of specific gene variants and specific social behaviors in different ethnic populations are not likely to yield meaningful information if environmental forces, for instance, gene–environmental interactions, are ignored. Life circumstances are critical for calculating a person's risk of domestic violence, maltreatment of others, harsh criminal activity, or any other socially illicit behavior. But this has not stopped some researchers from investigating the potential role of specific gene variants in antisocial activity, including juvenile delinquency and adult crime, for the purpose of comparing what they call the genetic risk of social deviance across various ethnic populations.

This is not to dismiss the valuable applications, especially in medicine, that can and sometimes do result from disease-related genetics research across ethnic populations. But there is less agreement over the benefits, if any, of comparing genes associated with social behaviors in diverse racial or ethnic groups. Genes impact both prosocial and antisocial behavior by regulating brain chemicals, or neurotransmitters, that control feelings, mood, and cognition. Recall that neurotransmitters like serotonin, dopamine, and norepinephrine send signals from the junctions between nerve cells, or neurons, to other neurons, triggering cascades of chemical and electrical activity that control nerve impulses. The messages, or nerve impulses, transmitted between neurons govern the action of muscles and hormones, and are an integral part of a person's built-in wiring. Genes lay down a blueprint for the "hardwiring" of the brain and central nervous system (CNS). However, the pliable connections between brain fibers, or the softwiring, and to some extent even an individual's seemingly permanent hardwiring, are influenced by not only genetics, but also numerous environmental and other nongenetic factors.[10]

Genes help shape both physical and psychological traits by guiding the action of neurotransmitters, which in turn influence the pattern and activity of brain circuitry, especially the nerve wiring in the prefrontal cortex—the key thinking center—as well as the emotional areas of the brain.[11] The brain is essential for cognition and behavior, but the role of genes in controlling brain function is sometimes as a "back seat driver." The concept of a "gene controlling or causing" a behavioral trait (or, for that matter, most physical traits) is simplistic, misleading, and often erroneous. As Philip Kitcher explained decades ago, when scientists claim they have identified a gene for a particular trait, they are actually saying that individuals who carry a certain allele—a gene variant on one strand of their DNA—will make a protein that ultimately results in the development of a certain trait in a "standard environment." But if that standard environment is altered, then the gene might be expressed differently. Nowhere is this truer than with regard to such psychological characteristics as temperament and self-control or social behaviors like aggressiveness. Personality traits are determined partly by multiple genes that interact with not only the multilayered environment, but also one another.[12]

In the past 25 years, scientists have identified numerous genes implicated in personality and a variety of social tendencies. Specific genetic variants, or versions, associated with aggression, particularly violent crime, include the cadherin gene, several dopamine-related and serotonin-related genes, and low-activity monoamine oxidase A, or MAOA, genes. (MAOA is an enzyme that degrades such neurotransmitters as serotonin, epinephrine, norepinephrine, and dopamine. If the MAOA enzyme is not functioning properly, the amount of neurotransmitter released into the synaptic junction—the area between two neurons—will increase, potentially causing a biochemical imbalance that could trigger aberrant thoughts and behaviors.) These "troublesome" gene variants have become prime targets of contention in the current nature–nurture debate concerning social behavior for one reason: Ethnic populations sometimes differ in their overall frequencies of these gene versions.[13] Hereditarian supporters applaud the search for ethnic population variance in genes linked with personality and intelligence. And yet, the same gene variant can sometimes have considerably divergent effects in different ethnic populations. This occurs partly because the same variant may interact differently with the same or even similar genetic and nongenetic forces, depending on the population and its environment.[14]

Twin Genes for Severe Aggression: The Finish Offenders' Example

In 2014, Scandinavian researchers reported that variants of two genes—cadherin 13, known as CDH13, and the low-activity MAOA gene were more common in a group of Finnish prisoners than in the general population of Finland. Almost 5 to 10 percent of extreme violent crime in that country was associated with offenders who carried these two gene variants. A link was found between violent offending and these two gene versions, regardless of the offenders' environmental circumstances—childhood maltreatment, a history of substance abuse, or an antisocial personality disorder. Although most people with this genetic duo do not commit severely violent crimes, the risk for violence was almost 13 times higher for the men who carried both gene variants. Yet, this combination of gene versions was absent in some individuals with a history of brutal criminal behavior, suggesting that other triggers, possibly environmental stimuli, must also be present for violence to manifest.[15]

CDH13 disrupts nerve cells within the brain's self-control centers that normally keep a lid on impulsive behavior. This gene variant is also linked with attention deficit hyperactivity disorder—ADHD. If CDH13 can activate a propensity for violent crime, it does so within the context of other influences, notably other genes, especially low-expressing MAOA. Low-activity versions of the MAOA gene yield limited amounts of MAOA, an enzyme that breaks down key brain chemicals involved in the normal functioning of the brain. People with reduced levels of MAOA have increased risk for depression, alcohol, illicit drug abuse, addiction, and, in some cases, personality disorders, particularly if they experienced maltreatment during childhood.[16] MAO enzyme inhibitors, or MAOIs, are a popular class of psychiatric medications that raise levels of neurotransmitters in the brain, thereby helping to reduce symptoms of depression and other negative emotional and mental states.

Research on CDH13 in people of African black descent is limited. But several scientists have found that some of the low-activity versions of the MAOA gene—often dubbed the "warrior gene"—are more common in violent criminals from various ethnic backgrounds. Still, a low level of the MAOA enzyme is not automatically a recipe for antisocial behavior, given that according to one study, almost one-third of Caucasian males possessed the low-expressing MAOA gene variant. Many more individuals are carriers of the low-activity MAOA gene variants than the percentage of the U.S. population arrested annually for violent crime.[17] Nonetheless, one MAOA gene variant that has extremely low activity is more common in African American men, a finding that has ignited one of the most heated controversies in the current nature–nurture debate about race and crime. We will return to our discussion of the MAOA gene, aggression, and ethnicity later in this chapter.

DOPAMINE: CHEMICAL NEXUS BETWEEN BRAIN, GENES, AND BEHAVIOR

Dopamine, a brain chemical released from the pleasure center in the frontal lobe of the brain, plays a major role in regulating movement, cognition, motivation, and desire.[18] The dopamine system is pivotally involved in mood and affective disorders, and may also contribute to ADHD—a condition affecting an estimated 10.4 percent of children five to 17 years of age in the United States.[19] As part of the biological machinery behind behaviors like learning and reward, several dopamine-related gene variants have been tied to increased risk for substance abuse, alcoholism, and other addictive behaviors.[20]

In the 1990s, Israeli researcher Richard Ebstein and colleagues discovered long and short versions of DRD4—the D4-dopamine receptor gene—which influences the action of dopamine on nerve cells. Scientists found that the long variant—popularly labeled the "adventure gene"—was associated with novelty-seeking in both Ashkenazi and non-Ashkenazi Jews.[21] The long version of DRD4 has been implicated in personality tendencies to seek out novel stimuli and environments, as well as in individual reactions to stress. And it is associated with various psychiatric conditions. Examples include ADHD; schizophrenia; and substance dependences, for instance, cocaine addiction.[22] The long gene form of DRD4 has been detected in thrill-seekers, such as hot-air balloon navigators, race-car drivers, and parachute jumpers—individuals who crave adventures and excitement. These people are often described as impulsive, fickle, excitable, quick-tempered, and extravagant, even though they are otherwise considered mentally healthy.[23]

After the discovery of the short and long DRD4 versions, researchers identified a specific variant of DRD4 characterized by seven repeat sequences in a small section of the DNA within this gene. Known as "7R," the repeated section of DNA has been linked with "wanderlust" and migrations. The 7R tandem—a strip of DNA correlating with what some researchers called "adventurous behavior"—was detected in 63 percent of several migratory populations of South American Indians. By contrast, the DRD4-7R gene variant was almost completely absent in such populations as the Han Chinese and Yemeni Jews—populations that historically have remained in the same geographic area for the last 30,000 years.[24] Still, the results from multiple investigations have produced conflicting findings, forcing some experts to question whether any DRD4 variant truly fits the criteria of a novelty-seeking gene.[25] Resolving this dilemma might require taking a deep dive beyond simplistic, binary links between genes and personality traits. After all, social behaviors are influenced by multiple genes, numerous environmental forces, complex interactions between genes and environment, and, in some cases, possibly epigenetic marks.

DRD4-7R, one of the most widely studied dopamine-linked gene variants associated with psychological traits, is reportedly more common in people with ADHD symptoms.[26] Whether a dopamine-related gene variant implicated in social behavior is advantageous may depend on an individual or population's environment. To illustrate this conundrum, University of Washington anthropologist Dan Eisenberg investigated social behaviors that correlated with the presence of DRD4-7R in the Ariaal tribe of Northern Kenya. The Ariaal were traditionally nomadic pastoralists, but today almost half of their 10,000 remaining members reside in settled villages. The sharp contrast between Ariaal nomads and settled villagers has been an open invitation for scientific scrutiny, especially from curious researchers intent on mapping a three-way interplay between personality-linked gene variants, culture, and society.[27]

When Eisenberg's team compared individuals with and without the DRD4-7R in the two groups of the Ariaal, they found that the traits that connected this gene variant varied, depending on the individual's lifestyle. Ariaal nomads carrying the 7R sequence tended to be physically stronger and better nourished than their counterparts who lacked this variant. Apparently, the 7R confers a distinctive adaptive advantage to people living a nomadic existence in an arid region like Northern Kenya. According Eisenberg's team, the limited attention span reportedly linked with the 7R variant might be beneficial in nonacademic settings where the ability to detect rapidly shifting changes in the ecological surrounding would be an asset for survival. But a short attention span—an ADHD-type orientation to the world—would not be well-suited for traditional school learning, whether in a Western classroom of formal schooling or a settled village.[28]

When the circumstances were reversed, a different association was seen between genetics and behavior. Compared to their kin lacking the 7R marker, the Ariaal with the 7R gene variant who led a sedentary existence had less strength and were not as well nourished. They were described as restless people who jumped from one thought to the next, or one idea to another, without a steadfast stream of thought. These traits, Eisenberg's team surmised, might be a welcomed asset in a society facing constant ecological shifts and sudden cultural upheaval. But these same tendencies might prove challenging in a settled environment, for example, a classroom where book learning is taught. In this case, racing thoughts and short attention span—the researchers' characterization of the typical DRD4-7R personality type—would not be a good fit. In Darwinian terms, the implication is that an ethnic population's genetic background might influence whether the 7R variant of the DRD4 is likely to be retained and passed on to future generations.[29]

The process by which a gene variant is transmitted from parents to offspring and maintained in future generations is adaptation, or fitness. It is an integral part of the microevolution of populations within a species—for our purposes, the microevolution of ethnic populations within the human species. As a biological phenomenon, fitness refers to reproductive success. And yet, the forces that help determine if a genetic trait like DRD4-7R will be favorable or unfavorable for a population are both biological and environmental—both genetic and cultural. Whether the genetic underpinnings—genes like DRD4-7R—of specific social behaviors are either universal across different ethnic populations or less common in only a few ethnic populations is not entirely clear. This conundrum speaks to one of the core controversies concerning the role of biology in population differences in multiple social patterns, including marriage and mating, child rearing, family wealth, educational achievement, aggression, altruism, violence, and benevolence.

There are multiple types of dopamine receptors, notably the DRD2 version, as well as the DRD4-7R just discussed. Consider recent findings reported for the Datoga, a multitribal people considered one of the oldest cultural groups of Tanzania. The Datoga are pastoralists who traditionally practice cattle husbandry, even though today many of them operate onion plantations. And while they face continued discrimination and marginalization from a variety of social sources, they have maintained their reputation as fierce warriors. Ironically, they are also known for their exceptional egalitarianism, especially in distributing household wealth to members of their community.[30]

For years, experts have known that two dopamine-related genes—the dopamine transporter gene (DAT1) and one of the dopamine receptor genes (DRD4)—contribute to ADHD and intellectual performance. Variants of both genes have been identified in adolescents and adults at increased risk for severe violent crime.[31] Another dopamine-related gene, COMT, inactivates the neurotransmitter dopamine, with some COMT gene variants pivotally involved in emotional processing, regulating dopamine levels, and possibly such cognitive functions as numerical counting.[32] Aware of the potential role of these three genes in triggering combative impulses, a team of Russian researchers led by anthropologist Marina Butovskaya examined hypothetical links between genes, aggression, and other social behaviors in several relatively isolated East African tribal societies.[33]

In one study, researchers calculated the prevalence of four variants of the dopamine-linked genes, DAT1, DRD2, and COMT, in a small sample of Datoga men from the same community. Butovskaya's team discovered that a version of DRD2 by itself—a single-gene effect—was associated with anger. By contrast, the combined influence of the DAT1 and DRD2 gene variants was tied to hostility. The DAT1 and COMT variants had an interactive effect on anger and hostility, but they did not influence physical aggression. One of the most significant findings was that the "risk variant of DAT1"—the 10-repeat form—alone seemed to predict which of the Datoga men were likely to be angry and hostile in their interpersonal interactions.[34] (The Datoga men in this study were also evaluated for links between a testosterone-related gene, physical aggression, and anger. These findings are discussed in chapter 14.)

Butovskaya and colleagues surmised that physical aggression was an adaptation, a sign of Darwinian fitness, in Datoga men because it enhanced their abilities as warriors and made them fearsome competitors. The researchers speculated that intense physical aggression and warrior skills might be prized attributes for traditional pastoralists in Tanzania, rendering them successful in securing mates and resulting in reproductive success by producing more children. But the investigators also concluded that this type of behavior, as an outgrowth of natural selection, particularly sexual selection, is not beneficial in Westernized societies.[35] The implications of their words extend well beyond the Datoga and dopamine genes, in particular. Were they issuing an ominous warning? Were they suggesting that even though the behavioral genetic makeup of certain populations—in this case black African pastoralists—may serve a valuable social function in their own indigenous environment, the same genetic profile is not advantageous and may even be harmful in modern societies?

In a different investigation, two-year-old children, primarily nonwhite ethnic minority kids who carried two copies of the DAT1 variants—one inherited from each parent—had an increased risk for uninhibited temperament leading to behavior problems if they had unresponsive mothers. And yet, the gene variants did not seem to have a damaging effect on a child unless a specific environmental condition—in this case, an emotionally absent mother—was also present.[36] Similarly, another study reported that children who carried two DAT1 variants and were reared in families with highly unresponsive mothers had a heightened risk for uninhibited temperament—bold, sociable, or outgoing actions—followed by disruptive behavior problems by two years of age.[37]

Certain behavior-linked gene variants, for instance, DAT1, appear to make children more vulnerable to child mistreatment, potentially compromising their normal impulse control and leaving them susceptible to antisocial behavior. Still, this might be a chicken-and-egg scenario in which an environmental influence elicits a potentially detrimental genetic response, which in turn promotes further environmental discord. The situation might, theoretically, be even worse than initially thought if multiple problematic gene variants are involved. As the authors pointed out, compared to youth reared in healthy homes, children who carried a larger number of "risk gene variants"—dopamine receptors D4, D2, DAT1, and COMT—are more likely to be impulsive if they are maltreated in their home environment.[38]

This is not a trivial observation; however, blacks are not the only targets of explorations of dopamine-related or other neurotransmitter-related genes tied to social behaviors. One research team found that the DAT1 variant occurs in disproportionately high rates of American adolescents, in general, labeled as antisocial.[39] The DAT1 risk variant predicted an increased number of negative interactions with the police, especially chronic encounters with the criminal justice system. To further complicate matters, just as several gene variants correlate with antisocial behavior, multiple forms of mistreatment may amplify or interact with these gene variants to affect gene expression.[40]

Maltreatment refers to a wide array of harmful behaviors, including emotional distress, neglect, physical abuse, and sexual abuse, which can lead to severe emotional problems when they occur at a young age. The scientific evidence supports what many people intuitively suspect. Children who carry multiple risk gene variants and also experience more than one type of maltreatment are less psychologically equipped to bounce back from adverse circumstances. These young people may be more prone to impulsive behavior than those who carry fewer risk gene variants or have less exposure to disturbing environmental conditions, or both.[41]

Pronature researchers often view various dopamine-related gene variants, like DRD4, as part of the biological underpinnings of such personality traits as extraversion and novelty-seeking.[42] Fortunately, few individuals possess most, let alone all, of the gene variants that to date have been linked with socially undesirable character traits—for example, impulsive temperament, addictive behavior, aggressiveness, and violence. But even for those who do possess two or three gene versions that reportedly increase the risk for inimical personality traits, environmental forces still exert a substantial, and sometimes buffering, impact. This finding should not go unheeded. Environmental circumstances and personal experiences, particularly when they occur early in life, may be the deciding factors that determine which social traits, for instance, the risk for mental illness, are dominant in a person.[43]

Some DRD2 and DRD4 gene variants, especially when combined with low-activity versions of the MAOA gene, have been implicated in aggression and criminal activity. In one study, variants of three genes—DAT1,

DRD2, and MAOA—interacted with environmental forces to influence delinquent behavior in young males.[44] And yet, the DRD2 variant that appeared to increase the likelihood of juvenile delinquency was neutralized if a boy or male adolescent had parents and close family connections who were positively involved in his life. While this finding may seem to support the pronurture position, ironically, it reinforces the idea of the supremacy of the gene. For a growing number of researchers today, not only does gene–environment interaction, or gene expression, or both, depend on environmental circumstances, but also environmental circumstances may depend on genetic tendencies.[45]

Concepts of the genetic activity—what a gene can and cannot do—are rapidly evolving. Consider recent evidence that paints a picture of the effects of DRD4 variants that contrasts with the conventional wisdom prevalent almost 20 years ago. The long version of DRD4 has historically been linked

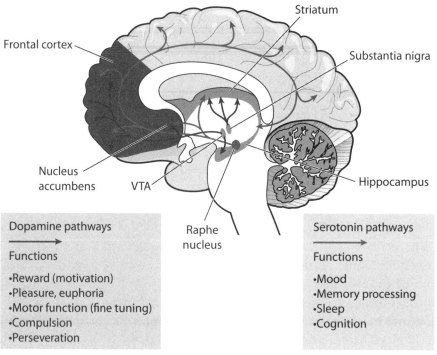

Dopamine and serotonin pathways in the brain.
SHUTTERSTOCK

with impaired attention, as in ADHD. But in a study led by cognitive neuroscientist Marissa Gorlick, the long variant was associated with good performance in work situations requiring selective attention to detail. This may mean that the effects of the DRD4 gene variant are plastic. The genetic propensity of DRD4 is apparently flexible enough to encourage either increased or decreased attention, depending on a person's needs in a specific environment.[46] Gorlick's team found that the attention of someone carrying the long DRD4 variant is likely to be heightened when he has to organize or categorize goal-directed priority items in a complex environment. He is more likely to have selective, or focused, attention when he has to complete an assigned task for his employer that demands a good working memory. But his attention might be reduced or even impaired if he is dealing only with minor items or if finishing the tasks at hand is not a priority or an urgent matter. For this reason, some researchers no longer categorize DRD4 as a risk gene for attention-related deficits. Instead, they see it as a plasticity gene that is pliable enough to promote enhanced selective attention in high-priority situations.[47]

SEROTONIN: MOOD CHEMICALS INSIDE THE BRAIN

The brain chemical serotonin, or 5-HT, helps nerve cells communicate by binding to receptors, or molecular docking stations, on the adjacent nerve cell. Serotonin is concentrated in the midbrain, where it controls mood, emotion, cognition, and motor function. This neurotransmitter also influences the action of hormones that affect such basic physiological drives as hunger and eating, sleep, sex drive, and aggression. Low serotonin levels have traditionally been linked with negative affect and personality traits—sadness, depression, aggression, and, in extreme cases, psychiatric mood and behavioral disorders and violence.[48] The serotonin system is governed by multiple genes that coordinate the regulation of the serotonin transporter, 5-HTT, a large molecule that inactivates serotonin, as well as myriad serotonin receptors. The function of the serotonin transporter is to carry serotonin to various parts of the brain, where it regulates nerve cell activity. This molecule is produced by the serotonin transporter gene, also known as the 5-HTT gene, and designated by the scientific abbreviated name SLC6A4.[49]

In the 1990s, Peter Lesch, a scientist at the National Institutes of Health (NIH), discovered two forms of the serotonin transporter gene—a short form called S and a long form labeled L, based on a key area within the gene known as the promoter region. The long and short versions of this gene determine how much serotonin is transported into nerve cells and repackaged for further use. Everyone inherits two gene variants for the 5-HTT gene, one from each parent, with each person having one of three possible combinations: two short variants, two long variants, or one long variant and one short variant.[50] An extra-long form of this gene has also been detected in research samples of African Americans, Japanese, and non-white Hispanic participants.[51]

In multiple studies, the short variant has been linked with reduced activity of 5-HTT and overall decreased efficiency of the serotonin transporter system. In some studies, people carrying one or two copies of the short gene version—the SS genotype—who were also exposed to disruptive environmental conditions, particularly during childhood, were more susceptible to psychological imbalances than those carrying only the long gene form. Individuals with the SS genotype were also more likely to experience a major depressive episode and increased anxiety than those without this gene variant.[52] In one study, the SS genotype was overrepresented in a group of white offenders, as well as a group of extremely violent Chinese criminals.[53] However, the serotonin transporter gene may account for only 5 percent of violent behavior, as its effect seems to depend on gene–environmental interactions.[54] By contrast, the presence of two long variants of the serotonin transporter gene—the LL genotype—may protect against depression, even in the face of environmental stress.

Some findings on the serotonin transporter gene across ethnic populations may seem at odds with evidence related to stress-related disorders, for example, essential hypertension, or high blood pressure. Case in point: Essential hypertension is a major medical problem in African Americans that can start as early as childhood, and perhaps even in the embryo.[55] According to medical researcher Redford Williams, serotonin genetics might contribute to disparities in rates of high blood pressure between black and white Americans. In 2001, Williams and colleagues at Duke University published a groundbreaking study on the association between serotonin transporter gene variants and high blood pressure. The long gene form predicted a higher risk for hypertension, a disease that affects proportionately more African Americans than whites.[56]

Almost 75 percent of East Asians, 44 percent of Europeans, and 20 percent of Africans have at least one short serotonin transporter gene variant.[57] Previous research revealed that more than 70 percent of Africans and African Americans have the long gene versions compared with 50 percent to 60 percent of whites. And less than 30 percent of people of Northern Asian descent (including those of Chinese and Japanese ancestry) carry the long gene variant.[58] In an early study, Williams's research team found that it was not the length of the gene variant that directly increased the likelihood of developing hypertension. Instead, the risk for this disease was tied to having an angry response to acute or immediate stress when confronted with an anger-provoking situation. People carrying one or two copies of the long gene variant tended to have a greater rise in blood pressure compared with those who had inherited the SS genotype.[59] When individuals from these two groups—SS and LL—sat quietly, there was no overall difference in the increase in blood pressure. But when they recalled an event or experience that made them angry, those with the LL combination overall showed a larger rise in blood pressure.[60]

Both the long and short variants of the serotonin transporter gene, then, appear to influence how a person reacts to stressful life events, especially aggravating events that can ignite an intense emotional, even hostile, reaction. To illustrate this effect, Williams noted that studies in Europe and Japan reveal that people with the long gene variant are more likely to be victims of heart attacks compared to their counterparts carrying two short variants. On the surface, these findings might suggest that African Americans on the whole are genetically more prone to reacting or overreacting to stress less effectively compared with ethnic populations in which the short variant is more prevalent. But the correlations are not cut and dry. Williams and associates found that the presence of either two short variants or two long variants of the serotonin transporter gene predicted different risks for hypertension, including both diastolic and systolic (e.g., hypertension), in different ethnic populations. The risk for this disease was higher in whites carrying the long variant, African Americans and Native Americans with two short variants, and Asians carrying the short variant.[61] Psychological stress can contribute to hypertension, but the LL genetic inheritance does not necessarily mitigate the risk for high blood pressure in white Americans. However, in African Americans, the LL genotype is linked with a *lower* rate of hypertension, whereas the short gene variant is associated with a higher prevalence.[62]

Studies on correlations between the serotonin transporter gene and personality traits have also produced contradictory results. As in the case of the dopamine-related genes discussed earlier, the same gene variants sometimes have variable effects on behavioral predispositions across ethnic populations. In some cases, individuals carrying the long variant may have an enhanced built-in cushion against stress, at least chronic stress. And yet, Williams's investigations revealed that African Americans, who are more likely to possess one or two long variants, may be at increased risk for anger. It is well known that anger can raise the likelihood of developing hypertension. Hypertension afflicts 43 percent of African American men and 45.7 percent of African American women. In recent years, the rates of hypertension have increased in low-income nations, including many African black societies, due to aging, urbanization, and an overall lack of good health. But even with elevated frequencies of hypertension, the overall rate for this disorder in four different Sub-Saharan black populations was 25.9 percent, well below the frequency in African Americans and even white Americans.[63]

Several factors may account for the uneven prevalence of hypertension across ethnic populations. First, there is no reason why the long gene variant could not protect against certain forms of stress—prolonged chronic stress, for instance—and yet not provide an adequate buffer against immediate stressors, or situations likely to invoke anger and frustration. Second, emerging studies reveal that the short and long variants of the serotonin transporter gene each have subtypes. Experts have discovered that one subtype of the long variants is associated with lower serotonin activity, causing it to essentially act like a short variant.[64] If serotonin transporter gene variants play a role in the onset of hypertension in some populations, then perhaps different environmental circumstances might explain why the frequencies of this disorder vary so widely in some genetically closely related populations.

On the other hand, if some researchers are correct, environmental forces seem to have only a negligible effect, if any, on the serotonin transporter gene in influencing personality traits. This is a profound notion because it suggests that this gene variant acts almost independently of nongenetic forces. The effects of serotonin transporter gene variants, whether short or long, may, to a large extent, depend on whether different segments of DNA within the same gene variant interact with one another. If they do, the effect may be similar to a gene–environmental interaction, which can affect both physical health and psychological traits.[65] In this sense, the serotonin transporter gene variant is distinct because its impact may not be based on traditional gene–environmental interactions, at least in its role of precipitating psychiatric symptoms. The genetics underlying depression, for instance, involves gene–gene interactions between different regions within the same gene, mimicking a gene–environmental effect.

These factors may help explain why the evidence for the role of the serotonin transporter gene in aggression and violence, including antisocial personality, is so fraught with contradictions. The short variant has typically been linked with psychopathy, but the long version, when combined with other gene variants and untoward environmental circumstances, may also be a risk factor for mental illness characterized by violent tendencies. Additionally, multiple epigenetic signals may influence the activity of gene variants linked with depression, potentially producing different actions in different individuals.[66]

THE MAOA GENE: RISK OR RESILIENCE?

Neurotransmitters comprise a core part of the biology underlying mood, arousal, emotions, and impulse control. Recall that monoamine oxidase A—MAOA—is a key enzyme that breaks down the neurotransmitters serotonin, epinephrine (adrenaline), norepinephrine (noradrenaline), and dopamine in the brain. When this enzyme is not functioning properly, serotonin levels in the brain are altered, possibly leading to emotional disturbance that, if coupled with severely disadvantageous or stressful environmental situations, seems to increase a person's risk for aggression and perhaps even violent crime. Early investigations conducted almost two decades ago showed that different variants of the MAOA gene can either dramatically lower or raise serotonin levels. In one of the most famous studies conducted on genes, brain, and behavior, NIH scientists Avshalom Caspi, Terrie Moffitt, and colleagues discovered compelling evidence for how the interaction between genetics and environment can shape a socially dangerous behavior. These researchers found that people with the weak gene variant—the short version that produces low or mild MAOA activity—had an increased risk for mental and behavioral imbalances later in life if they were severely mistreated during childhood. A key finding was that the risk was not present if a person did not experience severe mistreatment at a young age.[67] Confirmation soon followed, with other studies suggesting that individuals with the weak MAOA gene version seemed to be more prone to committing antisocial acts compared to those carrying the long MAOA gene variant—the version linked with strong MAOA activity.[68]

In 2006, Paul Chapman, a journalist, claimed that the weak MAOA gene variant might underlie the poor health and high rates of violent crime in the South Pacific Maori, the original Polynesian inhabitants of New Zealand. The Maori had notably greater frequencies of the low-activity MAOA variant than whites. The MAOA story ventured well beyond the Maori and the higher rates of an "aggression-provoking variant" in nonwhite men compared to white men.[69] As research continued, scientists identified additional variants of the MAOA gene, including some linked with impulsive aggression, that were categorized as low-activity, as opposed to the benign high-activity variants. Scientists at the National Institute of Mental Health eventually made a startling discovery. In the earlier research of Caspi's team, adult carriers of the weak MAOA gene variant who had been seriously abused as children were more likely to be violent and antisocial when they grew older compared to people with the intense MAOA gene variant who were not abused in childhood. Intense MAOA activity—linked with the short gene variant—apparently invokes a protective or favorable adaptive response, even if abuse has occurred. As new evidence accumulated, one study reported that brain scans of participants with the weak MAOA gene variant showed intense electrical activity in the amygdala and hippocampus—two of the primary brain structures involved in emotional arousal.[70]

Despite vehement protest from opponents of genetic determinism, the weak MAOA gene variant was soon dubbed the first "antisocial gene" in some scientific circles. Several experts cautioned that the weak MAOA gene variant had only a slight effect on provoking violent behavior, pointing out a central axiom in behavioral genetics: There is, as yet, no precise genetic trait for antisocial personality. Many concerned citizens, including some geneticists themselves, argued that environmental factors are crucial in determining impulsive violence in individuals. This has generally been true, regardless of whether a person has the mild MAOA gene variant or any other genetic trait allegedly linked with antisocial activity.[71]

But even if a single gene were not directly, let alone fully, responsible for criminal or violent behavior, certain genes, it seemed, waged an impact on shaping social traits. The issue, then, became a matter of determining how much of an influence the short MAOA gene variant had on indices of antisocial behavior in adolescent males who had experienced threatening psychosocial experiences. According to one report, genetics and psychosocial factors interact to increase the risk for criminal behavior in young males. But when researchers analyzed the data further, they found that psychosocial conditions were more influential than the weak MAOA gene variant for promoting violent and delinquent behavior.[72]

It soon became apparent that an MAOA-linked genetic tendency toward socially deviant behavior was more likely to become a "genetic fate" in young males if they faced harsh life circumstances. As evidence mounted, the converse also seemed to hold true. People who were born with the short MAOA gene version and reared in healthy, economically stable, and psychologically nurturing environments were not at increased risk for violence. Multiple studies on neurotransmitter-linked genes began to show that environmental conditions have a strong impact on determining whether the short MAOA gene variant expresses either low or high levels of MAOA.[73] Still, this revamped narrative did not conclude with an air-tight pronurture spin. As scientists delved more deeply into the intricacies of the low-activity MAOA gene variants, they discovered two distinct, alternative versions: the 2-repeat (2R) and 3-repeat (3R). Both variants contain fewer repeat sequences, such that they are now often lumped together as the weak-activity MAOA gene. (Another version called 5R has more repeats, but it is less active than the 4R variant.) The two classes of MAOA gene versions—the weak- and strong-activity versions—tend to correlate with distinctly different behavioral tendencies. Low-activity versions are linked with reduced levels of MAOA in the brain, allowing them to affect mood by changing serotonin levels.

One of these variants, commonly called 2R—is reportedly linked with severe hostility and aggression, notably brutal crime.[74] Some researchers have proposed that this variant might exert its effect without any intermediating effects from the environment. The "2R extreme warrior gene" would indeed be unique if it promotes violent impulses, independent of environmental influences. The notion of a potentially, intensely potent antisocial gene that could affect social behavior irrespective of environmental effects remains highly controversial. To further complicate matters, reports soon surfaced that the 2R variant of the MAOA gene occurs at a significantly higher rate in African American men than in white American men. This gene variant is relatively uncommon in humans overall but comparatively much more prevalent in African Americans. According to the Add Health database, rates of the 2R version of the MAOA gene are 0.03 percent in white males and 4.79 percent in African American males. Add Health contains data from the largest and most comprehensive longitudinal health survey of adolescents in the United States. The frequencies of the 2R variant were collected in 2008 and 2009, on separate samples of 3,356 white and 960 black young adult males.[75]

The current debate surrounding the warrior gene, particularly the "extreme warrior gene," was sparked by a discovery of a rare genetic disease almost 25 years ago. In 1993, a team led by Hans Brunner, a geneticist and medical doctor at Maastricht University in the Netherlands, along with a Harvard neurogeneticist, identified a mutant variant of the MAOA gene in 14 male relatives of an extended Dutch family.[76] The Dutch men spanned five generations, with many of them displaying extreme antisocial and violent tendencies. Since 1978, other researchers have investigated this family to understand why so many males in this group were so aggressive. It is now known that their disorder, eventually labeled Brunner's syndrome, is triggered by a mutation that decreases the expression of the MAOA gene, causing the variant to produce less of the MAOA enzyme. Without a sufficient amount of this enzyme to degrade key neurotransmitters, pivotal brain chemicals—serotonin, norepinephrine, and dopamine—can climb to excessive levels in the brain.[77]

An oversupply of these chemicals swimming around in the brain raises the risk for impulsive and erratic behaviors, including disrupted sleep disorder, extreme mood swings, heightened sex drive, anger, and violent tendencies. Early research also indicated that males with this condition had mild mental disability, with low IQs of about 85 points. In the mid-1990s, the idea that a gene could spark violent behavior was highly contentious. This was not unlike the raging debates about such hot-button topics as the U.S. federal government–sponsored Violence Initiative and the book *The Bell Curve*, which revived an age-old pronature claim that the ethnic divide in IQ is due partly to racial genetics. By the 2010s, however, mounting evidence from advanced brain imaging and behavioral genetics studies seemed to confirm that the low-expression version of the MAOA gene was indeed associated with an increased risk of aggressive and potentially brutal social behavior.[78]

Multiple studies have since confirmed the findings of Caspi's team. The interaction between low-activity MAOA gene variants and abusive social conditions at a young age increases the odds, at least in males, for offending, conduct problems, and hostility later in life.[79] Nonetheless, contradictory outcomes have been reported, and some research has produced equivocal results. In a study of male convicts in a correctional facility, low-activity MAOA gene variants were significantly linked with violent crime in whites, but the association in the black inmates showed a marginal trend that was not statistically significant. Researchers attributed the inconsistent results in the two ethnic groups to demographics, noting that most participants were white convicts.[80] Yet, other factors undoubtedly come into play when estimating the influence of the low-activity MAOA gene variants. Links between this variant and violent crime have been observed for offenders but not for the general population.[81]

Despite popular claims to the contrary, the low-activity MAOA variants are usually only associated with aggressive behavior in individuals, typically male, who have had damaging life experiences, especially at a young age. Permissive parenting, early life trauma, certain forms of maltreatment, and substance abuse are among the chief environmental factors that appear to interact with the low-activity MAOA gene variants. For many researchers, it is the interaction between genetics and environment, rather than a gene variant or environmental condition separately, that influences socially destructive behavior, including unbridled aggression and violent crime.[82]

Another study evaluated the effects of low-activity MAOA gene variants in participants diagnosed with a psychiatric condition.[83] When individuals who had experienced childhood physical abuse were excluded, the low-activity MAOA gene variants correlated with scores for antisocial personality in whites but not in blacks. The high-activity MAOA variant seemed to protect against antisocial personality in the whites who had not been physically abused during childhood. Yet, the high-activity variants correlated with slightly higher percentages traits for antisocial personality disorder compared with low-activity variants. This finding contradicted the predicted outcome in which MAOA gene variants with lower activity are sometimes linked with a higher incidence of antisocial personality disorder, especially in males victimized by physical abuse during childhood.[84]

As the authors pointed out, however, the level of childhood abuse, as well as the ethnicity of the participants, might account for this discrepancy. Other gene variants, in addition to various aspects of the social environment, may mediate MAOA activity and influence antisocial behavior. Also, assessing a male participant's antisocial tendencies may be problematic if a researcher has implicit bias. An African American adolescent or adult is statistically more likely to have an arrest record than a white offender who committed the same crime. But for some researchers involved in this study, black participants who had more arrests than their white counterparts might have been subjectively evaluated as having antisocial personality traits. As the authors noted, researchers may have been less inclined to make this designation for the white participants.[85]

Still, some researchers propose that because the MAOA-2R genotype is the most extreme of the "MAOA warrior genes," regardless of the social context, it poses a serious risk for social deviance. They claim MAOA-2R is sufficient by itself to result in antisocial or violent behavior without interacting with environmental forces. If true, this would imply that males carrying the 2R variant would have an exceedingly pronounced proclivity for antisocial activity, regardless of their environmental circumstances. As noted, the vast majority of MAOA-2R carriers are African American. (Research on males of African black ancestry other than African Americans is limited.) This gene variant has also been implicated in female antisocial activity, but the association for women is not as strong as in men. The correlation is reversed in females, with the high-activity variant considered the risk version.

The initial research on low-activity MAOA gene variants did not tease apart participants with the 2R version from those with other low-activity variants. Caspi's research focused on low-activity MAOA variants as a whole without distinguishing between various subtypes of variants that apparently carry different levels of increased risk for violence.[86] In 2008, researchers reported that two dopamine-linked genes and low expression MAOA gene variants, one of them MAOA-2R, were associated with a higher frequency of self-reported, serious and violent delinquency in male youth.[87] During the last few years, Kevin Beaver, a biosocial criminologist at Florida State University, has investigated the connection between the 2R variant and various antisocial activities in males, particularly African Americans. Beaver claims his research demonstrates a strong link between the MAOA-2R variant and violent behaviors, arrest, and incarceration.[88]

In one study, Beaver and colleagues evaluated shooting and stabbing behavior in males in the Add Health database. The MAOA-2R variant was carried by 0.1 percent of white males and 5.2 percent of African American males in the sample. Based on interviews that spanned about 15 years, 5.6 percent of the men in the entire sample, including both white and black men, reported shooting or stabbing another person at some point in the previous 12 months. The final analysis was performed on 133 African American men in the sample, of whom 6 percent carried MAOA-2R. Virtually no whites in this study were MAOA-2R carriers.[89] The results suggested that American males with this gene version were significantly more likely than those with other MAOA variants to be involved in shooting and stabbing incidents, and to have had multiple shooting and stabbing victims.[90] Still, Beaver has stated that the MAOA-2R variant by itself does not explain a significant amount of violent behavior in African American men. He has said he doubts that the MAOA-2R version can account for all of the variation—the heritability—between men who do and do not display severe antisocial traits.[91]

Beaver has acknowledged that gene–environment interaction influences social behaviors, even though he also has noted that some social behaviors could be shaped solely by environment or genetics. Although his research has focused largely on African American males, he has cautioned against prematurely speculating about possible ramifications of the role of MAOA-2R in criminal trends.[92] Nonetheless, some of his findings have apparently inspired several genetic determinists to interpret, or misinterpret, his narrative about MAOA-2R to mean that a single gene variant puts African American males at increased inborn risk for antisocial behavior, regardless of their environmental circumstances.

One study investigated the link between MAOA gene variants, punitive discipline from parents or caregivers, and antisocial behavior later in life in white and African American young men.[93] Most participants had either the low-expression 3-R or high-expression 4-R MAOA variants. The punitive activity included such corporal discipline as spanking or harsh actions like yelling at a young child. Punitive actions are not confined to physical or verbal abuse, but depending upon a person's psychological makeup and sensitivity, punitive discipline may be experienced as abusive to a young child. As the researchers expected, punitive discipline in young males carrying the low-activity 3R variant was associated with a heightened risk for antisocial behavior in both the white and black men. No such risk was seen in males who had the high-expression 4R variant, regardless of their ethnicity.[94]

The original sample in this study included five African American men with the 2R version who had been excluded from the initial analysis. Aware that the 2R version reportedly represents the highest-risk variant and is linked with very low MAOA activity, researchers reanalyzed the findings for the entire sample, including the five black males with the 2R version. The results remained unchanged. Like the men with the 3R version, those with the 2R variant exhibited an increased risk for antisocial behavior if they had experienced firm discipline and not simply extreme maltreatment as a child.[95] Although the sample was exceedingly small, the findings suggested that the decreased expression of the 2R gene variant is unlikely to be expressed unless adverse environmental triggers are also present. Additionally, this study revealed that the degree of each boy's misbehavior seemed to depend on his age when the punishment occurred. Apparently, there are sensitive periods during early childhood when males with the low MAOA variants are more susceptible to the behavioral ramifications of punitive discipline. This sensitivity can potentially translate into increased risk for antisocial behavior later in life.[96]

Another study of males carrying the MAOA-2R variant examined the link between their risk for delinquency and their relationship with their parents.[97] For participants who had an incarcerated parent, the risk for delinquency was reduced if they were close to their biological or social father but not their biological mother. The researchers called this a "double whammy" for at-risk youth with the 2R variant, since their parents created an environment in which the impact of otherwise small genetic influences on the young men were amplified. This example illustrates a gene–environment interaction in which the effect of each element depends on the strength of the other. In this study, risk factors had to be present in the family environment to trigger gene expression of the 2R variant, which in turn might increase the risk for delinquency and violence. These findings also suggested that the risk for unacceptable behavior in young men could be substantially moderated by socially supportive familial environments.[98]

Few experts or even the lay public doubt the negative effects that disruptive life circumstances, for instance, imprisoned parents, may have on their children. And yet, while some youth seem to be protected against a life of delinquency, even amid pernicious life circumstances, the correlations reported between the MAOA-2R "extreme warrior" gene variant and antisocial behavior do not necessarily reflect cause-and-effect relationships. The 2R variant potentially influences social behavior, but as the authors cautioned, multiple components of the social environment contribute to intergenerational delinquency: low socioeconomic status (SES), abusive parenting, low self-control, lack of supportive and caring parenting skills, family instability, friendships, and poor school and neighborhood environments.[99] Delinquency is shaped by the interplay of thousands of genetic and environmental influences.[100] Low-activity MAOA gene variants linked with antisocial behavior not only interact with environment, but also, in some cases, appear to condition a person's response to environmental triggers.

In phenotypic plasticity, the activity of one or more genes may be expressed differently, depending on the environment.[101] The characteristics of some MAOA gene variants are reminiscent of plasticity genes capable of altering their activity in response to variations in the environment. These genes are susceptible to epigenetic imprinting that can influence whether a disease or a behavioral tendency manifests.[102] In a seminal study, epigenetic signaling reduced the expression of low-activity MAOA gene variants, resulting in more severe posttraumatic stress disorder, or PTSD.[103] This shows that the activity of MAOA genes can be altered by epigenetic marks. More research is needed to determine if epigenetics specifically influences antisocial traits by impacting MAOA gene variants.[104]

Still, findings for the role of low-activity MAOA variants in aggression and violence are sometimes contradictory. In another study, both white and black adolescents, aged 12 to 19 years, who had experienced maltreatment from a parent or caregiver before age 12 were more likely to develop conduct disorder as a youth and, as an adult, engage in antisocial behavior or be convicted for violence.[105] This association fits a well-described pattern, but in this case, the low-activity MAOA gene variants did not mediate a link between maltreatment and antisocial behaviors in any of the participants.[106] The male youth carrying the presumed risky gene variant were not predisposed toward violence if they had not been grossly mistreated. Based on this evidence, researchers surmised that some of the links between low-activity 2R or 3R MAOA gene variants and violence might not be accurate. According to their assessment, some of the variants reported in other studies may have been false positives.[107]

BEYOND ANTISOCIAL GENES

Overall, the evidence suggests that some gene variants reportedly associated with antisocial and law-breaking activity do not, in and of themselves, increase the risk for offending or socially unacceptable behaviors. Instead, these gene variants appear to change an individual's susceptibility to either negative or positive environmental factors.[108] These variants, then, essentially function as plasticity genes.[109] Today, concepts of genetic liability and genetic assets are shifting. Changing perspectives on the behavior of the gene itself, notably gene expression, offer novel insights into complex interactions between not only nature and nurture, but also nature and nature—or different biological phenomena. The interplay between nature and nature includes, for example, intricate interconnections of genes with environmental triggers via epigenetic pathways that affect gene expression, which in turn can impact the plasticity of the brain. For this reason, some genes, once termed "risk genes," have been reconceptualized as "plasticity genes." Regardless of their name, a key feature of these genes is that they are adaptable in responding to, and interacting with, other biological and environmental influences—for example, other genes, social factors, environmental toxins, physical disease, and epigenetic signals.[110]

As scientists unravel the maze of highly complex and transformative properties of plasticity gene variants, they may move us closer to understanding why ethnic populations sometimes differ in their social behaviors. The key take-home message is that plasticity genes, as their name implies, do not inevitably result in fixed, immutable, and predictable traits. Rather, the expression of plasticity genes depends largely, and perhaps ultimately, on delicately orchestrated interactions between environmental and epigenetic processes. The effects of some plasticity genes potentially can change across generations and, with appropriate training or intervention, even remarkably transform in a person during his lifespan, perhaps partly as a result of epigenetic processes.[111]

In one study, teenage delinquency was associated with interactions between four different gene variants and specific environmental or experiential factors—family conflict, sexual abuse, and child–parent relationships. The variants included a low-activity version of the MAOA gene and a section of the serotonin transporter gene (not a complete gene variant) implicated in several psychiatric disorders. Yet, despite claims that some gene variants can provoke antisocial activity, supportive relationships with parents or caregivers seem to buffer children against future undesirable social behavior. Positive nurturing shields individuals, even those who have inherited presumably at-risk genotypes, against detrimental environmental effects, including family conflict and sexual abuse, during youth and later in life.[112] Myriad environmental forces play a preeminent and perhaps decisive role in dictating the trajectory of genetic influence on personality and social behavior, no matter how seemingly troublesome an alleged genetic propensity may be. Just as many leading experts contest the notion of genes as destiny, gene–environmental interactions do not predict fated life outcomes. By the same token, environmental conditions do not guarantee either good or bad predetermined life outcomes, although they can have major consequences for an individual and, in some cases, his family.

According to University of Cincinnati criminologist J. C. Barnes, there is no gene for criminal behavior because crime is a learned behavior. Still, Barnes has postulated that multiple genes—hundreds, perhaps thousands—can incrementally raise the probability of participating in a criminal act, even if that increase is only 1 percent. From his vantage, genes do count, and their effects on deviant behavior can be potentially momentous.[113] The combination of certain variants of dopamine gene variants, for example, DAT1 and DRD2, reportedly elevates the risk for serious and violent crime in adolescents and adults. In some studies, various versions of these genes, as well as the mutant serotonin transporter gene (5-HTT) and the low-activity MAOA gene variants, have been implicated in criminal behavior.[114] Additionally, some of these gene variants can apparently interact with certain versions of the testosterone-linked AR gene, as well as another mutant gene variant that degrades neurotransmitters, possibly leading to an increased likelihood of aggressiveness in men.[115] As previously noted, we will examine research on the effects of the AR gene on social behavior in males in chapter 14.

Individuals who inherit multiple-risk gene variants appear to have an elevated risk for antisocial behaviors, for instance, criminal acts. But almost inevitably, these same gene variants are also modified by a wide array of environmental influences, in some cases via epigenetic processes, which can strongly affect a person's social tendencies. These personality and social traits, in turn, affect his individual life outcomes. Today, studies on the genetic underpinnings of criminal activity are increasingly examining antisocial tendencies through the lens of gene–environment interactions that raise the risk for destructive social behaviors. Yet, multiple environmental influences, which themselves may interact with one another, are often assessed only superficially, at best. One study examined the transmission of crime across generations in a group of males in London. Individuals who were labeled, or classified, by the criminal justice system based on their offense and punishment, and also had one or more convicted parents, were more likely to be targeted by the criminal justice system. The labeling effect was seen only in the individuals with convicted parents, suggesting that these two environmental factors resulted in a cumulative disadvantage. These findings shed light on the role that bias from the criminal justice system plays in transmitting criminal activity from parents to children and on down the line.[116] The risk for criminal activity in people with this environmental double whammy—a convicted parent combined with a young person's own conviction—is akin to an exacerbated self-fulfilled prophesy. One risk is amplified by the other risk, with potentially negative ramifications more likely to manifest when both risks are present.

Oxytocin and the "Love Gene"

Oxytocin—a chemical that acts as both a neurotransmitter and a hormone—has two main functions in the human body: causing the womb to contract during childbirth and promoting milk ejection for breastfeeding.[117] Oxytocin has also been linked with social behavior, notably mother–infant bonding, and various attitudes and behaviors that foster social approval and friendship.[118] In popular culture, oxytocin has been coined the "love hormone," heralded for eliciting positive emotions, trust, altruism, and generosity. But this hormone is also associated with autism, as well as socially undesirable traits, ranging from envy and gloating to perhaps anger.[119]

The oxytocin receptor gene—OXTR—is responsible for producing the chemical receptors to which oxytocin attaches.[120] OXTR has gained increasing attention because it is closely aligned with human emotional responsiveness and social behavior. Unlike OXT—the gene that makes oxytocin—OXTR can be affected by epigenetic processes that, in turn, can alter the expression of oxytocin gene variants.[121] Emory University scientists Alicia Smith, Erica Smearman, and colleagues explored the association between control sites within oxytocin gene variants, single nucleotide polymorphisms (SNPs located within DNA), childhood abuse, and adult depression and anxiety symptoms in 393 African American adults.[122] The interaction of both individual SNPs and epigenetic marks with abuse predicted psychiatric symptoms in the participants. (Recall that epigenetic marks can either silence or amplify the activity of various genes.) In some cases, the genetic and epigenetic effects occurred independently, whereas in other cases, they overlapped in their interaction with abuse. The results pointed to a possible mechanism whereby epigenetic marks can facilitate interactions between genes, or DNA, and environment, specifically childhood abuse; however, as the authors noted, the connection they found between SNPs and the control region of genes may have been unique to the participants in their study or African Americans, in general.[123]

In other studies, individuals with one or two copies of the A version of OXTR exhibited less optimism, poorer parenting skills, and lower self-esteem than those without this variant. These people were more prone to having a negative physiological and psychological reaction to stress. By contrast, those carrying the G variant of OXTR displayed greater empathy in their attitudes and actions.[124] Multiple investigations have recently scrutinized the effects of epigenetics on the expression of OXTR variants. DNA methylation, one of the most common epigenetic processes, modifies the structure of DNA, which changes the activity of OXTR. One study evaluated social bonding patterns in low-income African American men and women residing in an urban community. Individuals who inherited two G variants and had experienced severe childhood abuse were more likely to have disorganized adult attachment—erratic and unpredictable bonding behavior in their relationships—compared to carriers of the A/A and A/G gene variants. But carriers of the A variants of the OXTR gene were resilient against the effects of extreme childhood adversity. The interaction between their epigenetic marks and DNA seemed to protect them against emotional imbalance and disorganized attachment in bonding with others.[125]

As this study revealed, understanding the social context, particularly the lifelong adverse influences of early childhood traumatic experiences, is imperative for assessing how susceptible different individuals are to their environment. The findings may be leveraged to identify a person's risk of, and resilience against, challenging life circumstances.[126] The results underscore the benefits that can come from integrative research encompassing nature and nurture in a broad sense and not limited to classic genetics. After all, epigenetic marks are inherited, along with DNA. And yet, epigenetic mechanisms are reversible because they unfold above the physical sequence of DNA chemicals. When epigenetic processes work in tandem with environmental processes to affect gene expression, the epigenetic mechanisms themselves can potentially change as a result of shifting environmental influences. Consider, for example, the effects of psychosocial interventions that encourage children and caregivers to develop mutually supportive relationships. This environmental approach can help break the intergenerational cycle of poverty, maltreatment, domestic violence, and, in some cases, poor cognitive performance. On a biological level, societally beneficial changes that occur in an individual's disposition and behavior can, throughout time, dissipate negative epigenetic markings that he has inherited. For certain genes variants, then, an adverse epigenetic profile, when reversed, will not be transmitted to his or her offspring. In animal studies, mothers have been found to pass on maternal skills to their daughters through various patterns of oxytocin receptor gene expression.[127] Some of these patterns have positive outcomes, while others have negative effects. If detrimental epigenetic signals are thwarted in an individual before these signals can be passed on to her offspring, then genes like OXTR are more likely to be expressed in a healthy manner. Lest we forget, epigenetic activity, for good or bad, is triggered by a wide array of environmental influences.

Still, some experts advise exercising caution before drawing conclusions about the merit, or lack thereof, of this type of research.[128] More details are needed to pinpoint unequivocal connections between epigenetics, OXTR, and various social and emotional behaviors across different ethnic populations. As we have learned, the same gene variant may have a different effect on social traits in different populations. Teasing apart the multilayered effects of multiple genes interacting with multiple environmental forces—as well as individuals' experiences of those interactions—is complicated. This applies to candidate genes, SNPs, tandem repeats, and genetic signatures implicated in human personality traits, psychiatric conditions, and social behaviors. It also pertains to innovative nature-plus-nurture research on human social life, which has now introduced epigenetics into the equation.

Still, not everyone who inherits multiple-risk gene variants and simultaneously experiences inopportune circumstances, especially during childhood, seriously breaks the law or becomes a hard-core criminal. Individuals who persevere rather than fall prey to a downward spiral of life adversities may be the exception, but they do exist. And yet, relatively few investigations probe the genetic makeup and environmental circumstances, along with relevant epigenetic processes, of people who surmount the odds seemingly stacked against them only to become respectful and productive members of society. Only a handful of studies have dissected the interplay and perhaps synergism between separate environmental and experiential factors that can interact with plasticity genes to buffer some individuals from detrimental behavior, even amid appalling life circumstances. This approach might prove fruitful for unveiling psychosocial and environmental clues about the influences that foster improved social outcomes for individuals exposed to serious hardship, no matter their genetic ancestry or SES.

Girl Growth

Sexual Maturation in Black and Brown Bodies

FEMALE PUBERTY: A CHANGE FOR THE AGES

IN 1997, WHEN THE UNIVERSITY OF NORTH CAROLINA'S MARCIA HERMAN-Giddens and colleagues published their groundbreaking multiracial study on female puberty in the United States, it sent a silent shock wave that rippled throughout the medical community. Their investigation of 17,077 American girls, ages three to 12 years, revealed that African American girls started sexual development on average 1.3 to 1.5 years earlier than their white counterparts.[1] The pace of puberty in both groups of girls was apparently faster than reported in previous eras, although findings from different studies have produced conflicting results on changes in the start of puberty during the last few decades and even centuries.[2]

Puberty refers to the physiological changes that occur during the sexual maturation of girls and boys, encompassing physical growth, as well as emotional, mental, and social development. This is the time when hormonal changes rapidly but predictably transform a child's body into an adult body capable of sexual reproduction. In males, puberty involves the formation of the testes, as well as secondary sexual traits not directly related to procreation. These traits include facial and bodily hair, deepening of the voice, enlargement of the penis, and an Adam's apple. In recent years, American boys have experienced testicular growth and secondary sexual characteristics an estimated six months to two years earlier than in the past.[3] A study of more than 4,100 boys in the United States, conducted between 2005 and 2010, found that African American boys on the whole enter puberty at a younger age than white or Hispanic American boys. The earliest stage of puberty occurred at age 10.14 years in non-Hispanic white boys, 10.4 years in Hispanic boys, and 9.14 years in non-Hispanic African American boys.[4]

In girls, physical maturation corresponds to key physical signs that include the start of menstruation, breast development, and the first signs of pubic hair, but the hallmark of female puberty is menarche—a girl's first menstrual period. In the late 1990s, the onset of menarche in the United States averaged 12.16 years for black girls and 12.88 years for white girls, revealing a significant decline compared with the mean age of 15 years for girls (most likely white girls) reported for the start of menses in the 1800s. Most astounding were the contrasting ages of this event across ethnic groups—a contrast, literally and figuratively, in black and white. While the average onset of menarche was slightly younger than nine years for African American girls, it was 10 to 10.5 years for white girls when Herman-Giddens's team published their study in 1997.[5]

The findings of Herman-Giddens and associates did not fall on totally deaf ears of the American public. Still, members of the news media treaded cautiously in how they interpreted evidence for seemingly inborn race-based differences in sexual maturation. After all, for some observers, earlier female puberty suggests an increased likelihood of precocious sexual activity and with it the potential for unintended, out-of-wedlock teenage pregnancy. Three years after the 1994 release of *The Bell Curve*, who in the mainstream media would have the audacity to contemplate such a possible connection, let alone explicitly state it in writing? Most journalists at that time were circumspect in their words and tone. But others did not hesitate to speculate that the study's results—faster sexual maturation in black girls compared with white girls in the United States—just might lie at the root of what some were calling innate race-based differences underlying a wide array of adolescent behavioral problems.

The diminishing age of sexual maturation in girls is not confined to the United States. Numerous other ethnic populations from geographically diverse regions of the world have experienced a rapid decrease in the average age of the first menses. Western and Northern European countries like Denmark have reported sharp drops in the age of menarche, from about 17 to 18 years in the mid-nineteenth century to 12 or 13 years in the 1970s. A similar pattern is also evident in non-Western societies, with declines of 17 years to 12 to 14 years of age reported during a 50-year period starting in 1920.[6]

In 1997, the genetics of human puberty was still in its infancy, but some behavioral geneticists were already convinced that genes, specifically population genetics—or "racial genetics," as it was known at the time—contributes substantially to ethnic population differences in menarche. Some researchers theorized that during the formation of the human species, different evolutionary driving forces in divergent populations had resulted in different biological and social traits related to reproduction. These traits were not confined to purely physiological features, as they supposedly ranged from the timing of puberty to tendencies toward monogamous partnerships to social practices of parental investment and child rearing. All of these topics are worthy of exploration, but the focus in this chapter is on the pace of puberty in female children and youth of African black descent compared with the timing of puberty in other ethnic populations.

If the cadence of puberty is substantially determined by genetics, then we should expect to see similar rhythms in the onset of various signs of sexual maturation in genetically related populations. Yet, patterns of pubertal timing sometimes vary significantly among ethnic populations that share common ancestry. Despite popular stereotypes of the precocious sexual nature of black girls, particularly African American girls, the earliest ages of female puberty reported worldwide do not occur in black girls, descended from enslaved Africans, who are born and reared in the United States. In an international survey published in the 1990s, the earliest age for first menses occurred at 12.4 years in girls of affluent families in Istanbul, Turkey. The age of menarche in girls from poor families in that city was 13.2 years.[7] In a survey published in 2001, the earliest mean ages for menarche were 12.0 years in Greece and Congo-Brazza, 12.2 years in Italy, 12.3 years in Thailand and Spain, and 12.4 years in Mexico.[8] The divergent genetic ancestries of the peoples populating these countries reveal that demographic factors account for some, perhaps much, of the discrepancy in the age of menarche across ethnic populations and nations.[9]

In an older study published in the 1970s, the age of menarche for girls in Nigeria varied slightly, depending on their families' socioeconomic status (SES). The onset of first menses was 13.8 years in an "illiterate-urban" community, compared with 13.3 years in university-educated girls and 13.2 years in "better-off females."[10] In a later survey, girls in Nigeria from middle-class backgrounds had a mean age for menarche of 12.22 years, an age close to the onset of menses in African American girls.[11] Since some Nigerians and black Americans share a fair amount of their DNA, population genetics conceivably could play a role in the timing of menarche in these two continental populations. However, the parallels in menarche between African American females and their counterparts in other black African ethnic groups, particularly in West African societies, have been inconsistent. In the 1990s, the mean age of menarche for black African girls in Cameroon was approximately 12.72 years for those from privileged backgrounds and 13.98 years for those living in suburban areas. But for Cameroonians inhabiting poor rural communities, the average age of menarche was 14.27 years, almost two full years later than the age of first menses in African American girls.[12]

Other survey findings show even higher ages for the start of menses in black African nations: 14.61 years in Cameroon, 14.78 in Somalia, and 15.21 in Tanzania. And in Haiti menarche began at a mean age of 15.37 years.[13] Interestingly, one of the highest ages of menarche reported in peer-reviewed studies occurred in postsecondary female adolescent students in Kenya, where the mean age was 14.4 years with a range of nine to 16.5 years.[14] In the late 1980s, girls in Papua New Guinea also tended to have delayed menses, with an average age of onset of 15.8 years.[15]

Some of these findings are based on studies two to three decades old, but the trends and underlying causes for these trends still exist today. Poor nutrition, usually associated with poverty, can postpone the start of menarche, at least in the developing world.[16] Genetics plays a role, but multiple demographic and other environmental factors, including social status, family finances, climate, and exposure to toxins, are known to affect the timing of menarche. But the wide range in the average age of menarche across societies and ethnic populations, in some cases across groups that share close ancestry, strongly suggests that the contribution of population genetics to the onset of menarche is limited. Environmental factors, including social, economic, and other demographic influences, offer a more parsimonious explanation. Environmental circumstances best account for the overall evidence demonstrating wide disparities in the age of menarche in different ethnic populations, both within the same country and across nations.[17]

FEMALE PUBERTY'S OTHER SIDE: BUDDING BREASTS
AND SPROUTING HAIR

In 1973, the median age of menarche in American girls, in general, was 12.77 years, with significantly earlier onset of menses seen in black girls. And yet, about 30 years later, the median age of the first menstruation in girls overall in the United States was 12.43 years, reflecting only a minuscule decline during this period of time.[18] Even the historical record throughout time casts doubt on claims that the timing of the start of puberty has decreased dramatically during the last few centuries. In one study, archeologists excavated the skeletal remains of adolescents from medieval England and examined various bones for clues about their onset of puberty. The canine teeth, as well as the shape of the neck and wrist bones, and fusion of multiple bones in the centuries-old skeletons, suggested that children in medieval England began the process of physical maturation between 10 and 12 years of age. This is approximately the same age range in which young people enter puberty in contemporary English society.[19]

These findings might mean that the controversy regarding declining ages in female puberty is merely hyperbole. It may well be, at least if the conspicuous ethnic gap in the timing of first menses in the United States is ignored. Menarche is often considered the cardinal event corresponding to female sexual maturation, but puberty in young women is also defined by the appearance of breast buds or sprouted hairs in the pubic region, or, in some cases, both types of physical changes. While the later phase of puberty is associated with menarche, the onset of breast development and pubic hair growth represent early markers of sexual development. Contradictory findings on pubertal timing across populations are not uncommon, partly because researchers sometimes base their conclusions on different signs of puberty. If experts fail to specify which criteria they are using when they aggregate findings from various studies, their estimates of the average age of the start of puberty across nationalities, cultural groups, or ethnic populations may be misleading and possibly even inaccurate.

Early signs of American female puberty, specifically the development of breasts and pubic hair, have decreased, in general, in the past several decades. According to the U.S. government's National Household Education Surveys Program, in operation from 1966 to 1970, on 12- to 13-year-old Americans, 94 percent of both black and white girls showed signs of breast bud formation around their nipples. Pubic hair was detected in 90 percent of the white girls and 95 percent of the black girls in this study.[20] For most girls, sexual hair usually starts sprouting in the pubic region and in the armpits about one to 1.5 years after breast buds have appeared.[21]

By the mid-1990s, when Herman-Giddens and colleagues published their landmark study, the age of initial breast development, or breast budding, was 8.9 years in African American girls and 10.0 years in white American girls. (Prior reports had placed the average age of breast growth in American females overall at 10 or 11 years.) Among three-year-old female toddlers, breast or pubic hair was seen in 1 percent of whites and 3 percent of blacks. But by eight years of age, 14.7 percent of white girls and 48.3 percent of African American girls had started developing breasts or pubic hair, or both.[22]

These ethnic differences in the age of incipient breast formation are striking, but assessing the onset of breast development and pubic hair may be more complicated than determining the age of menarche. Consider the following: When pediatricians at Baylor College were asked how they evaluated breast development in adolescent girls from various ethnic backgrounds, the doctors could not agree on which girls had fully developed breasts. In the last few years, breast development and pubic hair have been detected in increasingly large numbers of seven-year-old girls. In 2010, a team of pediatricians examined 1,239 girls from three cities in the United States and found that almost 15 percent of the girls overall had started developing breasts by seven years of age. Early breast development was seen in about 23 percent of African American girls, 15 percent of Latino American girls, and 10 percent of white girls. Compared to previous surveys, twice as many white girls in the more recent study had breast growth at this age.[23]

Many pediatricians have been puzzled about the exact causes for the rising rates of early sexual maturation in American girls, in general, irrespective of their ancestry. Excess weight worsened by the childhood obesity epidemic in the United States has long been suspected to be one of the main culprits. In societies in which people live the longest, female puberty usually commences at a younger age. Yet, this trend contradicts the pattern seen in African American girls. Compared with white girls, black girls in the United States on average have a shorter life expectancy, even though they start puberty earlier.

Regardless of longevity, obesity has been increasingly linked with the diminishing age of female puberty, especially in industrialized societies, but recently also in developing nations. In China, for instance, girls who experienced earlier menarche were more likely to be obese and have central obesity, or excessive fat around the abdomen, than girls who were not excessively overweight.[24] In Nigeria, excessive weight and obesity are presumably major contributors to an earlier age of puberty, with menarche, breast development, and pubic hair occurring sooner on average than in the past.[25]

Irrespective of a young girl's genetic ancestry or social circumstances, excessive weight or obesity may trigger first menses at a younger age because extra fat cells can act like the female sex hormone, estrogen. This hormone affects the development of such female secondary sexual characteristics as breast formation, pubic hair, and the menstrual cycle. Although controversial, a small but growing body of evidence suggests that chemicals in the environment, for example, bisphenol A (BPA), that disrupt normal endocrine function may prematurely hasten sexually maturity in some young girls. This cascade of events may lead to faster physical development, while also increasing a female's risk of breast cancer later in life.[26]

Overall, the results are conflicting, however, since BPA correlates with puberty in some studies but not others.[27] Nonetheless, concern about the timing of puberty has escalated in the medical community and beyond, and not simply as an academic exercise for assessing genetic versus environmental influences on the process of becoming an adult human. The tempo of sexual maturation holds important implications for two crucial life outcomes: health and social behavior. Girls who experience early puberty are at increased risk for developing a small stature, obesity, gynecologic cancers, and type 2 diabetes. By contrast, later maturing girls reportedly have a heightened risk for decreased fertility, stroke, and Alzheimer's disease.[28] Perhaps even more disconcerting, the potentially adverse risks of early menarche are not limited to physical conditions. Girls who experience precocious puberty are especially vulnerable to behavioral disturbances and more likely to suffer from mood disorders.[29] Females who began menarche before 11 years of age are at increased risk for multiple behavioral problems, for instance, conduct problems, delinquency, and substance use.[30]

GENETIC UNDERPINNINGS OF FEMALE PUBERTY

Amid the wealth of genetics data now available on a wide array of physical traits, combined with new insights into puberty-linked genes, some researchers have turned their attention to the possible role of DNA in pubertal timing in both girls and boys.[31] Genes undoubtedly contribute to sexual maturation, but the exact role of population genetics, as opposed to individual genetics, remains unclear. The timing of sexual maturation is 50 to 60 percent heritable, meaning that genes explain about 60 percent of the variation in the onset of physical maturation seen among members of the same sex in the same population.[32]

Some commentators have interpreted this finding to mean that heritability accounts for the wide global age range of the starting age of puberty in girls (eight to 13 years) and boys (nine to 14 years). This is not the case, however, because a heritability estimate pertains to a specific population at a particular point in time. The heritability of the hormonal and neurological changes triggering puberty, which influences sexual behavior, is not equivalent in every ethnic population.[33]

One research team discovered some degree of overlap in the SNPs—single nucleotide polymorphisms—linked with puberty in black and white American children. (Recall that SNPs are gene variants, or a section of DNA in a specific gene in which one chemical letter is substituted for another.) There were differences in the SNPs of the two ethnic populations, as some of the gene variants associated with menarche were specific for race and sex. Scientists identified 31 SNPs in white females that were strongly associated with either menarche or both menarche and growth indicators of adult height and body mass index (BMI)—a measure of body fat. Researchers later discovered that 15 of the 31 SNPs linked with puberty in the white females were also present in African American females.[34]

In another investigation, 42 SNPs were associated with the timing of menarche in white American women. Again, there was partial overlap in the SNPs observed in the two ethnic populations, with 25 of the gene variants significantly linked with menarche in black American females.[35] This might mean that the onset of menses in African American girls involves alternative genetic pathways that are nonexistent or inactive in white females. Scientists are still investigating whether any of the gene variants involved in the onset of menarche possibly also play a role in obesity. Compared to several other ethnic groups, non-Hispanic black women have higher rates of diseases associated with early physical maturation, including childhood obesity, metabolic syndrome, and type 2 diabetes. (As noted earlier, metabolic syndrome is a cluster of physical abnormalities, including hypertension, obesity, elevated triglyceride level, and low HDL, or "good" cholesterol level, that triggers increased risk for cardiovascular disease and type 2 diabetes.) But in a large meta-analysis of 15 studies on gene variants connected with the timing of menarche in more than 18,000 African American women, no single SNP had a significant effect on sexual development.[36]

The lack of a firm relationship between SNPs and the timing of menarche may seem to suggest that environmental factors are the sole driving force for ethnic population differences in the onset of menses. But according to University of Minnesota epidemiology professor Ellen Demerath, the lead investigator of this meta-analysis, the results of genome-wide association studies, or GWAS, which analyze large numbers of gene variants at a time, show this is not the case. As Demerath explained, the sensitivity for GWAS is lower in populations of African ancestry. Compared with Europeans and Asians, black Africans have more diversity in their gene variants, making it less likely that scientists can identify a small number of SNPs specifically linked with puberty in black females. Demerath's team was quick to point out that other types of gene variants not detected in this study may influence the onset of menarche. They noted that any "concealed gene variants" that may—or may not—influence the timing of puberty in black girls have yet to be discovered.[37]

ENVIRONMENTAL UNDERPINNINGS OF FEMALE PUBERTY

Environmental factors alone may not account for population discrepancies in the timing of puberty, but they may play a more robust role than previously thought. Yet, disagreement about the amount of environmental influence that affects the onset of puberty abounds, more intensely than ever. One reason for this is that some of the relevant aspects of the environment that may impact physical development are social and linked with education, family income, and social status or social class. Early maturing girls on average usually have a higher body weight. And girls reared in low-income homes tend to be overweight and mature earlier than their counterparts raised in middle- and upper-class homes. This is not surprising, given that poor families are less likely to have access to healthy foods, especially fresh produce, and more likely to live in neighborhoods where it may not always be feasible or safe to be physically active.[38] The situation may be even more precarious for African American girls, since black girls across the demographic spectrum are more inclined than white and Asian girls to be overweight, if not obese.[39]

Diet is an obvious potential culprit in population variation in the onset of female puberty, with nutrition possibly responsible for differences in the timing of female puberty throughout time and across societies. But also entering the equation are other environmental factors that affect plasticity—the capacity of the brain, nervous system, and some of the body's hormonal systems to be pliable and adaptive in response to environmental pressures during a person's growth and development.

Like emotional and cognitive development, puberty is influenced by quality of sleep, specifically deep sleep. Sound sleep, in general, is crucial for normal psychological growth during childhood and adolescence. But deep sleep, also known as delta sleep or stage-three nondream sleep, is mandatory for the production of reproductive hormones, which in turn are essential for puberty. For example, human growth hormone is produced during deep sleep, a time when physiological processes unfold to restore the brain's capacity for new learning to occur the next day.[40] This is especially salient, given that a shortage of deep sleep can interfere with the normal sexual maturation of both girls and boys, as well as impact their daily activities, notably academic performance in school.[41]

One of the most dramatic examples of social and developmental influences on menarche comes from studies on girls who were adopted from orphanages in India or Bangladesh and placed in high-income homes in Scandinavia. Girls adopted at older ages had remarkably early first menses, occurring as young as seven years of age, presumably because they had spent more time when very young living in stressful conditions at orphanages. For girls raised partially in orphanages, sexually maturation was accelerated, depending on their age of adoption. Girls adopted at older ages had spent more time in less favorable conditions, but as their life circumstances improved, so too were they more likely to enter puberty at an earlier age.[42] For this group of girls, puberty signaled a developmental response to environmental stressors.

How a person responds to various environmental circumstances, especially susceptibility to certain diseases and psychological well-being later in life, may be determined by environmental factors that affected her at a very young age. These patterned responses are called developmental programming effects. Traumatic events that threaten a fetus, newborn, or young child can influence a person's risk for such chronic diseases as diabetes and metabolic syndrome in adulthood. Harrowing experiences that occurred in utero or in early childhood may have a lifelong impact on the physical health of that person's offspring and theoretically on his descendants in future generations for decades to come. Many scientists now think that developmental programming effects—notably an increased tendency to develop life-threatening chronic diseases—are transmitted from parent to child through epigenetic inheritance, or hereditary changes that do not affect DNA.[43]

The impact of plasticity on the onset of puberty may well be intricately intertwined to some extent with epigenetics. Plasticity is not a totally genetically preprogrammed event, but instead a response conditioned by a host of early life circumstances, for instance, nutritional status, both in utero and during early childhood.[44] Environmental conditions, for example, improved SES, may have been partly responsible for the younger menarche in both middle-class Nigerians and black Americans discussed in an earlier section. The "nurture side" of nature–nurture may also explain why an increase in the standard of living in a rural population in the West African nation of Gambia was associated with a downward trend in the age of menarche—16.06 years in 1989 to 14.90 years in 2008. Income and SES, literacy, protein consumption, and daily intake of vegetable calories appear to influence the onset of menarche. Weight gain associated with adopting a more Western diet and lifestyle of sedentary activity in some non-Western societies may contribute to earlier menarche in some African ethnic groups, as well as in populations in industrialized societies.[45]

In the United States, multiple environmental factors, possibly acting synergistically, may be linked to the younger starting age of sexual development. During the last 70 years, nutrition has improved overall for Americans in general, but such adverse influences as hormones in meat and dairy products, environmental pollutants, and disease can alter the balance of the female endocrine system. Like BPA, chemicals like PCBs (polychlorinated biphenyls), found in insecticides and numerous household products, cause estrogen-like activity in the body, potentially increasing the pace of sexual development. Laboratory research offers a glimpse into some of the possible causes for this effect. BPA, mentioned previously, is used to make commercial plastic products, but when consumed unintentionally, it can act like estrogen inside the body. In preclinical studies, compared to unexposed animals, laboratory animals exposed to BPA not only weigh more soon after birth, but also reach puberty prematurely.[46]

The situation is not encouraging for humans either. As mounting evidence reveals, environmental estrogen-like compounds, for example, pesticides and BPA, are endocrine disruptors that can trigger breast development at a younger age in girls. The potentially adverse ramifications of early puberty are both physical and psychological, and include increased risk for low self-esteem, eating disorders, depression, and even suicide attempts, in addition to earlier sexual activity. Adding insult to injury, females who start breast growth at a young age have an elevated risk of developing breast and endometrial cancers when they reach adulthood.[47]

One of the most profound examples of the relationship between external estrogen-like chemicals and female sexual development comes from the groundbreaking research of Chandra Tiwary, a former chief of pediatric endocrinology at Brooke Army Medical Center in Fort Sam Houston, Texas. Aware that many hair products used by African Americans contain

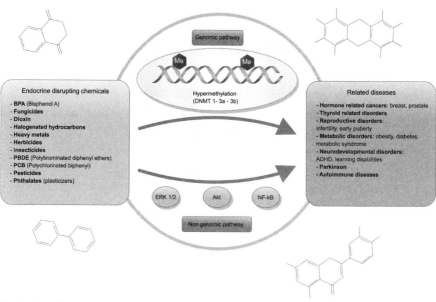

Endocrine disruptors.
SHUTTERSTOCK

estrogen-like hormones, as well as placenta, Tiwary hypothesized that hormone-containing cosmetic products might directly affect the physical maturation of children who use them. Most of these products are shampoos and treatments advertised as deep conditioners for dry, brittle hair.

In the late 1990s, Tiwary reported the effects of hair products on four black girls, age 14 months to about eight years old. All of them, even the baby, developed either pubic hair or breasts, or both, within just two to 24 months after being exposed to hair products containing estrogen or placenta. But this was only half the story. When the parents or caregivers stopped using these products on the girls, their breast development and pubic hair growth ceased within a few months.[48] The types of hair substances Tiwary evaluated have traditionally been used on black youth in the last several decades. According to Tiwary, almost 56 percent of 184 African American adults said they had used black hair products known to contain hormones or placenta on both themselves and their children. Approximately 64 percent of African Americans used these types of products compared with 7 percent of whites.[49] According to some experts who have examined popular shampoos marketed to African Americans, these products have sufficient amounts of estrogen to trigger precocious puberty in young preadolescent girls who use them. But spokespersons for the manufacturers of commercial hair products targeting black customers have disputed these claims, insisting there are no concrete links between estrogen-containing hair products and early puberty in African American girls.[50] In partial defense of their denial, other research suggests that accelerated pubic hair growth in African American girls may be related to higher levels of insulin-like growth factor.[51]

When Tiwary reviewed his survey results and realized that a sizeable proportion of African American parents use these products on their young daughters, he reckoned that ethnic hair products might stimulate precocious sexual development in female children who had not even reached adolescence. Tiwary suggested that the frequency of "sexual precocity"—meaning precocious physical development—in African Americans could be reduced simply by not using these hair products on children.[52] The empirical evidence from Tiwary's research demonstrated that in females, at least African American girls, population genetics in and of itself is probably not the primary, or at least not a direct, cause of ethnic differences in either the timing or the rate of sexual maturation.[53]

The reality is that multiple triggers can promote the onset of early sexual development. Tiwary's study on the estrogen-like effects of certain black hair products provided compelling evidence for the potential impact of an environmental factor—in this case a class of cosmetic products—on accelerating the physical development of African American female children. And while his conclusions did not go uncontested, his findings illustrate the possible impact of an environmental factor—in this case a class of cosmetic products—on both the direction and rate of development of physical traits, in this case physical traits related to puberty.[54]

Environmental forces, specifically commercial hair care substances containing estrogen-like compounds, appeared to play a role in the timing of different signs of puberty because the process of puberty itself is inextricably linked with plasticity. In keeping with the concept of biological plasticity, in some cases the developmental direction of a biological trait can be altered and potentially even reversed if the environment trigger is altered or removed. It cannot be overemphasized that the precocious physical development of the young black girls in Tiwary's case studies subsided once their parents stopped using certain hair products on their children.[55]

Many children, perhaps African American children, in particular, are exposed to environmental chemicals that contain estrogen-like compounds. Juxtaposed to the "Tiwary effect" of an environmental estrogenic chemical triggering accelerated sexual maturation in black American girls is empirical evidence demonstrating that in some African black ethnic populations, puberty starts at a younger average age compared to entry into puberty in African American girls. These findings, in concert, strongly suggest that the faster sexual maturation reported in African American girls (as well as its potential delay or reversal) may result largely from external or nongenetic exogenous factors that affect growth and development.

As some experts have admonished for almost three decades, a wide array of environmental factors—a high-fat diet, sedentary lifestyle, excess body fat, and exposure to environmental estrogens and various potential pollutants—may converge to substantially alter the pace of physical maturation in adolescents. Across ethnic groups, cultures, and countries, environmental influences specific to a particular population remain a formidable and perhaps decisive force in tipping the balance to either speed up or delay the timing of the onset of puberty.[56]

Tiwary's findings shed light on one potential source of black/white differences in the onset of American female puberty. On the other hand, it is not clear how much of the population difference in the timing of puberty can be explained by a single environmental factor like estrogen-like compounds absorbed through the skin. Earlier sexual maturation is sometimes considered an impetus for precocious sexual behavior, but the two concepts are distinctive. Sexual maturation—the full physical development of a person's reproductive system—is the biological benchmark of adulthood that results largely from developmental biology, including the plasticity of the hormonal system. By contrast, sexual activity is a social behavior shaped by learning, family upbringing, peer influence, and the wider cultural environment in which a child or adolescent is reared.

Connecting the dots between physical maturity at an earlier age and alleged sexual patterns in black adolescents, other ethnic minority youth, or even low-income (and increasingly all-income) white teenagers requires factoring in the contributions of multiple cultural and social factors, and calculating how these factors interact with an individual's biology, both genetic and nongenetic. Health, diet, nutrition, sleep patterns, parental upbringing (including family home and financial stability), social mores, neighborhood safety, and peer pressure individually and collectively have profound effects on adolescent behavior by interacting with and modifying genetic and epigenetic processes.

Group patterns of sexual behavior are formed within the context of culture, but even then, these patterns may change throughout time. As an ethnic or social group adapts to a Western middle-class lifestyle, the cognitive processes and social behaviors of its members tend to transform, regardless of their ancestry. This process is not instantaneous but usually takes several generations to occur. In African Americans, as in other ethnic groups in the United States, sexual norms correlate and tend to change with shifting socioeconomic and educational levels. The fact that this transformation occurs at all, even if it takes multiple generations, means that a group's sexual mores reflect its cultural norms and social status during a given historical moment in time.

Attributing an ethnic population's sexual norms (to whatever extent we can generalize about behavioral norms) to inborn population tendencies—race-based genetics—is not supported by current scientific evidence. To make this assertion is to trivialize the potentially far-reaching effects of environment, personal experience, and the impact of culture on human behavior. Generalizations of this sort misrepresent scientific research findings on the interlocking influence of human genetics, human social behavior, and human culture on human sexual activity. And they ignore the extraordinary human capacity of developmental flexibility in shaping social behaviors, sexual and nonsexual. Social behaviors in humans are determined by the interplay between environment, including ecological and social factors, and biology, which in turn comprises genes and gene expression, epigenetics, and plasticity.

Still, we cannot deny that sexual development may influence, even if only indirectly, overt sexual behavior in adolescent girls. Nor can we ignore national statistics showing that African American teen mothers have the highest rates of unwed pregnancies in the United States. In 2016, 39.8 percent of births in the United States were to unmarried women, down 1.2 percent from a peak in 2009. The rates of nonmarital births varied widely along ethnic lines, with rates of 12 percent for Asian mothers, 28.5 percent for non-Hispanic white mothers, and 52.6 percent for Hispanic mothers. As in past years, non-Hispanic black mothers, many of them unwed teens, had the highest rate of any single American ethnic population, at 69.8 percent. High rates of births to unmarried women also occurred in American Indian or Alaska Natives and Native Hawaiian or other Pacific Islander ethnic groups, with rates of 68.1 percent and 47.8 percent, respectively; however, as in the case of Asian women, the overall number of births to these minority populations was quite low.[57]

Regardless of a girl or young woman's ancestry, her risk for unintended pregnancy is related to several demographic factors, including her parents' and her own level of education, her family's economic status, and the behavioral norms of her cultural group and community. African American youth reared in neighborhoods or communities where early or multiple sexual liaisons are interwoven into the cultural fabric face an increased risk for unplanned pregnancy. Teenage mothers, in general, who become unintentionally pregnant generally have only a limited amount of formal schooling, often lacking the resources needed to comfortably raise their children in a middle-class environment. The increased likelihood of unintended teen pregnancy stems from sources that go well beyond peer pressure and the social values of young women within their social circle. More often than not, these young women are reared in a vicious cycle of poverty that leaves them—mothers and children, sometimes along with the fathers—on the margins of mainstream society. In families and social groups where domestic violence and corporal punishment represent the norm more than the exception, the cycle is often intergenerational.

ENVIRONMENT 2.0: THE COMPLEX CRUX OF NATURE PLUS NURTURE

In the United States, women who started puberty at an earlier age tend to be heavier and younger at the time of their first sexual intercourse, first pregnancy, and first childbirth. They also are more likely to have a lower level of educational attainment and work in lower-status occupations. As young adults, they reportedly have an increased risk for aggressive or delinquent behavior. These trends indicate that in the United States, early puberty is often associated with lower SES; however, this correlation, especially between SES and precocious menses, is not entirely consistent worldwide. As in the case of various traits influenced by gene–environmental interactions, different environmental pressures facing different populations may vary in how they affect the genes involved in the timing of puberty.

The environmental factors that can potentially trigger epigenetic influences to initiate menstruation early go well beyond nutrition and include social conditions of young girls' home life. In the United States, poverty, harsh parenting, and, worse yet, child maltreatment, being reared in a home with an absentee father, and poor parental investment, especially from a father figure, are correlated with an earlier onset of puberty.[58] The absence of a father in the home can be stressful for some girls. In one study, positive parental support was linked with delayed puberty.[59] American girls who grew up with parents, particularly fathers, present in their lives, were less likely to have started puberty while in the seventh grade. And yet, being raised by a stepfather tended to have an opposite effect, although the reasons for this seeming paradox are not known.[60]

According to some investigators, girls who experience precocious puberty are also at elevated risk for antisocial behavior, with a heightened tendency to display increased nonphysical aggression and relational aggression, with one example being social bullying to harm the reputation of a peer. (The standard used in most published studies is the average age of menarche in European

American females.) In some but not all studies, early maturing girls were also more likely than later-maturing females to develop friendships with girls considered deviant. These potentially harmful relationships usually faded when the girls reached 16 years of age, but these early maturing females still had a higher risk for delinquency and succumbing to stronger negative peer pressure compared with girls who experienced menarche at a later age.[61] To reduce risks for delinquency, some experts have recommended steering early maturing girls away from relationships with peers who participate in socially unacceptable activities. Girls who reached puberty at a younger age also seemed to be more sensitive to environmental circumstances, for instance, situations that put them at risk for law-breaking or other socially deviant behavior.[62]

One group of researchers found that the same set of gene variants that predisposed girls to an earlier onset of puberty also raised their susceptibility to delinquent behavior. The genes reportedly increased the girls' vulnerability to engaging in delinquent activity, again suggesting that early maturing girls may be more sensitive to a variety of environmental stressors.[63] For example, early maturing girls were more likely to become juvenile delinquents if they attended mixed-sex schools instead of same-sex schools. And girls who began puberty earlier were more likely to break the law if their mothers were poor nurturers and their parents did not monitor them.[64] For some researchers, these findings underscore the importance of gene–environmental interactions in increasing the risk for juvenile delinquency. Some experts claim that within a given population of at-risk girls, the heritability of delinquent behavior—the proportion of delinquency symptoms due to genetic variation within the group—may vary, depending upon each girl's age of first menses.[65]

Consider a study on the links between the timing of menarche and conduct disorder. For girls whose first period began at an average age of 12 to 13 years, 67 percent of the variance in conduct disorder symptoms was associated with genes, but only 33 percent was tied to the environment. Genetics reportedly had a larger impact than environment on variation in the antisocial behavior of the later-maturing girls, perhaps because these girls had escaped the social problems intertwined with precocious puberty. On the other hand, genetics was found to have less of an influence on differences in nonviolent delinquency or conduct in girls who started puberty before 11 years of age than in girls who matured "on time." Among girls who experienced early menarche at age 11 or younger, approximately 8 percent of the variance in conduct disorder symptoms was attributed to genetics, whereas 92 percent was due to environment.[66]

As these authors pointed out, these findings revealed that heritability is not static and that genetic tendencies may be suppressed or enhanced in different social conditions. And yet, although vulnerability genes seem to play a role in elevating the risk for delinquency, at least in early maturing girls, heritability estimates tell only part of the story. The precise genes and gene variants thought to be involved in both the timing of puberty and susceptibility to juvenile delinquency have not been fully identified.[67] Even if in the future, genes are discovered that definitively contribute to both traits, experts will face a momentous task. They will have to sort out exactly how these gene variants interact with a host of environmental factors to discourage law-breaking activity and, instead, foster proactive social behavior sanctioned in the wider society.

Do we have here a chicken-and-egg situation? Is the increased sensitivity of girls who experience "precocious puberty" influenced by genetics or more likely gene–environmental interactions, in turn making them more susceptible to environmental factors that could speed up puberty? Or is the early onset of puberty, conditioned partly but not totally by genetics, more likely to change a young girl's hormones and physiology such that she is at once more vulnerable to a wide gamut of environmental forces? The answer is not straightforward. The delicate interactions existing between genes and environment may ultimately be determined by the types of environmental encounters a girl experiences and her responses to those encounters.

THE EPIGENETICS OF FEMALE PUBERTY

Perhaps alternative explanations can help move us past a simplistic binary paradigm of nature versus nurture. Even if we shift our dichotomy of genes *versus* environment to a more integral and expansive model of genes *plus* environment, we still need an adequate concept to explain the complexities of how nature and nurture combined usher in sexual maturation in girls as they transform into young women. Epigenetics is neither a genetic nor an environmental phenomenon per se. But as an autonomous biological force that interacts with both environmental circumstances and other biological phenomena, it must be considered when weighing the probable influence of genes and environment on pubertal timing. This is the rationale for factoring epigenetic effects into any estimate of the impact of nature versus nurture on the timing of sexual maturation in an individual or a population at a particular point in time.

Adverse external and internal environments—poor diet, inadequate physical activity, exposure to toxic chemicals and pollutants, and numerous sources of psychosocial stress—trigger epigenetic changes that are known to have long-term effects on growth and development, metabolism, and health.[68] Although preliminary, some evidence indicates that interacting environmental factors influence epigenetic mechanisms, which in turn can impact gene expression. These changes can subsequently affect the plasticity of physiological traits, including hormonal and other bodily changes that culminate in puberty.

Genes are literally and figuratively blueprints that help guide, or at least influence, the development of traits—physical, psychological, and social. Traditionally, genetics has been viewed as fixed and immutable—a force that controls and reigns supreme over all other biological activity within a living thing. Genes are essential for life, but for decades scientists have recognized that gene expression, or the way a genetic trait manifests, can, at least for some traits, be profoundly altered through gene–environmental interactions, as well as interactions with other genes. Epigenetics has the capability to dramatically transform gene expression without transforming the DNA sequence. As a result, the physiological response that a particular gene variant was expected to trigger can be changed, possibly yielding a different response than predicted had there been no epigenetic signals. In humans, numerous gene variants, including some associated with sexual maturation, can potentially be modified by epigenetic processes.

Researchers at the University of Copenhagen discovered that while no single gene is responsible for the wide variation in the timing of puberty, epigenetics affects the activity of at least one of the genes associated with the onset of puberty. In short, factors other than genetics, particularly epigenetic signals, can slow or accelerate the onset of puberty, especially menarche.[69] In another study, scientists demonstrated that epigenetic changes, along with genetic and environmental factors, prompted the onset of puberty.[70] In females, epigenetic mechanisms can alter some of the physiological effects of estrogen, one of the major female reproductive hormones. Since estrogen is thought to play a crucial role in female obesity, epigenetic changes affecting estrogen activity may have a profound impact on initiating early sexual maturation, starting with breast development and hair growth in the pubic and axillary regions. This is not a trivial matter, as epigenetic changes occurring during puberty may have lifelong consequences, possibly contributing to chronic disease later in life.[71]

At any stage in the human life cycle, and certainly during the onset of puberty, subtle, less measurable, or even partly undetectable environmental factors can increase the risk for psychological and physiological imbalance in a person. It cannot be overstated that in young people especially, inadequate sleep, poor nutrition, environmental pollutants, and perhaps exposure to certain hair and cosmetic products are part of the same suite of potentially detrimental environmental forces that can be transformed via epigenetic mechanisms. As such, epigenetics acts as a bridge that mediates interactions between the environment and genes, producing lasting effects on both bodily and psychological traits. Harmful epigenetic effects can disturb the normal functions of the brain, nervous system, and hormones.

Just as epigenetics research has set the stage for revamping nature-plus-nurture perspectives on physical diseases, intelligence, and social behavior, it has also provided empirical evidence showing that physical maturation is not an exclusively automatic preprogrammed genetic event. The timing of puberty, particularly in young females, is not determined by a vague and scientifically questionable notion of racial ancestry, given the widely different average ages of menarche in multiple genetically related populations of African black descent. The timing of puberty is an individual developmental milestone that fits the classic definition of phenotypic plasticity—observable changes in a person's behavior, bodily traits, and physiology that emerge as she or he responds to new environmental pressures.

The environments of children transitioning to adolescence are both internal and external, encompassing psychological and social attributes, as well as physical qualities of water, air, and food, which impact health. Preadolescent youth who have a common ethnic and socioeconomic background, combined with a shared genetic ancestry, are likely to experience a similar pattern in the timing of the start of puberty. Young girls growing up in similar neighborhoods tend to have common experiences and environmental exposures, leaving them vulnerable to similar epigenetic effects. For this reason, complex interactions between genes, environment, and epigenetics combined offer a more logical explanation than population genetics alone for common trends in the onset of menarche in young females within the same ethnic population.

Genes are involved in the initiation and perhaps even the duration of sexual maturation. Again, however, the expression of those genes is conditioned by both biological and nonbiological factors, including the social and physical environment, as well as epigenetics, a nongenetic biological process. Like other biological and social traits shaped by phenotypic plasticity, the onset of puberty largely signals a young person's individual physical and psychological response to environmental conditions, whether positive or negative, or both. Sexual maturation, as a key phase of the human life cycle, is the outgrowth of a delicate balance between genes, environment, epigenetics, and plasticity. But in the case of female sexual maturation, the environment may well be the overarching force that guides the transition of girls becoming adult women in diverse societies throughout the world.

Testosterone Tales of Race and Aggression

RACIAL SUPREMACY AND THE TESTOSTERONE FLIP-FLOP

DURING THE PAST 30 YEARS, RESEARCH ON TESTOSTERONE, THE MOST MAS-
culine of human sex hormones, has straddled a thin line between the search
for a prostate cancer cure and the quest to find a biological black box for
social behavior.[1] Both the scientific literature and popular press are replete
with narratives linking testosterone with extreme aggression, violence,
antisocial behavior, and savage crime.[2] For more than a half-century, some
genetic determinists have claimed that this hormone is a major source of
race-based differences in social behaviors. Even amid a changing scientific
landscape, testosterone, for some hereditarian supporters, remains the holy
grail of modern racial science.[3]

In the 1980s, Carleton Coon created a firestorm in physical anthropology
that quickly spread throughout the Ivory Tower.[4] When Coon's small book,
Racial Adaptations, was published, many scientists vehemently protested
his overtly racialist views.[5] But his supporters commended him for crafting
what they called the most well-knit theories for racial origins and variation
ever penned. Still, the ratings and reputations of racialists rise and fall, even
within their own ranks. By today's scientific standards, much of Coon's text
has become obsolete. This book and his earlier work, *The Origin of the Races*,
are now barely acknowledged for more than their historical impact.[6] Even
contemporary racial scientists have challenged the merit of some of Coon's
claims and revamped, if not dismissed, some of his conclusions.

Like his successors—modern-day racialist scholars—Coon assembled an
assortment of facts and figures to prove the existence of a racial hierarchy. In

his scheme, whites sat confidently at the top, while blacks were predictably relegated to the bottom rung of his ladder. In the middle were Asians, who, in that era, were usually considered barbaric rather than savage, a category explicitly reserved for Africans. Coon presented scientific data to buttress his notion of a race-based gradient of inborn abilities. Ironically, his ideas about race and testosterone have been rejected by many modern racialist researchers. Intent on proving the dominance of white males, Coon noted that the weight of male testes is greater in white men than in other racial groups. (But he also said the ovaries are heavier in black females than in white females.)[7]

The brew of testosterone, Coon wrote, is virility. He concluded his short book with a homage to this hormone, praising testosterone as a core component of aggressiveness, territorial protection, and hoarding—the very traits, he boasted, that make white males superior to everyone else. From Coon's perspective, blacks were not the aggressors. In the mid-1900s, testosterone conveyed a totally different symbolic meaning for Coon than it does today for racialist scholars and their followers. He asserted that testosterone imbued European men with keen judgment, claiming that this hormone marked the history of "Caucasoids" throughout the world.[8]

During the last two decades, several vocal racial scholars have embraced testosterone as one of the preeminent biological drivers of ethnic behavioral differences, particularly aggressive and highly competitive behaviors.[9] When Lee Ellis and Helmut Nyborg's controversial study on testosterone levels in black and white men was published in 1992, it helped unleash an emerging racial science of that era based on supposed links between race, biology, and social behavior. Ellis and Nyborg concluded that black men overall, when compared to white men, have higher testosterone levels, predisposing them to aggression, violence, promiscuity, and numerous other unsavory tendencies. During the resurrection of racial science in the late 1900s, a movement that spawned present-day neo-racialism and its offshoot, the alt-right, several North American racialist scholars proposed that testosterone plays a key role in ethnic differences in multiple social behaviors, notably violent crime. J. Philippe Rushton, Edward Miller, Lee Ellis, Anthony Walsh, and Michael Levin are among the most vocal racialist scholars who have treated testosterone as a missing link in a race-based, evolutionary puzzle. They claim that this hormone explains what they perceive as interconnections between population genetics, reproduction, parenting, and law-abiding versus law-breaking activity in males.[10]

Rushton asserted that racial differences in testosterone levels show that antisocial behavioral traits like aggressiveness follow a low-to-high gradient from "yellow to white to black"—Northeast Asian to European to African. Many of his proponents still believe that his predicted race-based ranking holds for not only male sex hormones, but also multiple physical traits, with social ramifications. Rushton devised an elaborate, triracial, hierarchical model in which he compared various biological traits and social patterns supposedly observed in the world's three major continental populations, or racial groups: European whites, African blacks, and East Asians (e.g., people of Chinese and Japanese descent). Rushton claimed that East Asians have the lowest testosterone levels, evidenced by their delayed age of puberty and smaller genitals compared with both blacks and whites. He said these inborn racial traits explain why East Asians have a lower sex drive; less promiscuity; and the lowest rates of fertility, divorce, and crime of the world's major racial groups. Black Africans were characterized by biological and social traits diametrically opposed to those of East Asians. European whites were placed in between East Asians and blacks in this overly simplistic and scientifically dubious triad of the world's major continental populations.[11]

A key feature of Rushton's model was that higher testosterone levels in African descended men were responsible for what some racialists have called the "black male temperament"—an alleged predisposition that race-realists believe propels many blacks, particularly black males, to pursue multiple, morally reprehensible sexual relationships. The social behaviors that presumably follow, Rushton surmised, elicit what he termed the African marriage system and black promiscuity. To support his claims, Rushton repeatedly borrowed numerous findings from international studies, often conducted by other researchers, on personality and temperament. In his publications on race, he reached virtually the same conclusion: African blacks and their diaspora are less restrained and less quiescent than "Orientals" and whites.[12] In the 1990s, Edward Miller, an economics-professor-turned-racialist-writer, also proposed that higher testosterone levels in black men provide a biological explanation for a wide gamut of what he termed "African male personality traits." For Miller, these traits include a stronger sex drive, aggression and dominance, a highly sociable nature (even extroversion), impulsiveness, and sensation seeking.

TESTOSTERONE BIOLOGY—TURNING BOYS INTO MEN

Androgens comprise a group of male sex steroid hormones responsible for masculinization, triggering the physical and behavioral traits of male-hood, particularly the growth and development of the male reproductive system. These steroids regulate hair growth and sex drive in both men and women, but their levels and activity are notably higher in men. Testosterone is not the only androgen, but it is the one that has most fascinated researchers, as well as the public, intrigued with the role of sex steroids in aggression and violence. Male reproductive functions are closely interwoven with the activity of three steroids: testosterone and two other hormones, DHT (dihydrotestosterone) and estradiol. Two of these steroids—testosterone and DHT—are male hormones, or androgens, while the third is predominantly a female hormone.[13]

Of the three sex hormones in men, testosterone is the most abundant and the most important, because it guides the development of male sex characteristics, notably the male sex drive. When adolescent boys mature and physically turn into men, their reproductive organs transform, along with their secondary sexual traits. During puberty, testosterone controls male sex characteristics when levels of this hormone suddenly begin to climb, peaking at the onset of adulthood. During this period, DHT, rather than testosterone, regulates the development of male reproductive tissue, for instance, the external genitals and prostate. Yet, testosterone regulates almost all of the physical and behavioral characteristics associated with male puberty, including a deeper voice, aggressive attitude, and interest in the opposite sex, as well as the budding growth of a beard and pubic and chest hair. Additionally, testosterone increases muscle growth in male adolescents and plays a role in broadening the shoulders and triggering acne; however, as men age, their testosterone levels gradually diminish.[14]

Much of the knowledge about testosterone biology comes from research on the role of this hormone in prostate cancer. In the 1990s, scientists discovered that increased levels of testosterone, particularly its more active form, DHT, were linked with this potentially deadly cancer. Earlier research had shown that in the United States, African American males had the highest rates of prostate cancer and were more likely to die from it than men from other ethnic populations.[15] An early study revealed that mean testosterone levels were about 19 percent higher and free testosterone levels 21 percent higher in black men than in white men in the United States. At that time, it seemed obvious to some experts that testosterone was one of the main culprits of prostate cancer.[16]

The results of numerous studies demonstrated that African Americans have an exceptionally elevated risk for this disease. While some scientists focused on the risk for prostate cancer across ethnic groups, other experts explored ethnic population differences in testosterone levels. The growing evidence was widely applied to prostate cancer research, but it was also leveraged in emerging scientific studies of the roots of such social behaviors as aggression, violence, crime, and even parenting. Testosterone reportedly contributes to these social behaviors, making it a prime candidate for the biology underlying a litany of negative social traits in adolescent and adult males. And because men from different ethnic populations ostensibly varied in their testosterone levels, for some researchers this sex steroid soon symbolized a nexus between race, biology, and social behaviors—especially aggressive, violent, and socially deviant behaviors.

Testosterone: From Biology to Behavior

Whether esteemed or vilified—whether linked with disease or behavior—the traits associated with testosterone remain a source of ongoing debate. For some writers, inflated testosterone levels are tethered to the world's major problems, ranging from uncontrolled population explosions to a host of environmental problems, two of which are global warming and deforestation. The root of these problems, say some observers, is the testosterone-driven male quest for power.[17] Still, as other authors note, high levels of testosterone are not necessarily inherently bad, but instead, are acceptable, even desirable, in, for example, soldiers and athletes. When males compete with one another in sports or social encounters, the winners often show spikes in their testosterone levels. Conversely, some experts claim, society's male losers—whether in love, the job market, or game-playing—tend to have reduced testosterone levels.[18]

By the turn of the twenty-first century, the evidence for the lowest testosterone levels in men of Northeast Asian descent and the highest in males of African ancestry began to wane. One study compared morning and evening differences in testosterone in young adult African American and white men to identify possible links between male sex hormone levels and prostate cancer. Free testosterone levels were similar in both groups, with lower levels observed at 8:00 p.m. than at 8:00 a.m. for both groups—29.4 percent lower for African American men and 23.9 percent lower for white men[19]; however, in both the morning and evening, the African American men had significantly higher serum levels of total testosterone and SHBG, a protein that binds sex hormones. Compared to the white men in the study, the black men had a smaller waist circumference, a smaller waist-to-hip ratio, and a lower fasting insulin concentration. Researchers concluded that ethnic population differences in central body fat might account for the varying levels of circulating SHBG in black and white young adult males. SHBG has a major influence on testosterone concentrations because it stimulates another compound that, in turn, activates male hormone receptors.

In another study of 1,899 men ages 30 to 79, the black men had higher levels of DHT and a higher ratio of DHT to testosterone compared to Latino or white men. But there were no significant differences in testosterone concentration, or in one of its by-products, or in bioavailable testosterone between these three ethnic populations. While some studies reporting higher serum testosterone levels in black men compared to white men pertain to males younger than 41 years of age, a different trend occurs in older men. In another study, researchers compared serum testosterone levels in African American and white men older than 40 years of age who had undergone biopsy for prostate cancer. Both groups of men—those with and without prostate cancer—had comparable levels of serum testosterone.[20]

In the 1990s, sociologists Allan Mazur and Alan Booth found that testosterone exerts a different effect on behavior, notably sexual behavior, in boys during puberty than during adulthood.[21] In a study of 12- to 13-year-old boys, the age when puberty is under way, testosterone was linked with behavioral problems, including aggression, a need for dominance, or antisocial acts.[22] But the association between testosterone levels and behavior was not observed in a similar group of 15- to 16-year-old boys. Apparently, testosterone does not have a direct hormonal effect on the behavior of boys, but instead produces an indirect effect on young adolescent males as they go through puberty. The sexual and antisocial behavior of the boys was largely wedded to their changing social role during this period of life transition. The physical and sexual changes that male children experience during puberty can make this a rebellious time. The increased behavioral problems some-times seen in boys of this age group may be related more to having to adjust to physical changes than to testosterone levels per se. According to some authors, this may explain why some studies have failed to find a correlation between testosterone and aggression in normal 9- to 14-year-old boys.[23]

When boys have completed puberty at about 16 years of age, testos-terone in the bloodstream can produce physiological effects that trigger aggressive behavior, at least in some boys. But the results—even in men with criminal histories—are not entirely predictable. In a study on 692 prison inmates with an average age of 20 years, testosterone levels cor-related positively with the level of violence of the crime, and with dominant and confrontational acts.[24] But in other research on murderers, assaulters, and controls (who were not convicted of violent crimes), no significant differences were seen in testosterone levels between 13 violent offenders and 14 nonviolent offenders.[25] Although Mazur and Booth claimed that testosterone levels were higher in black men, subsequent studies have since revealed that black men have only a modestly, although statistically sig-nificantly, higher level of free testosterone compared to white men.[26] Free testosterone—the form that binds to androgen receptors—is involved in the development of male secondary sex characteristics. It is estimated to be 2.5 to 4.9 percent higher in black men but not large enough to explain African American men's increased risk of prostate cancer.

Testosterone levels can also vary depending on a man's age. In Ellis and Nyborg's study of 4,462 army veterans, aged 30 to 47 years, there were no ethnic differences in testosterone levels in veterans older than 37 years of age.[27] And in Mazur and Booth's landmark research, there was no significant black/white variation in testosterone levels among younger, college-educated veterans, perhaps because, the authors reasoned, youth who attend college are less likely to come from the inner-city. Based on this distinction, Mazur and Booth surmised that unusually high testosterone levels—levels signifi-cantly higher than those of whites—are observed only in younger African American veterans with little education. As these authors put it, younger black urban residents with minimal educational achievement are most likely to participate in an honor subculture—an environment that can trigger increased testosterone levels. Ethnic population differences in testosterone levels, they said, may reflect "higher defensive demands" facing young black adult men, at least in certain situations.[28] They suggested that ethnic variation in testosterone levels is caused by the social environment and not by racial genetics. As they explained, men in combative, inner-city environments are forced to compete for dominance and, as a result, spend much of their time and energy in daily life activities competing for respect, power, and influence in their community. But once they have attained this status, they feel they have to maintain an image of dominance to retain that standing.[29]

If young African American men residing in urban centers are continually required to assume a psychologically defensive posture, they may well be developmentally predisposed to higher testosterone levels (with levels that remain elevated) than young men in other social environments, regardless of their genetic ancestry. Most African American young men, like many Latino American young men, in America's low-income urban communities do not have the luxury of taking "time out" from the daily grind of trying to "make it" in an economically competitive, combative, and socially stressful world. If they could, then perhaps their hormones (which are "raging" in virtually all men of this age group) could be steered into balance. The blunt reality is that most male youth, particularly young men of color, reared in this type of social milieu, where they continually face strenuous pressures, can get caught up in a vicious cycle. Depending upon their own individual circumstances, they may unwittingly find themselves at increased risk for violence, crime, injury, and even premature death.[30]

Individual circumstances are tantamount to environment and experience—two factors now known to influence testosterone levels, even dramatically. For Mazur and Booth, testosterone is almost evenly distributed across social classes, except in the case of what they call "honor subcultures" like those of inner-city African American and Latino American adolescent boys and men.[31] In this case, testosterone is less likely to be increased in lower-income men unless these men are forced to constantly compete for dominance.[32] The implication of Mazur and Booth's analysis is that male testosterone levels are not fixed, but instead, malleable and not determined by genetics alone. Since testosterone levels can be affected by environment and experience, they are not a fated outcome of an inborn predisposition.[33]

Recent evidence has cast even more doubt on claims that testosterone levels are inevitably and significantly higher in black males universally. It may seem, then, that the debate about testosterone as a major cause of ethnic population differences in statistics on violent crime has been resolved. However, the biology of testosterone is only one aspect of an ever-growing, complex saga of race, violence, male sex hormones, and, more recently, genes linked with these hormones, as well as interactions between these genes and a host of environmental influences. The notion that testosterone or a related androgen—a male sex steroid—can explain black/white/Asian differences in aggression, violence, and antisocial acts persists in some circles, both scholarly and lay. Today, the racialist idea that testosterone is largely responsible for "bad black behavior" has been fine-tuned and revamped under the guise of other scientific evidence: the AR gene linked with the activity of androgen hormones, such as testosterone. This gene and its relevance to contemporary narratives about race-based differences in social behavior are explored in the next chapter.

Chapter Fourteen

The AR Gene

The "New Testosterone"

The AR Gene: Testosterone's Molecular Docking Station

Testosterone once seemed to be a likely candidate for prostate cancer, given that this disease occurs in disproportionately higher rates in men of African black descent, along with the assumption that black men consistently have higher levels of this male sex hormone. In the United States, African American men are almost 1.6 times more likely to develop prostate cancer than white men and 2.6 times more likely than Asian American men.[1] Although lifestyle factors contribute to the onset of this insidious cancer, genetic ancestry and family history also play paramount roles.[2] Recent evidence suggests that testosterone levels are not the main culprit in triggering prostate cancer, no matter a man's ancestral heritage.[3]

Scientists eventually realized that the risk and severity of prostate cancer were related to how a man's body responded to male sex hormones, or androgens, like testosterone. By the late 1990s, as growing scientific evidence countered the notion that black men do not have the highest testosterone levels, scientists discovered a relevant new finding. A variant of the androgen receptor gene—the AR gene—is associated with not only prostate cancer, but also more severe and potentially more deadly forms of this disease.[4] This gene variant is far more common in men of African black ancestry than in other ethnic populations.[5] Probing beyond the genetics of prostate cancer, some writers soon claimed that the AR gene variants commonly found in blacks are also linked with such socially unpalatable social traits as aggression, violence, crime, polygyny (marriage between a husband and multiple wives), and low parental investment, or essentially poor parenting skills.[6]

The actions of testosterone in the male body ultimately depend on the activity of androgen receptors—chemical slots in which testosterone molecules fit, similar to the way a key inserts into a keyhole. The AR gene gives the instructions, or a blueprint, for the body to manufacture a protein that stimulates the production of other genes sensitive to androgens. Testosterone, along with the male sex hormone DHT, binds to androgen receptors, forming a complex that attaches to DNA. In males, the androgen receptor can turn genes on or off as needed to guide sexual development, notably the growth of the prostate gland. Variations in AR gene activity can affect how a man reacts to androgens, or male sex hormones. Environmental factors can also affect testosterone levels, but the structure of the AR gene appears to be fixed at birth.[7]

THE AR GENE AND PROSTATE CANCER

One region of the AR gene—the CAG repeat—is a segment of DNA composed of three nucleotides, or three letters of the DNA alphabet, that repeats multiple times in the AR gene. The number of CAG repeats correlates closely with the activity of the androgen receptor, the docking station where androgens attach. Many studies have reported a negative, or inverse, correlation between the number of CAG repeats and sensitivity to androgen hormones like testosterone. Fewer CAG repeats usually lead to enhanced expression or activity of the androgen receptor, which in turn increases the action of circulating androgen hormones on target tissues in the body. By contrast, a larger number of CAG repeats, or a longer CAG segment, is often linked with decreased sensitivity to circulating androgens and lower testosterone levels. The CAG triplet size ranges from as few as six to about 39 repeats.[8]

In men both with and without prostate cancer, the lowest number of CAG repeats has been observed in African Americans, followed by an intermediate number in Caucasians, followed by more repeats in Hispanics, with the largest number occurring in Asians.[9] When this racial pattern of CAG repeats was first detected in the early 1990s, experts suddenly had a novel mechanism that might help explain the relatively higher incidence and higher death rates of prostate cancer in African American men. As the decade progressed, increasing evidence revealed that a lower number of AR gene CAG repeats, especially fewer than 18 repeats, predicted an increased risk of total prostate cancer compared to a reduced risk associated with more than 26 repeats.[10] By stimulating the androgenic activity of the receptors, short CAG repeat sequences appear to raise the risk for prostate cancer.[11] Early studies indicated that African American men were twice as likely as whites to have less than 20 CAG repeats.[12] And prostate cancer patients with fewer CAG repeats tended to develop an aggressive and more extensive type of this disease, including metastatic prostate cancer.[13] This eureka in prostate cancer research laid the groundwork for an innovative approach to studying this disease.

Numerous studies conducted in the last two decades have confirmed the ethnic trend in the number of CAG repeats. Recent evidence points to CAG repeat averages numbering 19 to 20 in African Americans, 21 to 22 in Caucasians, 22 to 23 in Asians, and 23 in Hispanics.[14] Ethnic populations, therefore, can vary in their levels of androgen activity and androgen sensitivity, which can manifest as differences in hormone levels or hormone activity. Higher androgen hormone levels can affect the risk and severity of some cancers, and possibly even responses to drugs prescribed to treat those cancers. The AR gene with fewer CAGs found in most black American males provided an explanation for the increased susceptibility to prostate cancer in this population.

This was not the only application of the new knowledge surrounding AR gene biology. Intrigued with the evolutionary implications of AR gene variants, some researchers conjectured that variable CAG repeat lengths represented adaptations that evolved in human populations in the distant past. Did the genetically encoded number of CAG repeats emerge through natural selection or, instead, through neutral selection? According to West Indian urological surgeon William Derval Aiken, African American men have a heightened responsiveness to testosterone and other androgens because this trait was selected in African black slaves who survived the transatlantic journey from Africa to the New World.[15] Aiken speculated that enhanced sensitivity to androgen receptors would have been advantageous for increasing the oxygen delivery of red blood cells, an ability that would have helped enslaved Africans tolerate the stifling close quarters of a slave ship. Aiken surmised that black males who were innately more responsive to androgens would have had an edge in staying alive during their involuntary voyage to the New World. Many of their descendants would have inherited this gene variant, making them more sensitive to testosterone than their black brethren in West Africa and Central Africa. If Aiken were correct, then purportedly elevated testosterone levels might contribute to the exceptional prowess of African Americans and West Indians in sprinting.[16] But it could also raise the risk for developing prostate cancer and dying from that disease.

Aiken's hypothetical scenario might have seemed feasible when investigations were launched on AR gene CAG length, because at that time the studies focused on African Americans. Subsequent studies have revealed that black Africans, like African Americans, generally have shorter AR gene CAG repeat lengths.[17] Given the overall shorter repeat CAG length of the AR gene in African black men, including West African men in Africa, it appears that this trait did indeed originally evolve in Sub-Saharan African populations.

Most studies have shown a positive correlation between AR-CAG repeat length and prostate cancer, but contradictory findings have also been reported.[18] In the Flint Men's Health Study, there was no link between AR-CAG repeat length and prostate cancer risk in a sample of African American men.[19] One possible reason for this inconsistency is that scientists employ different techniques for analyzing the activity of AR-CAG repeats. To further confuse the issue, another segment of the AR has a different set of repeats—the GGN triplet code—even though the CAG triplet has received the most attention. The two types of repeats appear to be related, but it is not known if their lengths have a combined effect on increasing the risk for prostate cancer.[20]

In some studies, men who had fewer CAG repeats were more likely to have higher levels of upper body strength, self-reported competitiveness, and an enhanced testosterone response to potential mates. Still, some experts suspect that testosterone levels rather than AR-CAG repeat length or another hormone might be responsible for masculinization.[21] In light of the wide range of social behaviors associated with testosterone, it was only one small step from examining the role of androgens in prostate cancer to exploring its contributions to human behavior. The potential relationship between CAG repeat length and physical attributes of manliness—traits often associated with aggression and sexual proficiency—has made the CAG triad a prime target for the biological underpinnings of male-related social behaviors.

The AR Gene: From Cancer to Crime

Wittingly or unwittingly, the discovery of AR gene variants opened the door to a revised perspective on alleged links between race and male sex hormones. For some pronature advocates, variation in a pivotal gene related to androgens—testosterone and similar hormones—could explain ethnic population differences in such social traits as aggressiveness.[22] One early study reported that shorter repeat triplets in the AR gene were associated with conduct disorders and impulsive behaviors like attention deficit hyperactivity disorder (ADHD), oppositional defiant disorders, and Tourette's syndrome.[23] Relatively shorter AR-CAG repeat lengths in the AR gene of nonblacks have been associated with a wide array of adverse personality traits and destructive social patterns. In a study conducted on prisoners on the Indian subcontinent, shorter AR-CAG repeat lengths were more common in male rapists and murderers than in control subjects. Compared to men in the control group (who had an average of 21.19 repeats), the rapists had a shorter repeat length (an average of 18.44 repeats), whereas the murderers had even fewer repeats (an average of 17.59 repeats).[24] This study was one of the first to provide empirical evidence for a strong correlation between the number of AR-CAG repeats and antisocial or criminal activity.[25]

Nonetheless, there are multiple notable exceptions to this trend, suggesting that correlations between AR-CAG repeat length and antisocial behavior do not necessarily reflect a cause-and-effect relationship. Canadian male students with fewer CAG repeats showed a decreased tendency toward physical aggression and anger.[26] Similarly, when rage-filled criminals were compared with noncriminals in China, researchers could not find a link between the size of the AR repeat and violent convicts. The scientists who conducted this study concluded that the AR-CAG repeat length is not a major contributor to male violent criminal activity.[27] And yet, shorter AR gene repeats of 17 or less were more likely to occur in violent criminals than in Chinese males who led normal, noncriminal lives. The short repeats were found in 7.5 percent of the violent convicts versus 1.9 percent of the noncriminals.[28]

The AR Gene and Social Traits: The Backdrop

Today, overtly, antiblack racialist research has been largely supplanted by seemingly more objective and scientifically robust, politically savvy, and culturally competent studies in behavioral genetics and evolutionary psychology.

To grasp the complexities of ethnic population differences in social patterns, some biological determinists have turned to life history theory—a branch of evolutionary biology intended to predict and explain patterns of growth and reproduction, and such behaviors as mating and parenting. Life history theory is used to calculate the trade-off between the number of offspring versus the amount of parental care across species—and potentially across populations within the same species.[29] It was initially devised to shed light on different reproductive strategies in various species, but Canadian racial scholar J. Phillipe Rushton reframed it, labeling it the differential model of r/K life strategies, so that it could be applicable to human populations.[30] Following on the heels of Richard Herrnstein and Charles Murray's book *The Bell Curve*, Rushton's differential r/K model helped precipitate the resurrection of racial science in the 1990s.[31]

Rushton's model focused largely on the trade-off between reproductive success—the number of progenies—and parental investment in offspring. Depending on which outcome dominated, a species or population could be classified as either fast or slow life history strategists. In both the original model and Rushton's rendition designed specifically for human populations, fast life history strategists are said to inhabit unstable ecological environments, whereas slow life history strategists usually occupy stable environments. Rushton claimed that a population of fast strategists gives birth to large numbers of offspring but provides only limited parental investment for these children. Slow strategists characteristically produced small numbers of offspring but invested increased parental care into raising their progeny. The longevity of slow life history strategists is typically greater because they take longer to mature into adulthood. Inevitably centering on issues of race, Rushton's model of r/K life strategies was used to compare and contrast the world's primary racial groups (a classification rejected by most experts), based on their variable reproductive strategies. Interwoven into this model were population differences in a range of physical traits, including hormones (particularly testosterone), the volume of the adult brain, the size of the male genitalia, and the psychology of mating.[32]

Since the mid-1990s, many leading scientists and academics have disputed Rushton's rendition of r/K life strategies, challenging every aspect of it. The vast majority of anthropologists contest the notion that the human species can be divided into three main continental populations, or racial groups—Africans, Asians, and Europeans.[33] Testosterone was pivotal in Rushton's model of r/K life history. And he based many of his claims regarding race-based differences on his prediction that levels of this hormone are highest in African black males, intermediate in whites, and lowest in East Asians. In the last chapter we learned that there are far too many inconsistencies for this trend to be considered viable across continental populations or ethnic populations. Obesity and inadequate nutrition, in concert with numerous other environmental factors, can dramatically alter average testosterone levels in the males of an ethnic population. Also, average testosterone concentrations change with age, a finding that may explain some of the discrepancies seen when comparing levels of this hormone in different populations.[34] The evidence overwhelmingly demonstrates that testosterone levels can fluctuate, largely in response to shifting environmental influences.[35]

Despite contradictory evidence for ethnic population trends in male testosterone levels, several recent studies appear, at least at first glance, to provide some support for Rushton's assertions. Almost 20 years after the scholarly demise of Rushton's armchair hypothesis of r/K life strategies, some of his pronature protégés remain convinced that male sex hormones are an integral part of the biology that influences social relations. To be fair, some preliminary evidence seems to support their position. Testosterone reportedly mediates individual differences in personality in men, which in turn can impact how well people in a society cooperate with one another.[36] In some studies, fathers who raise children with their partners were found to have lower testosterone levels than unpartnered males who were thought to be more competitive when seeking mates.[37] In a study of men in the West African nation of Senegal, higher testosterone levels were linked with greater extraversion, less parenting, and an increased tendency to be polygynous.[38] In other research, men in the United States with high testosterone levels had larger testes, were less inclined to parent their children, and displayed weak signs of brain activity when shown photos of their babies.[39]

THE AR GENE: AGGRESSION, POLYGAMY, AND PARENTING

With testosterone losing ground as the indisputable Holy Grail of contemporary racial science and relinquishing its status as the grand differentiator of race-based superiority and inferiority, what better biological device to replace it than the AR gene? For starters, the genetic machinery that fuels the activity of androgens like testosterone varies overall across racial groups. During the last decade, a growing number of studies have examined the connection between testosterone and AR-CAG repeat lengths in men from different ethnic populations, especially black populations, and correlated repeat length with a variety of social behaviors. These behaviors include such reproductive practices as monogamy and child-rearing, individual and interpersonal physical aggression, number of sexual partners, courtship habits, and variation in partnering and paternal roles.[40]

In emerging research on nature versus nurture in shaping ethnic population differences in social traits, the length of CAG triplets may signal the "new testosterone": a preordained biological code that, depending on how many CAG repeats a man possesses, presumably can be used to justify relegating him to a position of either racial inferiority or racial superiority. Amid continuing efforts to correlate sex hormones with social traits, the AR-CAG repeat has replaced testosterone in some scholarly circles. Several proponents of Rushton's Differential K theory have turned to population patterns of the AR-CAG repeats to support their claims about links between testosterone, social deviance, and race. In a recent study, researchers (including Irish racialist psychologist Richard Lynn) examined differences in the AR gene, androgenic hair, rate of prostate cancer, and sex frequency and number of sex partners across various countries.[41]

The data reportedly confirmed the authors' predictions in some, although not all, of the outcomes they analyzed. The lowest levels of androgens were found in East Asians. However, the results for Sub-Saharan Africans were inconsistent, not unlike the findings of numerous other studies comparing population differences in testosterone levels. And the data on individual nations throughout black Africa in this study were limited. Researchers lumped together generalized data about testosterone levels in West Africans and then extrapolated from those data to produce values for specific African countries.[42] They failed to find a significant link between the length of the AR-CAG repeats and the incidence of prostate cancer or the number of sexual partners. The highest rate of prostate cancer occurred in Caucasians, a finding that seemingly goes against the global trends in other research. The authors noted that the correlations they found had a large effect size but acknowledged that the effect sizes failed to reach statistical significance. In the end, there was no firm evidence to support their hypothesis. But that did not prevent them from concluding that their findings partially validated the Differential K theory.

For Lynn and colleagues, differential life history theory—with its implications of black Africans purportedly prone to nonmonogamous male–female bonding and poor parental care—held up. Their paucity of supportive evidence or, worse yet, the glaring counterevidence that turned their position on its head appeared to matter only minimally, if at all. Instead, these writers merely recycled the views of other researchers, as if to blindly accept the premise—at times indistinguishable from their premature conclusion—that shorter CAG repeat sequences of the AR gene in men are genetic triggers for socially undesirable behaviors. This type of circular reasoning may be appropriate for marketing pitches, but it does little for fashioning a scientific argument grounded in replicable and robust empirical evidence.

Other research has also reported links between AR gene CAG repeat length and the amount of effort a man devoted to seeking a mate and taking care of his children. In one study, the results were precisely what investigators predicted. Men with a shorter AR-CAG repeat length had more upper body strength and also scored higher on tests for prestige and dominance than men with a longer AR-CAG repeat length.[43] Prestige and dominance are psychological traits that can be measured, to some extent. Like physical masculinity, they are presumably related to mating effort—the time and energy a person, in this case a sexually active adult male, spends on finding a desirable mate. The authors speculated that in harsh environments like Northern Europe and Northeast Asia, longer CAG repeats would have been advantageous because they would have resulted in a diminished androgenic response. This supposedly would have fostered increased parental investment. But in more predictable environments, for example, Africa, researchers proposed, men may have invested relatively more energy into mating than survival. The authors surmised that shorter AR-CAG repeats would have been beneficial for males inhabiting tropical regions because this genetic trait would have encouraged an androgenic response, reportedly along with higher testosterone levels.[44] However, as we learned in the last chapter, researchers disagree among themselves as to whether Africa or Europe had the more unpredictable environment.[45]

To some extent, this hypothesis is simply a reinvention of Rushton's r/K theory of life history traits, which some writers have used to contrast human populations in terms of social behaviors—sexual relationships, parenting styles, aggressive behavior, and violent crime. But contrary to researchers' expectations, a smaller number of AR-CAG repeats did not correlate with a tendency for short-term mating relationships, as their model would have predicted. There were no differences in total scores on sexual behavioral surveys between the men with fewer CAG repeats and those with more CAG repeats. And levels of testosterone were not associated with behavioral outcomes. On the other hand, the length of the CAG repeat sequence was associated with body strength and status seeking but not with the men's overall self-reported mating strategy. The scientists proposed that the shorter CAG sequences might be beneficial for male–male competition for long-term mates.[46]

To support this assertion, they cited previous research demonstrating links between shorter CAG repeats and an increased risk for criminal behavior, particularly violent crime. They speculated that criminally inclined men with shorter CAG triplets may be more likely to compete intensely with other men for female sexual partners. Extrapolating even further, they surmised that AR gene variants—the shorter versus longer CAG repeat triplets—represented a trade-off between mating and survival effort. A core feature of their hypothesis was that men with shorter AR-CAG sequences—the most common type of variants in males of African black descent—are more likely to aggressively compete for mates. And yet, researchers presented no scientific evidence to validate a key element of this assertion: the notion that males with a smaller number of AR-CAG repeats tend to be more promiscuous and more inclined to engage in short-term relationships.[47]

As the authors correctly noted, studies have produced mixed results for links between the length of CAG repeats and antisocial personality traits. [48] Shorter CAG repeat lengths have been linked with greater sperm production, but it is not clear if the relationship is causal.[49] In fact, some research suggests that gene variants for AR-CAG repeats do not have a critical impact on male infertility.[50] In separate studies on men from Nigeria, Germany, India, and Tunisia, no positive correlation was found between the number of CAG repeats and sperm concentration and sperm motility.[51]

Like some of the other researchers working in this field, the authors relied on a self-reported questionnaire called the sociosexual orientation inventory—SOI. This survey is designed to measure a person's willingness to engage in sex without closeness or commitment (and determine to some degree his or her desire for sex with multiple partners).[52] The participants in this study comprised 149 undergraduate college men who were given partial course credit for their involvement. All of them were self-reported heterosexuals, and the ethnic composition of the sample was 64 percent Caucasian, 21 percent Latino, 14 percent Asian, and 1 percent African American.[53] It is not clear if the results would have been noticeably different if the men had been more representative of a real-world population in urban centers of the United States. Also, it is difficult to determine if research participants answer questions with equal honesty in home interviews and a clinical or laboratory setting or social psychology lab.[54]

The findings from several recent studies seem to suggest that the number of AR gene CAG repeats is linked with not only male aggression, but also male reproductive patterns and parenting styles. Since a man's testosterone level correlates with his AR gene CAG length in some studies, testosterone is thought to mediate a man's partnering and child-rearing styles. For some observers, this putative association has crucial social ramifications for ethnic population differences in some aspects of family structure. Despite contradictory findings, the presence of relatively short CAG repeat lengths has been associated with higher androgen receptor activity in males of African black descent. The implication here is that fewer AR gene CAG repeats, sometimes accompanied by higher testosterone levels, result in a particular pattern of male social traits, including excess aggression, polygamy, and poor parenting. For some commentators, notably scientifically oriented race-realists, these character traits are suitable for small, preliterate societies but not advantageous for fostering civil and constructive behavior in large-scale modern Western societies.

Exactly how inherited tendencies supposedly linked with AR gene CAG repeat length manifest in terms of social behavior is the focus of a current wave of interdisciplinary research. Consider recent studies on the Hadza and Datoga, two indigenous traditional tribal societies of Northern Tanzania that have limited contact with Western civilization. These two social groups live in relatively close proximity to one another, but they differ in some of their social patterns, including their sexual and reproductive behaviors.[55] The Hadza, a traditional hunter-gatherer society, are genetically isolated from other living populations, even though they are culturally similar to the Khoisan of Southern Africa. The Hadza language is remotely related to the

Khoisan tongue. (The Khoisan are the indigenous peoples of Southern Africa who were historically known by the derogatory name "Bushmen.") When a group of researchers studied Hadza men, they found no correlation between the number of AR-CAG repeats and aggression. A tribe of only about 1,000 living members, the Hadza generally have limited contact with the Western world. Approximately 300 Eastern Hadza of Tanzania still maintain a pristine hunter-gatherer lifestyle, sharing the egalitarianism and communalism that defines such small-scale, peaceful societies as the Khoisan speaking San of Namibia.[56]

The Hadza have no central leader and have never been recognized as warriors, instead seemingly preferring to avoid aggression. They usually form stable monogamous bonds with their mates and invest considerable amounts of time and energy in rearing their children. Hadza men have been described as having low levels of aggressiveness, but according to some researchers, this characterization might be due to the small number of Hadza males evaluated in research studies.[57]

Located in Northern Tanzania, the Datoga are traditional, seminomadic, polygamous pastoralists that practice cattle husbandry, an occupation requiring young men to be guardians and warriors.[58] With almost 50,000 to 100,000 members, the Datoga far outnumber the Hadza.[59] Today, the Datoga remain marginalized and stigmatized, continually subjected to political discrimination, but the males have maintained their reputation as fierce warriors. Their culture demands them to be aggressive to guard against theft of their cattle and be poised for combat against other males from neighboring societies.[60]

These two East African traditional populations also contrast in their marriage patterns. With multiple wives as the norm, the Datoga men tend to have more children and reportedly invest less time and energy in their children compared with the Hadza. In keeping with their roles as warriors, the researchers noted, the Datoga devote time to male–male competitions, both within and outside their immediate social circle, and use corporal punishment on their children.[61]

Most Hadza and Datoga males have a "medium number" of AR gene CAG repeats, ranging from 20 to 25, but the frequency of gene variants of specific lengths differs among individual males. The Datoga are more likely to have greater variability in CAG repeat length, with more of them likely to have higher frequencies of either fewer repeats (less than 17) or more repeats (more than 26). Gene variants with 21 and 22 CAG repeats were found in about 50 percent of the Hadza men versus 70 percent of the Datoga men.[62] One study reported that the Datoga men with fewer CAG repeats were more likely to display physical aggression, as well as negative emotions like anger and hostility, than men with a greater number of CAG repeats.[63] Still, researchers who rely on CAG repeat length as a yardstick of male reproductive behavior have yet to explain why both the Hadza and Datoga have a shorter CAG sequence—one that supposedly predicts mating and child-rearing practices thought to run against the values of Western society. One investigator proposed that the lack of aggressiveness in the Hadza society might mean that genetic tendencies (presumably toward violence, of course) manifest differently in small-scale societies.

Aggression refers to any behavior directed toward another person for the purpose of harming that person. But researchers do not always differentiate between reactive aggression and proactive aggression when they report links between testosterone and aggression.[64] This may be one reason why discrepancies persist in published studies. Reactive aggression is defensive and accompanied by anger, impulsive and offensive or rude behavior, and emotional instability. Proactive aggression, by contrast, is a means to an end rather than a response to being provoked or antagonized. Proactive aggression is used to attain a goal or acquire a resource, for instance, winning a tournament or sports competition.

Further confounding the situation, testosterone levels are not static, but instead they fluctuate, often rapidly, in relation to competitive interactions, especially between males.[65] Populations may vary considerably in their cultural attitudes toward aggression.[66] Depending on one's perspective, such an act as a soldier or warrior going into battle, or a father inflicting harsh corporal punishment on his child, could be viewed as either reactive aggression or proactive aggression. Some forms of aggression may be socially sanctioned and valued in certain societies and ethnic populations, for example, small-scale pastoralist communities, in which these activities are considered necessary for the common good. As such, an ethnic group's cultural belief system and social setting are part of its environment that influences the average testosterone levels of men in that group.[67]

The Ariaal are a rural, small-scale African population of pastoralists in Kenya. Some of the Ariaal are nomadic, whereas others have recently shifted to settled cattle farming, in which they trade animal products like meat and hides with their neighbors.[68] The traditional Ariaal nomads usually rely on a subsistence diet that leaves them undernourished with thin bodies characterized by a low percentage of body fat. In undernourished males like the Ariaal, testosterone may have an increased influence on overall body size and muscle mass during puberty.[69] Research on aggression in the Ariaal is limited, but polygyny is widely practiced in this ethnic group, with men tending to have more wives as they age. One study found that Ariaal men are inclined to be aloof in their relationships with their wives, also providing only minimal direct care for their children. As the researchers predicted, among 20- to 39-year-old males, testosterone levels were lower in monogamously married men than in unmarried men. But contrary to the researchers' expectations, testosterone levels were not higher in polygynous married men than in monogamous married husbands. As the authors explained, differences in testosterone may be linked with mating effort in young Ariaal men, but in older men political networks and wealth—signs of higher social status—may correlate better with marital status.

In another study, the length of CAG repeats in the Ariaal averaged 22.6, ranging from 15 to 34. There was no difference in the average length of CAG repeats (about 23) between the settled and nomadic Ariaal men 20 years of age or older.[70] This length is noticeably longer than the average of 16.7 to 17 CAG repeats reported in most other African black populations.[71] Still, the findings were not totally unique, as the mean size of the Ariaal's CAG repeats corresponded to the upper limit observed in some black males of the Ivory Coast.[72] These CAG repeat lengths were reported in a different study, but if CAG repeat length and testosterone levels are indeed strong predictors of the mating behavior of men, the Ariaal presumably would tend to be more monogamous. There are various reasons for why no strong three-way correlation was observed between CAG repeat length, testosterone levels, and monogamy or polygamy in the Ariaal. The simplest explanation is that social conditions can change a man's testosterone levels, which tend to vary anyway with age and the time of day.[73]

The results of several other studies also run contrary to some researchers' predicted pattern of low testosterone correlated with monogamous marriage. In Kenyan Swahili men, no differences in testosterone concentrations were seen between unmarried men and married, monogamous men. In this study, the highest testosterone levels occurred in men with two wives.[74] Testosterone levels can be substantially altered by multiple environmental forces, notably diet and infectious diseases, regardless of the length of the AR gene CAG. And environment exposures, especially during childhood, can affect the sensitivity of the AR gene to testosterone, as well as testosterone levels in males of different populations in different ways.[75]

Some studies have failed to consistently find a strong correlation between testosterone levels and paternal parenting. Researchers examined a large sample of 683 Filipino men during a four-year period to determine if their AR gene CAG repeat length was linked with their parenting style.[76] The men who were less involved in raising their children fit two distinctly different profiles: those who had fewer CAG repeats with an elevated testosterone level and those who had a greater number of CAG repeats length with a lower testosterone level. In contrast to what the researchers expected, the men with the most stable partnerships and strongest investment in fathering had intermediate levels of androgen activity. [77] It is unclear if the correlation between an increased number of CAG repeats and lower androgen sensitivity, or less responsiveness to testosterone, holds up universally. It may apply to men in industrialized nations but not necessarily to those in non-Western societies.[78]

One of the potentially worst attributes of testosterone is its possible contribution to criminal, antisocial activity. Despite evidence showing that testosterone levels are higher in people with aggressive behavior, for example, criminal violence, a meta-analysis reported a weak correlation of only 0.08 between testosterone and aggression.[79] And, as previously noted, minimal concentrations of testosterone do not correlate with AR-CAG repeat length in every study. These contradictory findings have yet to be explained.[80] This does not mean that elevated testosterone levels do not contribute to extreme aggression and violence. They undoubtedly do, but so too do other factors, namely environmental influences.

BEYOND THE GENE-EQUALS-BEHAVIOR EQUATION
The more experts explore the influence of AR gene CAG repeat length on male social behaviors, the more they are likely to encounter inconsistent results. There are several reasons for this. The CAG repeat length in the AR gene generally varies across populations, with short CAGs occurring in as many as 75 percent of African Americans compared with 49 percent

of Asians. Earlier, we noted that the size of the CAG repeats may work in tandem with other genes and gene products, including another repeat triplet within the androgen receptor gene.[81] Moreover, exactly how CAG repeats are divided into long and short sequences can vary from one study to the next. Some authors consider fewer than 20 as "short" and more than 20 as "long."[82] How CAG repeat lengths are categorized, then, may be in the eye of the scientist. Whether a shorter AR-CAG repeat length—combined with its supposed effect on enhancing testosterone activity—is positively correlated with aggressive behavior may depend on a male's stage of development during puberty or adulthood.[83] Without an appreciation of the subtle effects of AR-CAG genetics on testosterone, the precise impact of differently sized AR gene variants on social behaviors remains inconclusive.

This is not to deny the importance of androgen research, as it has clearly advanced our understanding of this pivotal hormone in relation to prostate cancer and, to some degree, social patterns. But the biological underpinnings of complex social behaviors tied to reproduction and fathering encompass more than genetics. They also include the plasticity of the brain and such hormones as androgens.[84] Emotional aggression is triggered by hormones like testosterone and cortisol, such neurotransmitters as serotonin and dopamine, and the combined action of these chemical mediators on the prefrontal region of the brain's frontal lobe. Testosterone can activate hostile emotional responses, whereas cortisol, popularly dubbed the stress hormone, and serotonin may counteract the effects of testosterone. These actions are part of a cascade of biological influences that, in interaction with environmental forces, help guide the formation and expression of social traits.

Some evidence suggests that the interplay between the AR gene and cortisol can predict whether a man will bond in a monogamous relationship with a woman or, instead, have polygamous relationships. But knowing the primary biological players involved in hypothetical scenarios of the origin of polygamy or personal aggression or fathering tells only part of the story, at best. Multiple interlocking pathways in the brain contribute to social behaviors, but these pathways can vary across individuals and populations. The amygdala and hypothalamus—two key emotional centers in the brain's limbic system—are also activated during aggressive situations.[85]

Cortisol acts on the prefrontal region of the brain's cognitive control center to restrain impulsive tendencies. It seems to inhibit testosterone responses that can lead to aggression and violence.[86] In one study, a lower number of CAG repeats and a lower baseline cortisol level each correlated with larger testosterone responses in men while they interacted with women as potential mates.[87] Other research suggests that baseline testosterone levels are linked with mating behaviors but only in men with relatively low baseline cortisol levels.[88] And serotonin blocking receptors also influence the degree of impulsive aggressive behaviors.[89] Bonded partnering and paternal investment, therefore, are, to some extent, shaped by the orchestrated interplay of neurotransmitters and hormones,[90] possibly including the actions of oxytocin, often coined the "love hormone," on gene expression.[91]

In chapter 11, we discussed research showing that the activity of oxytocin in some social relationships is influenced by epigenetics.[92] The brain chemical dopamine is associated with motivation to earn a reward or experience pleasure. In some studies, DAT1, the dopamine active transporter 1 gene, was found to be a primary genetic marker of antisocial behavior in adolescent delinquents.[93] Some dopamine-linked genes are associated with aggressive behavior, including hostility, physical aggression, certain mental disorders, alcoholism, and drug addiction.[94] The biology of such human social traits as aggression, anger, hostility, and perhaps parenting behavior involves reciprocal action between steroid hormones, particularly androgens like testosterone, and multiple genes, especially dopamine- and serotonin-linked genes.[95] As Russian anthropologist Marina Butovskaya, who has studied East African pastoralists, put it, for scientists to understand the impact of the AR genes, they will have to assess the impact of genes, not in isolation, but as parts of a "multifactorial miasma" that encompasses other genes, as well as the environment.[96]

As Butovskaya explained, certain genes may reveal an influence on social behavior only when coupled with certain environments.[97] Therein lies a key dilemma. Conducting scientific studies in which environmental conditions can be fully controlled is not only difficult, but also virtually impossible. If experts cannot quantify the effects of multiple, interacting environmental forces that in turn interact with biology to shape social traits, they may fall short in mapping the precise interplay between nature and nurture. And this does not even take into account the many problems that can arise if investigators' professional or personal biases compromise their ability to formulate research questions, design scientific studies, collect data, and objectively interpret the short-term and long-term significance of their findings.

AR GENE CAGS: THE EPIGENETIC CONNECTION

For some people who think personality traits, including aggression and violence, are shaped largely, if not predominantly, by genetics, short AR gene CAG repeats have become the latest installment of a genetic product earmarked to relegate those who possess it to an indelible status of mental, social, and moral inferiority. For some readers, the idea that shorter AR-CAG repeat length is linked with an increased tendency toward aggressiveness, polygamous mating, and low parental care may seem solid enough to demonstrate a cause-and-effect relationship. But how, then, do they explain notable anomalies and exceptions to this correlation? After all, equating a short AR-CAG repeat length with an innate propensity toward disruptive social behaviors has disturbing inferences. It implies that males of African black descent are hardwired for atrocious social traits. This faulty perception, when allowed to trickle down from some sectors of the Ivory Tower to mainstream audiences, has serious ramifications for not only black men and their families, but also global society, at large.

Whether AR gene CAG repeats of variable lengths evolved in human populations as adaptations through natural selection, notably sexual selection, or, instead, a neutral process, for example, genetic drift, is beyond our present scope. But even if AR-CAG repeats arose through natural selection, we now know that numerous traits, particularly social traits, are not inherently fixed and unalterable. We can no longer gauge the impact of a gene variant simply by its presence. Rather, we must examine its expression—its manifest activity, which ultimately sets the stage for the impact of genes and other biological processes on emotions, cognition, and behavior. Multiple factors impact the biological activity of the AR gene, including the activity between different regions within the AR gene, as well as its interactions with other gene variants, with a wide array of environmental factors and epigenetics—non-DNA related inheritance.[98] In the AR gene, epigenetic mechanisms can occur alongside genetic mutations to trigger a complex cascade of events leading to prostate cancer.[99] Not surprisingly, the AR gene is currently being examined in relation to multiple epigenetic processes that affect this gene and possibly contribute to more aggressive forms of prostate cancer in black men.[100]

When geneticist Joshua Galanter and colleagues examined epigenetic mechanisms associated with specific genes in a diverse population of Latinos, they made an unexpected discovery. Different ethnic groups falling under the broad umbrella of the Hispanic population in the United States varied in epigenetic imprints at genes influenced by social and environmental exposures. The DNA ancestry of the participants in the study explained much of the link between ethnicity and the epigenetic signals, but social, economic, cultural, and environmental exposures were also responsible for part of the association.[101] These researchers were not studying the AR gene, but their results underscore a crucial observation: Environmental factors may account for different epigenetic patterns seen in different ethnic populations. Diseases and other traits that contrast across populations are shaped by not only genetic ancestry, but also various environmental forces, for instance, the social milieu. Environmental influences can trigger epigenetic signals that can transform the activity, or expression, of various genes, which in turn impacts how a particular trait will manifest.

Since the length of AR-CAG repeats is genetically determined, the size of the CAG repeats is presumably fixed. But the activity of the AR gene can potentially be modified by epigenetic signals that in turn can alter the effects of CAG repeats on AR gene expression. Epigenetic mechanisms are stimulated at least partly in response to environmental pressures that vary for individuals and ethnic populations, as a whole. Whether interactions between environmental conditions, epigenetics, and AR gene CAG repeats that shape gene expression have a measurable impact on social behaviors remains to be seen. But if epigenetics can influence gene activity associated with AR repeat lengths and AR-CAG repeats can affect social behavior, perhaps indirectly, then epigenetics might help explain why men with AR-CAG repeats of similar length, regardless of their ancestral genetics and geographical origin, can behave so differently. Diverse environmental influences, including nutrition and cultural practices, then, may affect the expression of the AR gene and its impact on personality and other social traits. Lest we forget, environment plays a pivotal role in this process.

Epilogue

AS WE HAVE REPEATEDLY SEEN, GENE–ENVIRONMENT EFFECTS FOR SOCIAL traits can vary across ancestral populations. Various genes or sets of genes are sometimes linked with distinct social behaviors in some ethnic populations but not in others. Conversely, the same genes may be associated with different social behaviors across different populations. This is one of several key reasons why, in the last two decades, the concept of genetics has shifted from a model in which genes and gene activity, or gene expression, were considered static to a model in which genes themselves are now viewed as dynamic and, in some cases, pliable in their interactions with other genes and environmental forces.[1] Even when a gene variant, or multiple variants, in some individuals is linked with an increased risk of alcoholism, drug addiction, violent crime, or other socially aberrant behavior, that genetic risk requires multiple environmental triggers and at times interactions with other genes to become a reality.

The science surrounding the nature–nurture controversy about the underlying causes of racial differences in social outcomes is unlikely to be fully resolved in the foreseeable future. Ironically, the more experts learn about links between human gene variants and social behaviors, the more weight many of them ascribe to the role that environment plays in directing biological processes, particularly gene expression. To some degree, then, insofar as social traits are concerned, environment is not only an active player in interactions between nature and nurture, but perhaps also the lynchpin of gene–environmental interactions. This statement is technically not one of environmental determinism, if only because it recognizes the salience of both genes and environment. And yet, it challenges the hereditarian position—the notion that heredity contributes substantially to human mental and social traits. Philosophy of science professor Jonathan Kaplan makes a compelling case for why the hereditarian doctrine, when applied to the black/white gap in IQ, cannot be proven. The reason, Kaplan says, is because experts have failed to tease apart the influence of DNA from the impact of the environment on intelligence performance.[2]

Kaplan succeeds in showing that there are multiple loopholes in hereditarian assertions that the black/white divide in IQ (and presumably in other measures of cognitive performance, e.g., scholastic achievement) is substantially due to genetic ancestry.[3] To date, no one has proven that all ethnic populations have comparable cognitive abilities. Instead, what Kaplan describes is a different outcome revealing that ethnic populations are not necessarily unequal in their capacity to perform cognitive tasks. Like several other scholars, he effectively dismantles the "evidence" on which hereditarians rely to build their case for genetically driven, or innate, ethnic population differences in intelligence performance.

Intelligence performance, assessed through IQ scores, is the focus of numerous nature–nurture explorations of ethnic or racial differences in behavioral traits. But this score may or may not be equivalent to an individual's inherent cognitive abilities. The idea of an "average" level of intrinsic intelligence becomes even more problematic when applied to ethnic populations because, as in the case of individuals, intellectual performance may not be tantamount to intrinsic mental ability. Intelligence performance provides only a snapshot of a person's aptitude for demonstrating a particular skill set or level of knowledge at a given point in time. Depending on a host of environmental and experiential influences—including the test-taking situation, nutritional status, quantity and quality of sleep, and family dynamics—her performance may not be an accurate barometer of various modes of intelligence.

Nonetheless, environment, even in all its multifaceted and layered forms, cannot account for all of the ethnic disparities in social behaviors and life outcomes that influence the ethnic achievement gap. But it offers a starting point for appreciating the complexities of interlocking social, biotic, and ecological factors that, either separately or combined, intersect with gene variants to determine individual social behavior. Individual social traits, collectively, can influence the overall social patterns, or perceived patterns, of an ethnic population. Many social behaviors of any ethnic group are culturally rewarding and beneficial to both the group and society, at large. But some social patterns are harmful, diminishing quality of life on many fronts. Today, as scientific revelations about nature and nurture unfold, the possibility of early psychosocial and educational interventions to help people avoid disturbing life outcomes may finally be within reach. Still, there is no panacea, and there will always be individuals prone to disruptive behavior, aggression, and violence. Just as environment assumes a pivotal position in the equation of nature-*plus*-nurture equals behavior, biology too contributes to behavior, sometimes including a heightened risk of detrimental behaviors. But in this context, the relevant biology is not confined to the action of genes in a vacuum.

The biology behind personality and social life is dynamic and mutable, continually changing in response to both long-term and immediate environmental influences. This view of human nature supersedes the inflexible biological determinism that historically has been used to justify unproven claims of race-based genetic determinism. It speaks to a more realistic and encompassing vantage of the subtle interplay between biology and culture—between genes, epigenetics, environment, and plasticity—in shaping human cognition and social behaviors. Numerous environmental factors affect gene expression, with some of them having a profound effect on whether a trait related to a disease, physical growth, or personality manifests. It is through plasticity—flexibility in individual growth and development—that a person's inherited tendencies toward a physical or psychological characteristic can shift, at times dramatically.

Epigenetics offers a unique window into the biology tethered to intergenerational patterns that, in worst-case scenarios, seem to sustain vicious cycles of poverty and underachievement, all the while deepening the ethnic achievement divide. Some skeptics complain that epigenetics signals the last resort of egalitarians in their perennial quest to find nebulous, non-DNA influences on social behavior. These doubters see epigenetics as an excuse, an apology of sorts, for the inherent power of genetics in determining personality and social traits. For them, epigenetics is merely a contrived emblem of nurture, wishfully intended to transform the activity of otherwise unchangeable horrific genetic traits, which, they claim, carry known risks for aggression and violence. Among the most contentious risk gene variants are the shorter length of CAG repeats in the AR gene, the low-activity MAOA gene, the low-activity short variant of the serotonin transporter promoter gene, and the nine-repeat version of the dopamine active transporter 1 gene.[4]

To some extent, epigenetics represents an existential threat to the primacy of a gene—virtually any gene—once thought to be immutable and the source of a predestined mindset and personality. And here caution is warranted, lest, without sufficient scrutiny, as we uncritically embrace epigenetics, whether symbolically or empirically, as the great equalizer of human populations and societies. It may indeed end up moving us closer to a more level-headed understanding of the lack of a level playing field—imbalances in the nurture effect across ethnic populations. But epigenetics research is still quite young.

Even scholars outside the hereditarian tradition and opposed to it have expressed concern about the possibly destructive effects of epigenetics. Speaking about the emerging field of biological plasticity and epigenetics in the postgenomic era, geography professors Becky Mansfield and Julie Guthman critiqued these biological concepts in relation to environments, genes, bodies, and health.[5] They suggested that research in this area may have negative consequences for some people because of its implications for female reproduction and race. Epigenetics, they argue, is promoting a new biological ontology of race. Epigenetics does this, we are told, because it intensifies racial normalization, even while rejecting genetic determinism. Instead, it revolves around concepts of abnormality and improvement, especially regarding the female body and the fetus during the course of future generations. For Mansfield and Guthman, epigenetics is not overtly racist. And yet, the optimization of life that it potentially allows is founded on principles of racial normalization. From this vantage point, epigenetics focuses on the pathological, such that anyone whose body strays from normalcy is physiologically deviant and therefore requires corrective intervention. As they put it, the goal of epigenetic therapies is to eliminate the abnormalities of bodies that differ from the norm and reinvent these differences from being born. In this way, they noted, epigenetics seeks to improve human life.[6] But in the process, epigenetics propagates a rather odd type of scientific racism.

Regarding human social behavior, the pro and con perspectives of epigenetics showcase its benefits and limitations. To some degree, the validity of this emerging field of biology may depend on the specific social trait being discussed. The data-driven reality is that, in contrast to environmental factors that have no or only minimal influence on genuinely fixed genetic traits, for instance, certain inherited diseases, some epigenetic markers can redirect gene expression. In short, epigenetic signals do matter, and their impact can be momentous. They can potentially transform how genes behave in triggering, exacerbating, or possibly mitigating various physical diseases. And now, it seems, epigenetics can also affect some social traits, especially when adverse environmental effects are involved.[7] As we have learned from multiple examples of recent research, nongenetic biological processes, namely epigenetics and plasticity, function in tandem with genes and environment. Epigenetics can instruct certain genes to switch on or off, to turn the volume up or down, so to speak, thereby affecting gene expression, or the degree to which a specific genetic tendency is likely to manifest. This series of events can have a dramatic impact on some personality traits and social tendencies, which in turn can propel a person toward or away from unsavory life outcomes. With epigenetics mediating some of the effects of both genes and environment, and especially influencing gene-environmental interactions, claims that genetics, notably population genetics, contribute to ethnic population disparities in, for example, crime rates are not supported by the bulk of current scientific findings.

In no way does this condone the antisocial actions of anyone, as people are ultimately personally responsible for their actions. But even for individuals who presumably have a heightened genetic risk for offensive proclivities, environmental conditions, for instance, family dynamics, wield a major impact on their life outcomes, favorable or unfavorable. This may mean that a person's likelihood of falling prey to disturbing behavior is perhaps best explained not in terms of an ethnic population's susceptibility to the adversities of nature plus nurture. Rather, the risk for harmful social behavior may be tethered to a family's vulnerability to intersecting genetic liability, environmental adversity, and, perhaps most profoundly, epigenetic mediating forces—an intersection that for some individuals in that family might result in suboptimal softwiring of the brain.

And here we must tread lightly as we embark on a potential slippery slope. This is not intended to scapegoat families by targeting those who have been marginalized by poverty, encounters with the criminal justice system, or other aggravating conditions. But it is to recognize the value of assessing group differences partially in relation to clan genomics, a concept used to determine disease risk. Clan genetics is based on the idea that genetic variation inherited from a person's parents and recent relatives may provide more information than gene variants from distant ancestors about his likelihood for developing specific diseases.[8] In the context of personality traits, including the risk for unwelcome social behavior, the level of the clan, as opposed to the level of continental races or ethnic populations, offers a unique perspective. It potentially provides a more precise lens through which to envision subtle and less recognized interactions between genes, epigenetics, and environment in directing gene expression and influencing plasticity.

Research findings across disciplines and throughout decades, if not centuries and millennia, seem to support the familiar adage that a person's life outcomes—the good, the bad, and the in-between—can largely be traced back to his family background, to his family's life circumstances. These include not only social and ecological conditions, but also an individual's inherited sensitivities to various elements of his environment. If a person has an increased risk for problematic personality traits, that risk may be connected more to the way he reacts to stressful and distressing life circumstances than an inborn predisposition to commit dreadful acts. And while genetic ancestry might influence a person's chance of turning to violence, that ancestry—or package of inherited behavioral tendencies—most likely involves epigenetic patterns, as well as genetic propensities. As noted, epigenetic signals are transmitted across generations within a family. This may be one reason why, despite popular racial stereotypes, members of the same ethnic population sometimes, perhaps frequently, differ remarkably in their motivation, cognitive performance, educational attainment, and a variety of social outcomes, both acceptable and socially shunned.

Similar social patterns that occur in individuals and families within the same ethnic population are just as likely, and in some cases more likely, to result from shared or similar environmental conditions than genetically inherited inclinations. Individuals of the same ethnic population residing in the same general geographical and cultural region, for example, the United States, do not all share the same level of educational attainment, income, housing, healthcare, and other indices of socioeconomic status, or SES. As such, the diversity of their environmental situations could just as easily account for much, and some experts say virtually all, of the differences in many of their social behaviors.

If, on the other hand, race-based genetics genuinely accounted for variable social patterns and life outcomes across ethnic populations, we would not witness the incredible diversity of social outcomes evidenced in individuals and, especially, families within a particular ethnic group, for instance, African Americans. Black Americans, in particular, are, quite simply, not homogeneous. In large-scale, modern, Western societies, individuals and families within the same ethnic population that share an overall common or strongly overlapping genetic heritage sometimes reside in diverse social and ecological environments. The way in which nature plus nurture affects the lives of these individuals and their families is at times strikingly different. Despite common genetic ancestry, the social patterns observed within an ethnic population can vary tremendously. Consider the following oversimplified example. A person's inclination toward, for instance, either law-abiding or law-breaking social behaviors is largely shaped by multiple interactions between, on the one hand, his experience of, or vulnerability to, his environment and, on the other hand, the intertwined genetics and epigenetics of his family ancestry. This concept of family ancestry is akin to clan genomics, which was just discussed.[9]

Clan genomics offers novel insights into patterns of disease transmission,[10] but it may also shed light on the interactive gene–environmental underpinnings that contribute to the intergenerational ethnic achievement gap. This emerging field of genetics research may someday provide a more realistic window on the role of Nature—biological inheritance that encompasses but also ventures beyond genetics—in conditioning social traits. Still, tales of clans and genes should not be misconstrued to legitimize blaming families for "poor genes" that supposedly produce "at-risk individuals" starting in the womb. Rather, clan genomics offers a laser-sharp focus for honing in on why some individuals, within some families, are likely to have intensified negative reactions to environmental stressors. Clan genomics may help to explain why they are less able than other individuals, or sometimes other families, to surmount those adversities. The key point here is that clans, or extended families, may provide a more scientifically accurate level than population genetics—race-based genetics—for assessing the role of ancestral inheritance in a wide array of social traits. Even then, however, the role of ancestry in shaping social traits—whether at the level of the individual, clan, or ethnic population, or, for that matter, the entire human species—becomes measurable and meaningful only within the context of a broad range of overlapping environmental forces and personal life experiences.

NOTES

Prologue

1. L. Moody Simms, "Philip Alexander Bruce and the Negro Problem, 1884–1930," *Virginia Magazine of History and Biography* 75, no. 3 (1967): 349–62, https://www.jstor.org/stable/4247325.

2. David R. Williams, Naomi Priest, and Norman B. Anderson, "Understanding Associations among Race, Socioeconomic Status, and Health: Patterns and Prospects," *Health Psychology* 35, no. 4 (2016): 407–11, https://doi.org/10.1037/hea0000242.

3. Williams, Priest, and Anderson, "Understanding Associations among Race, Socioeconomic Status, and Health."

4. "2017 Kids Count Data Book: State Trends in Child Well-Being," *Annie E. Casey Foundation*, 2017, accessed January 15, 2019, https://www.aecf.org/m/resourcedoc/aecf-2017kids
countdatabook.pdf.

5. J. Philippe Rushton, *Race, Evolution, and Behavior: A Life History Perspective*, 3rd ed. (Port Huron, MI: Charles Darwin Research Institute, 2000).

Chapter 1. Nature Versus Nurture: Redrawing the Battle Lines

1. L. N. Magner, "The Rediscovery of Mendel's Laws of Heredity Science," in *Science and Its Times: Understanding the Social Significance of Scientific Discovery: 1900–1949*, ed. Josh Lauer and Neil Schlager (Detroit: Gale Group, 2000), 94–95; Hub Zwart, *Understanding Nature*, vol. 13, International Library of Environmental, Agricultural and Food Ethics (Dordrecht: Springer Netherlands, 2008), https://doi.org/10.1007/978-1-4020-6492-0.

2. John C. Loehlin, "History of Behavior Genetics," in *Handbook of Behavior Genetics*, ed. Yong-Kyu Kim (New York: Springer New York, 2009), 3–11, https://doi.org/10.1007/978-0-387-76727-7_1.

3. Loehlin, "History of Behavior Genetics."

4. Loehlin, "History of Behavior Genetics."

5. Lee Ward, *John Locke and Modern Life* (Cambridge, UK: Cambridge University Press, 2010), https://doi.org/10.1017/CBO9780511761461.

6. James Farr, "I. 'So Vile and Miserable an Estate' the Problem of Slavery in Locke's Political Thought," *Political Theory* 14, no. 2 (May 1986): 263–89, https://doi.org/10.1177/0090591786014002005; Sarah X. Pemberton, *Locke's Political Thought and the Oceans: Pirates, Slaves, and Sailors* (Lanham, MD: Lexington, 2017).

7. Holly Brewer, "Slavery, Sovereignty, and 'Inheritable Blood': Reconsidering John Locke and the Origins of American Slavery," *American Historical Review* 122, no. 4 (October 1, 2017): 1,038–78, https://doi.org/10.1093/ahr/122.4.1038.

8. Theresa Richardson, "John Locke and the Myth of Race in America: Demythologizing the Paradoxes of the Enlightenment as Visited in the Present," *Philosophical Studies in Education* 42 (2011): 101–12, https://eric.ed.gov/?id=EJ960330.

9. K. Malik, "On Enlightenment's Race Problem," *Pandaemonium* (blog), 2013, accessed January 15, 2019, https://kenanmalik.com/2013/02/13/on-the-enlightenments-race-problem/.

10. T. Fowler, *Locke*, vol. 11 (n.p.: Palala Press, 2016).

11. Stanley Harrold, *American Abolitionists*, Seminar Studies in History (Harlow, England, and New York: Longman, 2001).

12. G. E. Allen, "Essays on Science and Society: Is a New Eugenics Afoot?" *Science* 294, no. 5,540 (October 5, 2001): 59–61, https://doi.org/10.1126/science.1066325.

13. Loehlin, "History of Behavior Genetics."

14. Charles Darwin and Ernst Mayr, *On the Origin of Species*, print., a facs. of the first ed. (Cambridge, MA: Harvard University Press, 2003).

15. D. Burbridge, "Francis Galton on Twins, Heredity, and Social Class," *British Journal for the History of Science* 34, no. 122, pt. 3 (September 2001): 323–40; Gil Norman, "Genes, Experience, and What Makes Us Human," *Great Debate*, April 14, 2003, accessed January 15, 2019, http://www.thegreatdebate.org.uk/Gillreview1.html; G. Radick, "A Life of Sir Francis Galton," *Heredity* 89, no. 4 (October 2002): 328, https://doi.org/10.1038/sj.hdy.6800125.

16. Norman, "Genes, Experience, and What Makes Us Human."

17. Norman, "Genes, Experience, and What Makes Us Human"; Brendan O'Flaherty and Jill Shapiro, "Apes, Essences, and Races: What Natural Scientists Believed about Human Variation, 1700–1900," in *Race, Liberalism, and Economics*, ed. David Colander, Robert Prasch, and Sheth Falguni (Ann Arbor: University of Michigan Press, 2004), 21–55, http://www.jstor.org/stable/10.3998/mpub.17696.5.

18. John P. Jackson and Nadine M. Weidman, *Race, Racism, and Science: Social Impact and Interaction*, Science and Society Series (New Brunswick, NJ: Rutgers University Press, 2006).

19. Jackson and Weidman, *Race, Racism, and Science*; Robert J. Richards, *The Tragic Sense of Life: Ernst Haeckel and the Struggle over Evolutionary Thought* (Chicago and London: University of Chicago Press, 2009).

20. Elleke Boehmer, *Colonial and Postcolonial Literature: Migrant Metaphors*, 2nd ed. (Oxford and New York: Oxford University Press, 2005).

21. M. W. Berger, "A New Take on the Nineteenth-Century Skull Collection of Samuel Morton," *Penn Today*, October 4, 2018, accessed January 15, 2019, https://penntoday.upenn.edu/news/new-take-on-infamous-Morton-skulls.

22. Jackson and Weidman, *Race, Racism, and Science*; J. Poskett, "Victorian Phrenology: 'To Find a Good Wife, You Have to Ascertain She Has a Good Head,'" *History Extra*, December 24, 2015, accessed January 15, 2019, https://www.historyextra.com/period/victorian/victorian-phrenology-to-find-a-good-wife-you-have-to-ascertain-she-has-a-good-head/.

23. O'Flaherty and Shapiro, "Apes, Essences, and Races."

24. Jackson and Weidman, *Race, Racism, and Science*.

25. Darwin and Mayr, *On the Origin of Species*.

26. Audrey Smedley and Brian D. Smedley, *Race in North America: Origin and Evolution of a Worldview*, 4th ed. (Boulder, CO: Westview Press, 2012).

27. Smedley and Smedley, *Race in North America*.

28. Madison Grant, *The Passing of the Great Race, or, The Racial Basis of European History* (N.p.: Ostara Publications, 2016); Joseph L. Graves, *Biological Theories of Race beyond the Millennium*, vol. 1 (Oxford, UK: Oxford University Press, 2018), https://doi.org/10.1093/oso/9780190465285.003.0002.

29. O'Flaherty and Shapiro, "Apes, Essences, and Races."

30. O'Flaherty and Shapiro, "Apes, Essences, and Races."

31. N. W. Gillham, "Sir Francis Galton and the Birth of Eugenics," *Annual Review of Genetics* 35, no. 1 (2001): 83–101, https://doi.org/10.1146/annurev.genet.35.102401.090055.

32. Hans-Jörg Rheinberger, Staffan Müller-Wille, and Robert Meunier, "Gene," in *Stanford Encyclopedia of Philosophy*, ed. Edward N. Zalta (Stanford, CA: Metaphysics Research Lab, Stanford University, 2015), https://plato.stanford.edu/archives/spr2015/entries/gene/.

33. Stephen Jay Gould, *The Mismeasure of Man*, rev. and expanded, with a new introduction (New York: W. W. Norton, 2008).

34. J. Philippe Rushton, "Race, Intelligence, and the Brain: The Errors and Omissions of the 'Revised' Edition of S. J. Gould's *The Mismeasure of Man* (1996)," *Personality and Individual Differences* 23, no. 1 (1997): 169–80, https://www.sciencedirect.com/science/article/abs/pii/S0191886997809841; Mark Snyderman and Richard J. Herrnstein, "Intelligence Tests and the Immigration Act of 1924," *American Psychologist* 38 (September 1983): 986–95, http://dx.doi.org/10.1037/0003-066X.38.9.986; Steven A. Gelb, Garland E. Allen, Andrew Futterman, and Barry A. Mehler, "Rewriting Mental Testing History: The View from the *American Psychologist*," originally published in *Sage Race Relations Abstracts* 11, no. 2 (May 1986): 18–31.

35. Allen, "Essays on Science and Society"; Gould, *The Mismeasure of Man*.

36. Allen, "Essays on Science and Society"; Steven Selden, *Inheriting Shame: The Story of Eugenics and Racism in America*, Advances in Contemporary Educational Thought Series (New York: Teachers College Press, 1999).

37. Norman, "Genes, Experience, and What Makes Us Human."

38. John C. Malone, "Did John B. Watson Really 'Found' Behaviorism?" *Behavior Analyst* 37, no. 1 (May 2014): 1–12, https://doi.org/10.1007/s40614-014-0004-3.

39. John B. Watson, *Behaviorism* (New York: Norton), 82.

40. Franz Boas, "New Evidence in Regard to the Instability of Human Types," *Proceedings of the National Academy of Sciences of the United States of America* 2, no. 12 (1916): 713–18, http://www.jstor.org/stable/83642.

41. D. J. Galton and C. J. Galton, "Francis Galton: And Eugenics Today," *Journal of Medical Ethics* 24, no. 2 (April 1998): 99–105.

42. G. Lindzey et al., "Behavioral Genetics," *Annual Review of Psychology* 22, no. 1 (January 1971): 39–94, https://doi.org/10.1146/annurev.ps.22.020171.000351.

43. Ralph J. Greenspan, "The Origins of Behavioral Genetics," *Current Biology* 18, no. 5 (March 11, 2008): R192–98, https://doi.org/10.1016/j.cub.2008.01.015.

44. Loehlin, "History of Behavior Genetics."

45. Arthur Jensen, "How Much Can We Boost IQ and Scholastic Achievement?" *Harvard Educational Review* 39, no. 1 (April 1969): 1–123, https://doi.org/10.17763/haer.39.1.l3u15956627424k7.

46. Jensen, "How Much Can We Boost IQ and Scholastic Achievement?"

47. Steven Pinker, *The Blank Slate: The Modern Denial of Human Nature* (New York: Viking, 2002).

48. Jensen, "How Much Can We Boost IQ and Scholastic Achievement?"

49. Alondra Yvette Oubré, *Race, Genes, and Ability: Rethinking Ethnic Differences*, vol. 1 (Woodland Hills, CA: BTI Press, 2011).

50. Pinker, *The Blank Slate*.

51. Christine R. Harris and Harold E. Pashler, "Evolution and Human Emotions," *Psychological Inquiry* 6, no. 1 (January 1995): 44–46, https://doi.org/10.1207/s15327965pli0601_6.

52. National Research Council, *Measuring Racial Discrimination* (Washington, DC: National Academies Press, 2004), https://doi.org/10.17226/10887.

53. Sarah E. Ali-Khan et al., "The Use of Race, Ethnicity, and Ancestry in Human Genetic Research," *HUGO Journal* 5, no. 1–4 (December 2011): 47–63, https://doi.org/10.1007/s11568-011-9154-5.

54. Graves, *Biological Theories of Race beyond the Millennium*.

55. Alan R. Templeton, "Biological Races in Humans," *Studies in History and Philosophy of Science Part C: Studies in History and Philosophy of Biological and Biomedical Sciences* 44, no. 3 (September 2013): 262–71, https://doi.org/10.1016/j.shpsc.2013.04.010.

56. Graves, *Biological Theories of Race beyond the Millennium*.

57. Joseph L. Graves, "Evolutionary versus Racial Medicine: Why It Matters," in *Race and the Genetic Revolution: Science, Myth, and Culture*, ed. Sheldon Krimsky and Kathleen Sloan (New York: Columbia University Press, 2011), 142–70; Koffi N. Maglo, Tesfaye B. Mersha, and Lisa J. Martin, "Population Genomics and the Statistical Values of Race: An Interdisciplinary Perspective on the Biological Classification of Human Populations and Implications for Clinical Genetic Epidemiological Research," *Frontiers in Genetics* 7, no. 22 (February 17, 2016): 2, https://doi.org/10.3389/fgene.2016.00022; Alan R. Templeton, "Human Races: A Genetic and Evolutionary Perspective," *American Anthropologist* 100, no. 3 (September 1998): 632–50, https://doi.org/10.1525/aa.1998.100.3.632.

58. Graves, *Biological Theories of Race beyond the Millennium*.

59. Templeton, "Biological Races in Humans."

60. Helen Tilley, "Racial Science, Geopolitics, and Empires: Paradoxes of Power," *Isis: An International Review Devoted to the History of Science and Its Cultural Influences* 105, no. 4 (December 2014): 773–81.

61. Oubré, *Race, Genes, and Ability*.

62. Peter H. Schuck, "Racism and Racialism Are Different," *Huffpost: The Blog* (blog), December 22, 2014, accessed January 15, 2019, https://www.huffingtonpost.com/peter-h-schuck/racism-and-racialism-are-_b_6368010.html.

63. John Wakeley et al., "The Discovery of Single-Nucleotide Polymorphisms—and Inferences about Human Demographic History," *American Journal of Human Genetics* 69, no. 6 (December 2001): 1,332–47, https://doi.org/10.1086/324521.

64. Jensen, "How Much Can We Boost IQ and Scholastic Achievement?"

65. Charles J. Lumsden and Edward O. Wilson, *Genes, Mind, and Culture: The Coevolutionary Process*, 25th anniversary ed. (Hackensack, NJ: World Scientific, 2005).

66. Dorothy Tennov, "The Public Image of Sociobiology and Evolution," in *Research in Biopolitics*, ed. Albert Somit and Steven A. Peterson (Stamford, CT: JAI Press, 1998), 119–36.

67. David A. Mrazek, "Psychiatric Pharmacogenomic Testing in Clinical Practice," *Dialogues in Clinical Neuroscience* 12, no. 1 (2010): 69–76.

68. Mary O'Hara and Pamela Duncan, "Why 'Big Pharma' Stopped Searching for the Next Prozac," *Guardian*, January 27, 2016, accessed January 25, 2019, https://www.theguardian.com/society/2016/jan/27/prozac-next-psychiatric-wonder-drug-research-medicine-mental-illness; Naomi R. Wray et al., "Genome-Wide Association Analyses Identify Forty-Four Risk Variants and Refine the Genetic Architecture of Major Depression," *Nature Genetics* 50, no. 5 (May 2018): 668–81, https://doi.org/10.1038/s41588-018-0090-3.

69. Dean Hamer and Peter Copeland, *Living with Our Genes: Why They Matter More Than You Think* (New York: Anchor Books, 1999).

70. Leon J. Kamin, Richard C. Lewontin, and Steven Rose, *Not in Our Genes: Biology, Ideology, and Human Nature* (Chicago: Haymarket Books, 2017); Richard C. Lewontin, *Human Diversity*, 1st ed., Scientific American Library Series 2 (New York: Scientific American Library, 1995).

71. Julio Licinio, "Ten Years of Molecular Psychiatry," *Molecular Psychiatry* 10, no. 1 (January 2005): 1–2, https://doi.org/10.1038/sj.mp.4001626.

72. Patrick F. Sullivan et al., "Psychiatric Genomics: An Update and an Agenda," *American Journal of Psychiatry* 175, no. 1 (January 1, 2018): 15–27, https://doi.org/10.1176/appi.ajp.2017.17030283.

73. Tennov, "The Public Image of Sociobiology and Evolution."

Chapter 2. Darwin's Legacy: Selection and Adaptation

1. Theodosius Dobzhansky, "Nothing in Biology Makes Sense Except in the Light of Evolution," *American Biology Teacher* 35, no. 3 (1973): 125, https://doi.org/10.2307/4444260.

2. F. J. Ayala, "Darwin and the Scientific Method," *Proceedings of the National Academy of Sciences* 106, supplement_1 (June 16, 2009): 10,033–39, https://doi.org/10.1073/pnas.0901404106.

3. Craig Holdrege, "The Giraffe's Short Neck: Why Evolutionary Thought Needs a Holistic Foundation," *Nature Institute*, 2003, accessed December 21, 2018, http://natureinstitute.org/pub/ic/ic10/giraffe.htm.

4. Kostas Kampourakis and Vasso Zogza, "Students' Preconceptions about Evolution: How Accurate Is the Characterization as 'Lamarckian' When Considering the History of Evolutionary Thought?" *Science and Education* 16, no. 3–5 (February 27, 2007): 393–422, https://doi.org/10.1007/s11191-006-9019-9.

5. Heidelise Als, "Lamarck, Darwin, and the Science of NIDCAP: Epigenetics in the NICU," *Developmental Observer: The Official Newsletter of the NIDCAP, Federation International*, 2011, accessed December 21, 2018, http://nidcap.org/wp-content/uploads/2013/11/NIDCAP-DO-Vol4-No2.pdf; Richard W. Burkhardt, "Lamarck, Evolution, and the Inheritance of Acquired Characters," *Genetics* 194, no. 4 (August 2013): 793–805, https://doi.org/10.1534/genetics.113.151852; Ilya Gadjev, "Nature and Nurture: Lamarck's Legacy," *Biological Journal of the Linnean Society* 114, no. 1 (January 2015): 242–47, https://doi.org/10.1111/bij.12439.

6. Burkhardt, "Lamarck, Evolution, and the Inheritance of Acquired Characters"; Gadjev, "Nature and Nurture."

7. Frank J. Sulloway, "The Evolution of Charles Darwin," *Smithsonian.com*, December 2005, accessed December 21, 2018, https://www.smithsonianmag.com/science-nature/the-evolution-of-charles-darwin-110234034/.

8. "Galápagos Islands," *UNESCO World Heritage Centre*, accessed December 21, 2018, http://whc.unesco.org/en/list/1.

9. Dennis J. Geist et al., "A Paleogeographic Model of the Galápagos Islands and Biogeographical and Evolutionary Implications," in *Geophysical Monograph Series*, ed. Karen S. Harpp et al. (Hoboken, NJ: John Wiley & Sons, 2014), 145–66, https://doi.org/10.1002/9781118852538.ch8; Sulloway, "The Evolution of Charles Darwin."

10. Michelle Starr, "A New Bird Species Has Evolved on Galápagos, and Scientists Watched It Happen," *Sciencealert*, November 24, 2017, accessed December 21, 2018, https://www.sciencealert.com/darwin-s-finches-evolve-into-new-species-in-real-time-two-generations-galapagos.

11. Stephen Molnar, *Human Variation: Races, Types, and Ethnic Groups*, 6th ed. (Upper Saddle River, NJ: Pearson Prentice Hall, 2006).

12. Molnar, *Human Variation*.

13. Ranajit Chakraborty, "Analysis of Genetic Structure of Populations: Meaning, Methods, and Implications," in *Human Population Genetics*, ed. Partha P. Majumder (Boston, MA: Springer US, 1993), 189–206, https://doi.org/10.1007/978-1-4615-2970-5_14.

14. Greg Breining, "Are Darwin's Finches One Species or Many?" *Discover Magazine*, February 26, 2015, accessed December 21, 2018, http://discovermagazine.com/2015/april/2-species-stuck-in-neutral.

15. Brian J. Small, "Darwin's Finches," *Surfbirds*, accessed December 21, 2018, http://www.surfbirds.com/Features/darwinfinches/darwinfinches.html.

16. Arkhat Abzhanov, "Evolution of Beak Shapes in Darwin's Finches," *Serious Science*, June 23, 2015, accessed December 21, 2018, http://serious-science.org/evolution-of-beak-shapes-in-darwins-finches-3123; Small, "Darwin's Finches."

17. Small, "Darwin's Finches."

18. Abzhanov, "Evolution of Beak Shapes in Darwin's Finches."

19. Small, "Darwin's Finches."

20. A. W. F. Edwards, "More on the Too-Good-to-Be-True Paradox and Gregor Mendel," *Journal of Heredity* 77, no. 2 (March 1986): 138, https://doi.org/10.1093/oxfordjournals.jhered.a110192; M. Simunek, U. Hoßfeld, and V. Wissemann, "'Rediscovery' Revised: The Cooperation of Erich and Armin von Tschermak-Seysenegg in the Context of the 'Rediscovery' of Mendel's Laws in 1899–1901," *Plant Biology* 13, no. 6 (November 2011): 835–41, https://doi.org/10.1111/j.1438-8677.2011.00491.x.

21. Mauricio De Castro, "Johann Gregor Mendel: Paragon of Experimental Science," *Molecular Genetics and Genomic Medicine* 4, no. 1 (January 2016): 3–8, https://doi.org/10.1002/mgg3.199.

22. De Castro, "Johann Gregor Mendel."

23. "Understanding Human Genetic Variation," *National Institutes of Health*, 2007, accessed December 21, 2018, https://www.ncbi.nlm.nih.gov/books/NBK20363/.

24. De Castro, "Johann Gregor Mendel"; Aysha Divan and Janice Royds, *Molecular Biology: A Very Short Introduction*, 1st ed., Very Short Introductions 485 (Oxford, UK: Oxford University Press, 2016).

25. Simunek, Hoßfeld, and Wissemann, "'Rediscovery' Revised."

26. Simunek, Hoßfeld, and Wissemann, "'Rediscovery' Revised."

27. Vertika Singh and Kiran Singh, "Modern Synthesis," in *Encyclopedia of Animal Cognition and Behavior*, ed. Jennifer Vonk and Todd Shackelford (Cham: Springer International, 2018), 1–5, https://doi.org/10.1007/978-3-319-47829-6_203-1.

28. "Understanding Human Genetic Variation," *National Institutes of Health*, 2007, accessed December 21, 2018, https://www.ncbi.nlm.nih.gov/books/NBK20363/.

29. Jan Annigan, "Physiological Roles of Protein," *SFGate*, December 27, 2018, accessed December 21, 2018, https://healthyeating.sfgate.com/physiological-roles-protein-6182 .html.

30. Genetics Home Reference, "What Is Noncoding DNA?" *National Institutes of Health U.S. National Library of Medicine*, accessed December 18, 2018, https://ghr.nlm.nih.gov/ primer/basics/noncodingdna.

31. Molnar, *Human Variation*.

32. Molnar, *Human Variation*.

33. Charles Darwin and Ernst Mayr, *On the Origin of Species*, print, a facs. of the first ed. (Cambridge, MA: Harvard University Press, 2003); Molnar, *Human Variation*.

34. Robert Jurmain, Lynn Kilgore, Wenda Trevathan, and Russell L. Ciochon. *Introduction to Physical Anthropology* (Boston: Cengage Learning, 2014).

35. Molnar, *Human Variation*.

36. Laura Cooling, "Blood Groups in Infection and Host Susceptibility," *Clinical Microbiology Reviews* 28, no. 3 (July 2015): 801–70, https://doi.org/10.1128/CMR.00109-14; Molnar, *Human Variation*.

37. Alexandre Courtiol, Felix C. Tropf, and Melinda C. Mills, "When Genes and Environment Disagree: Making Sense of Trends in Recent Human Evolution," *Proceedings of the National Academy of Sciences of the United States of America* 113, no. 28 (December 2016): 7,693–95, https://doi.org/10.1073/pnas.1608532113.

38. "Natural Selection Has Strongly Influenced Recent Human Evolution, Study Finds," *Science Daily*, October 23, 2005, accessed December 21, 2018, https://www.sciencedaily.com/ releases/2005/10/051023115936.htm.

39. "Natural Selection Has Strongly Influenced Recent Human Evolution, Study Finds."

40. Molnar, *Human Variation*.

41. Molnar, *Human Variation*.

42. Raelia Lew et al., "Tay-Sachs Disease: Current Perspectives from Australia," *Application of Clinical Genetics* 8 (January19, 2015): 19–25, https://doi.org/10.2147/TACG.S49628; Molnar, *Human Variation*.

43. Molnar, *Human Variation*.

44. Carlos D. Bustamante et al., "Natural Selection on Protein-Coding Genes in the Human Genome," *Nature* 437, no. 7,062 (October 20, 2005): 1,153–57, https://doi.org/ 10.1038/nature04240.

45. James D. Watson, *The Double Helix: A Personal Account of the Discovery of the Structure of DNA* (New York: Scribner, 2001).

46. U.S. Department of Health and Human Services, National Institutes of Health, and National Institute of General Medical Sciences, *The New Genetics*, NIH Publication No.10-66 (Washington, DC: National Institutes of Health, 2010), www.nigms.nih.gov.

47. U.S. Department of Health and Human Services, National Institutes of Health, and National Institute of General Medical Sciences, *The New Genetics*.

48. U.S. Department of Health and Human Services, National Institutes of Health, and National Institute of General Medical Sciences, *The New Genetics*.

49. U.S. Department of Health and Human Services, National Institutes of Health, and National Institute of General Medical Sciences, *The New Genetics*.

50. L. Loewe and W. G. Hill, "The Population Genetics of Mutations: Good, Bad, and Indifferent," *Philosophical Transactions of the Royal Society B: Biological Sciences* 365, no. 1,544 (April 27, 2010): 1,153–67, https://doi.org/10.1098/rstb.2009.0317.

51. M. W. Nachman and S. L. Crowell, "Estimate of the Mutation Rate per Nucleotide in Humans," *Genetics* 156, no. 1 (September 2000): 297–304.

52. Nachman and Crowell, "Estimate of the Mutation Rate per Nucleotide in Humans."

53. Kat McGowan, "Most Mutations in the Human Genome Are Recent and Probably Harmful," *DiscoverMagazine.com*, December 17, 2013, accessed December 21, 2018, http://discovermagazine.com/2013/julyaug/07-most-mutations-in-the-human-genome-are-recent-and-probably-harmful.

54. Loewe and Hill, "The Population Genetics of Mutations."

55. Loewe and Hill, "The Population Genetics of Mutations."

56. Loewe and Hill, "The Population Genetics of Mutations."

57. Loewe and Hill, "The Population Genetics of Mutations."

58. Loewe and Hill, "The Population Genetics of Mutations."

59. Roshan Karki et al., "Defining 'Mutation' and 'Polymorphism' in the Era of Personal Genomics," *BMC Medical Genomics* 8, no. 1 (December 2015): 31–38, https://doi.org/10.1186/s12920-015-0115-z.

60. Karki et al., "Defining 'Mutation' and 'Polymorphism' in the Era of Personal Genomics."

61. Genetics Home Reference, "What Are Single Nucleotide Polymorphisms (SNPs)?" *National Institutes of Health U.S. National Library of Medicine*, accessed December 18, 2018, https://ghr.nlm.nih.gov/primer/genomicresearch/snp.

62. Genetics Home Reference, "What Are Single Nucleotide Polymorphisms (SNPs)?"

63. Monica Singh et al., "SNP–SNP Interactions within APOE Gene Influence Plasma Lipids in Postmenopausal Osteoporosis," *Rheumatology International* 31, no. 3 (March 2011): 421–23, https://doi.org/10.1007/s00296-010-1449-7.

64. Genetics Home Reference, "What Are Single Nucleotide Polymorphisms (SNPs)?"

65. "IGSR and the 1000 Genomes Project," *International Genome Sample Resource*, December 17, 2018, accessed December 21, 2018, http://www.internationalgenome.org/; "1000 Genomes Project," *National Human Genome Research Institute*, October 4, 2017, accessed December 21, 2018, https://www.genome.gov/27528684/1000-genomes-project/.

66. Wojciech Branicki, Urszula Brudnik, and Anna Wojas-Pelc, "Interactions between HERC2, OCA2, and MC1R May Influence Human Pigmentation Phenotype," *Annals of Human Genetics* 73, no. 2 (March 2009): 160–70, https://doi.org/10.1111/j.1469-1809.2009.00504.x.

67. Charlotte Huppertz et al., "The Dopaminergic Reward System and Leisure Time Exercise Behavior: A Candidate Allele Study," *BioMed Research International* 2014, article 591717 (2014), https://doi.org/10.1155/2014/591717.

68. Mark A. Bellgrove et al., "DRD4 Gene Variants and Sustained Attention in Attention Deficit Hyperactivity Disorder (ADHD): Effects of Associated Alleles at the VNTR and -521 SNP," *American Journal of Medical Genetics: Part B, Neuropsychiatric Genetics: The Official Publication of the International Society of Psychiatric Genetics* 136B, no. 1 (July 5,

2005): 81–86, https://doi.org/10.1002/ajmg.b.30193; Guang Guo and Yuying Tong, "Age at First Sexual Intercourse, Genes, and Social Context: Evidence from Twins and the Dopamine D4 Receptor Gene," *Demography* 43, no. 4 (November 2006): 747–69.

69. Bellgrove et al., "DRD4 Gene Variants and Sustained Attention in Attention Deficit Hyperactivity Disorder (ADHD)"; Guo and Tong, "Age at First Sexual Intercourse, Genes, and Social Context."

70. Austin de Rubira, Lea Georges, and Lars Fehren-Schmitz, "Ancient DNA Reveals That the Variability of the DRD4 -521 C/T SNP Associated with Novelty-Seeking Behavior Is Influenced by Selection in Western South American Populations," *Adaptive Human Behavior and Physiology* 2, no. 1 (March 2016): 77–91, https://doi.org/10.1007/s40750-015-0033-5.

71. Brian Krans, "ADHD and Evolution: Did ADHD Help Keep Humans Alive?" *Healthline.com*, October 12, 2017, accessed December 21, 2018, https://www.healthline.com/health/adhd/evolution#1.

72. Luis B. Barreiro et al., "Natural Selection Has Driven Population Differentiation in Modern Humans," *Nature Genetics* 40, no. 3 (March 2008): 340–45, https://doi.org/10.1038/ng.78.

73. Loewe and Hill, "The Population Genetics of Mutations."

74. Ziada Ayorech et al., "Publication Trends over 55 Years of Behavioral Genetic Research," *Behavior Genetics* 46, no. 5 (2016): 603–7, https://doi.org/10.1007/s10519-016-9786-2.

Chapter 3. Darwinian Medicine

1. Richard Milner, *The Encyclopedia of Evolution: Humanity's Search for Its Origins*, 1st Owl book ed. (New York: H. Holt and Co., 1993); Stephen Molnar, *Human Variation: Races, Types, and Ethnic Groups*, 6th ed. (Upper Saddle River, NJ: Pearson Prentice Hall, 2006); Christopher Stringer and Clive Gamble, *In Search of the Neanderthals: Solving the Puzzle of Human Origins*, 1st paperback ed. (London: Thames and Hudson, 1994).

2. Luis B. Barreiro et al., "Natural Selection Has Driven Population Differentiation in Modern Humans," *Nature Genetics* 40, no. 3 (March 2008): 340–45, https://doi.org/10.1038/ng.78; Dong-Dong Wu and Ya-Ping Zhang, "Different Level of Population Differentiation among Human Genes," *BMC Evolutionary Biology* 11 (January 14, 2011): 16, https://doi.org/10.1186/1471-2148-11-16.

3. Randolph M. Nesse, "Darwinian Medicine," *Encyclopedia Britannica*, September 27, 2013, accessed December 21, 2018, https://www.britannica.com/science/Darwinian-medicine; Randolph M. Nesse, "How Is Darwinian Medicine Useful?" *Western Journal of Medicine* 174, no. 5 (May 2001): 358–60; Randolph M. Nesse and George C. Williams, *Why We Get Sick: The New Science of Darwinian Medicine*, 1st ed. (New York: Times Books, 1994).

4. Nesse, "How Is Darwinian Medicine Useful?"

5. Nesse, "How Is Darwinian Medicine Useful?"

6. "International HapMap Project," National Human Genome Research Institute, May 1, 2012, accessed December 21, 2018, https://www.genome.gov/10001688/international-hapmap-project/.

7. "International HapMap Project."

8. J. Flint et al., "Why Are Some Genetic Diseases Common? Distinguishing Selection from Other Processes by Molecular Analysis of Globin Gene Variants," *Human Genetics* 91, no. 2 (March 1993): 91–117; Molnar, *Human Variation*.

9. Molnar, *Human Variation*.

10. Genetics Home Reference, "What Are Single Nucleotide Polymorphisms (SNPs)?" *National Institutes of Health U.S. National Library of Medicine*, December 18, 2018, accessed December 21, 2018, https://ghr.nlm.nih.gov/primer/genomicresearch/snp; Peter M. Visscher, "Challenges in Understanding Common Disease," *Genome Medicine* 9, no. 1 (December 2017), https://doi.org/10.1186/s13073-017-0506-1.

11. I. C. Gray, D. A. Campbell, and N. K. Spurr, "Single Nucleotide Polymorphisms as Tools in Human Genetics," *Human Molecular Genetics* 9, no. 16 (October 2000): 2,403–8; Visscher, "Challenges in Understanding Common Disease."

12. "Genome-Wide Association Studies Fact Sheet," *National Human Genome Research Institute*, August 27, 2015, accessed December 21, 2018, https://www.genome.gov/20019523/genomewide-association-studies-fact-sheet/.

13. Genetics Home Reference, "What Are Single Nucleotide Polymorphisms (SNPs)?"

14. Eléonore Toufektchan and Franck Toledo, "The Guardian of the Genome Revisited: P53 Downregulates Genes Required for Telomere Maintenance, DNA Repair, and Centromere Structure," *Cancers* 10, no. 5 (May 6, 2018): 135, https://doi.org/10.3390/cancers10050135.

15. P. D. P. Pharoah, N. E. Day, and C. Caldas, "Somatic Mutations in the P53 Gene and Prognosis in Breast Cancer: A Meta-Analysis," *British Journal of Cancer* 80, no. 12 (August 1999): 1,968–73, https://doi.org/10.1038/sj.bjc.6690628.

16. Jorge Zeron-Medina et al., "A Polymorphic P53 Response Element in KIT Ligand Influences Cancer Risk and Has Undergone Natural Selection," *Cell* 155, no. 2 (October 2013): 410–22, https://doi.org/10.1016/j.cell.2013.09.017.

17. Zeron-Medina et al., "A Polymorphic P53 Response Element in KIT Ligand Influences Cancer Risk and Has Undergone Natural Selection."

18. Zeron-Medina et al., "A Polymorphic P53 Response Element in KIT Ligand Influences Cancer Risk and Has Undergone Natural Selection."

19. Elizabeth Norton, "Testicular Cancer May Be the Price Paid for Sun Protection," *ScienceMag*, October 10, 2013, accessed December 21, 2018, https://www.sciencemag.org/news/2013/10/testicular-cancer-may-be-price-paid-sun-protection; Zeron-Medina et al., "A Polymorphic P53 Response Element in KIT Ligand Influences Cancer Risk and Has Undergone Natural Selection."

20. Zeron-Medina et al., "A Polymorphic P53 Response Element in KIT Ligand Influences Cancer Risk and Has Undergone Natural Selection."

21. Bob Grant, "Testicular-Skin Cancer Tradeoff," *Scientist Magazine*, October 14, 2013, accessed December 21, 2018, https://www.the-scientist.com/the-nutshell/testicular-skin-cancer-tradeoff-38561; Norton, "Testicular Cancer May Be the Price Paid for Sun Protection"; Zeron-Medina et al., "A Polymorphic P53 Response Element in KIT Ligand Influences Cancer Risk and Has Undergone Natural Selection."

22. Oluwatoyin Olatundun Ilesanmi, "Pathological Basis of Symptoms and Crises in Sickle Cell Disorder: Implications for Counseling and Psychotherapy," *Hematology Reports* 2, no. 1 (January 26, 2010): e2, https://doi.org/10.4081/hr.2010.e2.

23. Graham R. Serjeant, "Evolving Locally Appropriate Models of Care for Indian Sickle Cell Disease," *Indian Journal of Medical Research* 143, no. 4 (April 2016): 405–13, https://doi.org/10.4103/0971-5916.184282.

24. Ann McElroy and Patricia K. Townsend, *Medical Anthropology in Ecological Perspective*, 6th ed. (Boulder, CO: Westview Press, 2014); Robert W. Snow and Judy A. Omumbo, "Malaria," in *Disease and Mortality in Sub-Saharan Africa*, ed. Dean T. Jamison et al., 2nd ed. (Washington, DC: World Bank, 2006), 195–214, http://www.ncbi.nlm.nih.gov/books/NBK2286/.

25. R. Carter and K. N. Mendis, "Evolutionary and Historical Aspects of the Burden of Malaria," *Clinical Microbiology Reviews* 15, no. 4 (October 1, 2002): 564–94, https://doi.org/10.1128/CMR.15.4.564-594.2002.

26. Frank B. Livingstone, "Anthropological Implications of Sickle Cell Gene Distribution in West Africa," *American Anthropologist* 60, no. 3 (June 1958): 533–62, https://doi.org/10.1525/aa.1958.60.3.02a00110.

27. Bob Carter and Simon M Dyson, "Territory, Ancestry, and Descent: The Politics of Sickle Cell Disease," *Sociology* 45, no. 6 (December 2011): 963–76, https://doi.org/10.1177/0038038511416159.

28. Bruce M. Hannon and Matthias Ruth, *Dynamic Modeling of Diseases and Pests* (New York: Springer, 2009).

29. Snow and Omumbo, "Malaria."

30. H. Franklin Bunn, "The Triumph of Good over Evil: Protection by the Sickle Gene against Malaria," *Blood* 121, no. 1 (January 3, 2013): 20–25, https://doi.org/10.1182/blood-2012-08-449397.

31. Nalina Eggert, "Tracing Sickle Cell Back to One Child, 7,300 Years Ago," *BBC News*, March 13, 2018, accessed December 21, 2018, https://www.bbc.com/news/world-africa-43373247; Daniel Shriner and Charles N. Rotimi, "Whole-Genome-Sequence-Based-Haplotypes Reveal Single Origin of the Sickle Allele during the Holocene Wet Phase," *American Journal of Human Genetics* 102, no. 4 (May 2018): 547–56, https://doi.org/10.1016/j.ajhg.2018.02.003.

32. Joseph E Maakaron, "Sickle Cell Anemia: Practice Essentials, Background, Genetics," *Medscape*, September 4, 2018, accessed December 21, 2018, https://emedicine.medscape.com/article/205926-overview#showall; "Learning about Sickle Cell Disease," *National Human Genome Research Institute*, May 9, 2016, accessed December 21, 2018, https://www.genome.gov/10001219/learning-about-sickle-cell-disease/; Prithu Sundd, Mark T. Gladwin, and Enrico M. Novelli, "Pathophysiology of Sickle Cell Disease," *Annual Review of Pathology* 14 (October 2018): 263–92, https://doi.org/10.1146/annurev-pathmechdis-012418-012838.

33. "Learning about Sickle Cell Disease."

34. "Learning about Sickle Cell Disease."

35. Livingstone, "Anthropological Implications of Sickle Cell Gene Distribution in West Africa"; S. L. Wiesenfeld, "Sickle-Cell Trait in Human Biological and Cultural Evolution: Development of Agriculture, Causing Increased Malaria, Is Bound to Gene-Pool Changes Causing Malaria Reduction," *Science* 157, no. 3,793 (September 8, 1967): 1,134–40.

36. Pascale Gerbault et al., "Evolution of Lactase Persistence: An Example of Human Niche Construction," *Philosophical Transactions of the Royal Society of London: Series B, Biological Sciences* 366, no. 1,566 (March 27, 2011): 863–77, https://doi.org/10.1098/rstb.2010.0268; R. J. Grand et al., "Changing Genes; Losing Lactase," *Gut* 52, no. 5 (May 2003): 617–19.

37. Gerbault et al., "Evolution of Lactase Persistence."

38. Gerbault et al., "Evolution of Lactase Persistence."

39. Sarah A. Tishkoff et al., "Convergent Adaptation of Human Lactase Persistence in Africa and Europe," *Nature Genetics* 39, no. 1 (January 2007): 31–40, https://doi.org/10.1038/ng1946.

40. Tishkoff et al., "Convergent Adaptation of Human Lactase Persistence in Africa and Europe."

41. Anke Liebert et al., "Worldwide Distributions of Lactase Persistence Alleles and the Complex Effects of Recombination and Selection," *Human Genetics* 136, no. 11–12 (2017): 1,445–53, https://doi.org/10.1007/s00439-017-1847-y; Tishkoff et al., "Convergent Adaptation of Human Lactase Persistence in Africa and Europe."

42. Liebert et al., "Worldwide Distributions of Lactase Persistence Alleles and the Complex Effects of Recombination and Selection."

43. Gerbault et al., "Evolution of Lactase Persistence"; Liebert et al., "Worldwide Distributions of Lactase Persistence Alleles and the Complex Effects of Recombination and Selection"; Tishkoff et al., "Convergent Adaptation of Human Lactase Persistence in Africa and Europe."

44. Gerbault et al., "Evolution of Lactase Persistence."

45. "Tay-Sachs Disease," *SNPedia.com*, November 7, 2017, accessed December 21, 2018, https://snpedia.com/index.php/Tay-Sachs_disease.

46. Amos Frisch et al., "Origin and Spread of the 1278insTATC Mutation Causing Tay-Sachs Disease in Ashkenazi Jews: Genetic Drift as a Robust and Parsimonious Hypothesis," *Human Genetics* 114, no. 4 (March 2004): 366–76, https://doi.org/10.1007/s00439-003-1072-8; A. G. Knudson, "Founder Effect in Tay-Sachs Disease," *American Journal of Human Genetics* 25, no. 1 (January 1973): 108.

47. Eran Elhaik, "The Missing Link of Jewish European Ancestry: Contrasting the Rhineland and the Khazarian Hypotheses," *Genome Biology and Evolution* 5, no. 1 (January 2013): 61–74, https://doi.org/10.1093/gbe/evs119; Frisch et al., "Origin and Spread of the 1278insTATC Mutation Causing Tay-Sachs Disease in Ashkenazi Jews."

48. Lise Dubois et al., "Genetic and Environmental Contributions to Weight, Height, and BMI from Birth to 19 Years of Age: An International Study of over 12,000 Twin Pairs," *PloS One* 7, no. 2 (2012): e30153, https://doi.org/10.1371/journal.pone.0030153; Molnar, *Human Variation*.

49. Dubois et al., "Genetic and Environmental Contributions to Weight, Height, and BMI from Birth to 19 Years of Age."

50. Soo Heon Kwak and Kyong Soo Park, "Recent Progress in Genetic and Epigenetic Research on Type 2 Diabetes," *Experimental and Molecular Medicine* 48 (March 11, 2016): e220, https://doi.org/10.1038/emm.2016.7.

51. Lisa A. Urry et al., *Campbell Biology*, 11th ed. (New York: Pearson Education, 2017).

52. Blanca M. Herrera, Sarah Keildson, and Cecilia M. Lindgren, "Genetics and Epigenetics of Obesity," *Maturitas* 69, no. 1 (May 2011): 41–49, https://doi.org/10.1016/j.maturitas.2011.02.018.

53. Herrera, Keildson, and Lindgren, "Genetics and Epigenetics of Obesity."

54. Howard Slomko, Hye J. Heo, and Francine H. Einstein, "Minireview: Epigenetics of Obesity and Diabetes in Humans," *Endocrinology* 153, no. 3 (March 2012): 1,025–30, https://doi.org/10.1210/en.2011-1759.

55. Slomko, Heo, and Einstein, "Minireview."

56. Slomko, Heo, and Einstein, "Minireview."

57. Raquel Patricia Ataide Lima et al., "The Role of Epigenetics in the Etiology of Obesity: A Review," *Journal of Clinical Epigenetics* 3, no. 4 (2017), https://doi.org/10.21767/2472 -1158.100075; Caroline Hohensee, Tricia Varela, and Dustin Harris, "Child Obesity and Epigenetics," in *Epigenetics, the Environment, and Children's Health across Lifespans*, ed. David Hollar (Cham: Springer International, 2016), 335–43, https://doi.org/10.1007/978-3-319 -25325-1_13.

58. Slomko, Heo, and Einstein, "Minireview," 1,025.

59. K. Norrgard, "Genetic Variation and Disease: GWAS.," *Nature Education* 1, no. 1 (2008): 87.

60. "Genome-Wide Association Studies Fact Sheet"; John S. Witte, "Genome-Wide Association Studies and Beyond," *Annual Review of Public Health* 31, no. 1 (March 2010): 9–20, https://doi.org/10.1146/annurev.publhealth.012809.103723.

61. James R. Lupski et al., "Clan Genomics and the Complex Architecture of Human Disease," *Cell* 147, no. 1 (September 30, 2011): 32–43, https://doi.org/10.1016/j.cell.2011.09.008.

Chapter 4. From Bodies to Behavior: Population Genetics and Social Life

1. Manfred Göthert, "Serotonin Discovery and Stepwise Disclosure of 5-HT Receptor Complexity over Four Decades. Part I. General Background and Discovery of Serotonin as a Basis for 5-HT Receptor Identification," *Pharmacological Reports: PR* 65, no. 4 (2013): 771–86.

2. "Ergot Alkaloids and Derivatives," *DrugBank*, accessed January 4, 2019, https://www .drugbank.ca/categories/DBCAT000606.

3. Yojiro Muneoka and Masahiro Matsuura, "Effects of the Molluscan Neuropeptide FMRFamide and the Related Opioid Peptide YGGFMRFamide on Mytilus Muscle," *Comparative Biochemistry and Physiology Part C: Comparative Pharmacology* 81, no. 1 (January 1985): 61–70, https://doi.org/10.1016/0742-8413(85)90092-1; Yojiro Muneoka and Betty M. Twarog, "Neuromuscular Transmission and Excitation–Contraction Coupling in Molluscan Muscle," in *Mollusca*, ed. A. L. M. Saleuddin and Karl M. Wilbur (Amsterdam: Elsevier, 1983), 35–76, https://doi.org/10.1016/B978-0-12-751404-8.50010-4.

4. P. M. Whitaker-Azmitia, "The Discovery of Serotonin and Its Role in Neuroscience," *Neuropsychopharmacology: Official Publication of the American College of Neuropsychopharmacology* 21, no. 2 suppl. (August 1999): 2S–8S, https://doi.org/10.1016/S0893-133X(99)00031-7.

5. J. J. López-Ibor, "The Involvement of Serotonin in Psychiatric Disorders and Behavior," *British Journal of Psychiatry* 153, no. S3 (September 1988): 26–39.

6. Harvey Lodish et al., "Neurotransmitters, Synapses, and Impulse Transmission," in *Molecular Cell Biology*, 4th ed. (New York: W. H. Freeman, 2000), 67, https://www.ncbi.nlm .nih.gov/books/NBK21521/.

7. Lodish et al., "Neurotransmitters, Synapses, and Impulse Transmission."

8. K. P. Lesch et al., "Serotonin (5-HT) Receptor, 5-HT Transporter, and G Protein-Effector Expression: Implications for Depression," *Pharmacology and Toxicology* 71, suppl. 1 (1992): 49–60; K. P. Lesch et al., "Regional Brain Expression of Serotonin Transporter MRNA and Its Regulation by Reuptake Inhibiting Antidepressants," *Brain Research: Molecular Brain Research* 17, no. 1–2 (January 1993): 31–35; A. Weizman and R. Weizman, "Serotonin Transporter Polymorphism and Response to SSRIs in Major Depression and Relevance to Anxiety Disorders and Substance Abuse," *Pharmacogenomics* 1, no. 3 (August 2000): 335–41, https://doi.org/10.1517/14622416.1.3.335.

9. Lodish et al., "Neurotransmitters, Synapses, and Impulse Transmission."

10. J. Gelernter et al., "Genetic Association between Dopamine Transporter Protein Alleles and Cocaine-Induced Paranoia," *Neuropsychopharmacology: Official Publication of the American College of Neuropsychopharmacology* 11, no. 3 (November 1994): 195–200, https://doi.org/10.1038/sj.npp.1380106; Lodish et al., "Neurotransmitters, Synapses, and Impulse Transmission"; Radek Ptácek, Hana Kuzelová, and George B. Stefano, "Dopamine D4 Receptor Gene DRD4 and Its Association with Psychiatric Disorders," *Medical Science Monitor: International Medical Journal of Experimental and Clinical Research* 17, no. 9 (September 2011): RA215–220.

11. H. Kunugi et al., "Serotonin Transporter Gene Polymorphisms: Ethnic Difference and Possible Association with Bipolar Affective Disorder," *Molecular Psychiatry* 2, no. 6 (November 1997): 457–62; Chee Hong Ng et al., "Serotonin Transporter Polymorphisms and Clinical Response to Sertraline across Ethnicities," *Progress in Neuro-Psychopharmacology and Biological Psychiatry* 30, no. 5 (July 2006): 953–57, https://doi.org/10.1016/j.pnpbp.2006.02.015.

12. Joan Y. Chiao and Katherine D. Blizinsky, "Culture-Gene Coevolution of Individualism-Collectivism and the Serotonin Transporter Gene," *Proceedings: Biological Sciences* 277, no. 1,681 (February 22, 2010): 529–37, https://doi.org/10.1098/rspb.2009.1650.

13. Avshalom Caspi et al., "Genetic Sensitivity to the Environment: The Case of the Serotonin Transporter Gene and Its Implications for Studying Complex Diseases and Traits," *American Journal of Psychiatry* 167, no. 5 (May 2010): 509–27, https://doi.org/10.1176/appi.ajp.2010.09101452; Chiao and Blizinsky, "Culture-Gene Coevolution of Individualism-Collectivism and the Serotonin Transporter Gene"; G. R. Twitchell et al., "Serotonin Transporter Promoter Polymorphism Genotype Is Associated with Behavioral Disinhibition and Negative Affect in Children of Alcoholics," *Alcoholism, Clinical, and Experimental Research* 25, no. 7 (July 2001): 953–59.

14. Meeshanthini Vijayendran et al., "The Relationship of the Serotonin Transporter (SLC6A4) Extra Long Variant to Gene Expression in an African American Sample," *American Journal of Medical Genetics: Part B, Neuropsychiatric Genetics: The Official Publication of the International Society of Psychiatric Genetics* 159B, no. 5 (July 2012): 611–12, https://doi.org/10.1002/ajmg.b.32054.

15. Chiao and Blizinsky, "Culture-Gene Coevolution of Individualism-Collectivism and the Serotonin Transporter Gene."

16. Chiao and Blizinsky, "Culture-Gene Coevolution of Individualism-Collectivism and the Serotonin Transporter Gene"; M. Loi, L. Del Savio, and E. Stupka, "Social Epigenetics and Equality of Opportunity," *Public Health Ethics* 6, no. 2 (July 1, 2013): 142–53, https://doi.org/10.1093/phe/pht019.

17. Shreekantiah Umesh and Shamshul Haque Nizamie, "Genetics in Psychiatry," *Indian Journal of Human Genetics* 20, no. 2 (April 2014): 120–28, https://doi.org/10.4103/0971-6866.142845.

18. Sridhar Prathikanti and Daniel R. Weinberger, "Psychiatric Genetics—the New Era: Genetic Research and Some Clinical Implications," *British Medical Bulletin* 73–74, no. 1 (January 1, 2005): 107–22, https://doi.org/10.1093/bmb/ldh055.

19. Ralph J. Greenspan, "The Origins of Behavioral Genetics," *Current Biology* 18, no. 5 (March 11, 2008): R192–98, https://doi.org/10.1016/j.cub.2008.01.015.

20. Greenspan, "The Origins of Behavioral Genetics"; Aaron Panofsky, *Misbehaving Science: Controversy and the Development of Behavior Genetics* (Chicago and London: University of Chicago Press, 2014).

21. Kerstin Rohde et al., "Genetics and Epigenetics in Obesity," *Metabolism: Clinical and Experimental* 92 (March 2019), 37–50, https://doi.org/10.1016/j.metabol.2018.10.007.

22. Thorhildur Halldorsdottir and Elisabeth B. Binder, "Gene × Environment Interactions: From Molecular Mechanisms to Behavior," *Annual Review of Psychology* 68 (January 3, 2017): 215–41, https://doi.org/10.1146/annurev-psych-010416-044053; Panofsky, *Misbehaving Science*.

23. Oliver Krüger, Peter Korsten, and Joseph I. Hoffman, "The Rise of Behavioral Genetics and the Transition to Behavioral Genomics and Beyond," in *APA Handbook of Comparative Psychology: Basic Concepts, Methods, Neural Substrate, and Behavior*, ed. Josep Call et al. (Washington, DC: American Psychological Association, 2017), 365–79, https://doi.org/10.1037/0000011-018.

24. Panofsky, *Misbehaving Science*.

25. Gene E. Robinson, Russell D. Fernald, and David F. Clayton, "Genes and Social Behavior," *Science* 322, no. 5,903 (November 7, 2008): 896–900, https://doi.org/10.1126/science.1159277; Peter Wade and Yasuko I. Takezawa, "Race: Hereditarian Ideology and European Constructions of Race," in *Encyclopedia Britannica*, October 4, 2017, accessed January 24, 2019, https://www.britannica.com/topic/race-human; Anthony Walsh, *Science Wars: Politics, Gender, and Race* (New Brunswick, NJ: Transaction, 2013).

26. Michael Hansen et al., "The 2018 Brown Center Report on American Education: How Well Are American Students Learning?" *Brown Center on Educational Policy at Brookings*, June 2018.

27. Hansen et al., "The 2018 Brown Center Report on American Education."

28. Greg J. Duncan and Richard J. Murnane, eds., *Whither Opportunity? Rising Inequality, Schools, and Children's Life Chances* (New York and Chicago: Russell Sage Foundation/Spencer Foundation, 2011); Hansen et al., "The 2018 Brown Center Report on American Education."

29. Hansen et al., "The 2018 Brown Center Report on American Education."

30. V. P. Franklin, "The Tests Are Written for the Dogs: The Journal of Negro Education, African American Children, and the Intelligence Testing Movement in Historical Perspective," *Journal of Negro Education* 76, no. 3 (2007): 216–29.

31. J. Philippe Rushton and Arthur R. Jensen, "Thirty Years of Research on Race Differences in Cognitive Ability," *Psychology, Public Policy, and Law* 11, no. 2 (2005): 235–94, https://doi.org/10.1037/1076-8971.11.2.235.

32. Alondra Oubré, *Race, Genes, and Ability: Rethinking Ethnic Differences*, vol. 2 (Woodland Hills, CA: BTI Press, 2011).

33. Walsh, *Science Wars*.

34. J. Philippe Rushton, *Race, Evolution, and Behavior: A Life History Perspective*, 3rd ed. (Puerto Ayora: Charles Darwin Research Institute, 2000); Walsh, *Science Wars*.

35. Halldorsdottir and Binder, "Gene × Environment Interactions."

36. Robinson, Fernald, and Clayton, "Genes and Social Behavior."

37. Jenny Tung and Yoav Gilad, "Social Environmental Effects on Gene Regulation," *Cellular and Molecular Life Sciences* 70, no. 22 (November 2013): 4,323–39, https://doi.org/10.1007/s00018-013-1357-6.

38. Ronald L. Simons, Steven R. H. Beach, and Ashley B. Barr, "Differential Susceptibility to Context: A Promising Model of the Interplay of Genes and the Social Environment," in *Advances in Group Processes*, vol. 29, ed. Will Kalkhoff, Shane R. Thye, and Edward J.

Lawler (Bingley, UK: Emerald Group, 2012), 139–63, https://doi.org/10.1108/S0882
-6145(2012)0000029008.

39. Halldorsdottir and Binder, "Gene × Environment Interactions"; Robinson, Fernald,
and Clayton, "Genes and Social Behavior"; Tung and Gilad, "Social Environmental Effects
on Gene Regulation."

40. Avshalom Caspi et al., "Role of Genotype in the Cycle of Violence in Maltreated
Children," *Science* 297, no. 5,582 (August 2, 2002): 851–54, https://doi.org/10.1126/science
.1072290.

41. Caspi et al., "Role of Genotype in the Cycle of Violence in Maltreated Children."

42. Avshalom Caspi et al., "Influence of Life Stress on Depression: Moderation by a
Polymorphism in the 5-HTT Gene," *Science* 301, no. 5,631 (July 18, 2003): 386–89, https://
doi.org/10.1126/science.1083968.

43. Caspi et al., "Role of Genotype in the Cycle of Violence in Maltreated Children."

44. Robinson, Fernald, and Clayton, "Genes and Social Behavior"; Simons, Beach, and
Barr, "Differential Susceptibility to Context."

45. Tinca J. C. Polderman et al., "Meta-Analysis of the Heritability of Human Traits
Based on 50 Years of Twin Studies," *Nature Genetics* 47, no. 7 (July 2015): 702–9, https://
doi.org/10.1038/ng.3285.

46. Peter M. Visscher et al., "10 Years of GWAS Discovery: Biology, Function, and
Translation," *American Journal of Human Genetics* 101, no. 1 (July 6, 2017): 5–22, https://
doi.org/10.1016/j.ajhg.2017.06.005.

47. Simons, Beach, and Barr, "Differential Susceptibility to Context."

48. Walsh, *Science Wars*.

49. Daniel Frías-Lasserre, Cristian A. Villagra, and Carlos Guerrero-Bosagna, "Stress in
the Educational System as a Potential Source of Epigenetic Influences on Children's Development and Behavior," *Frontiers in Behavioral Neuroscience* 12 (2018): 143, https://doi.org/
10.3389/fnbeh.2018.00143; Jon K. Oxford et al., "Endocrine and Aggressive Responses
to Competition Are Moderated by Contest Outcome, Gender, Individual versus Team
Competition, and Implicit Motives," ed. Etsuro Ito, *PLoS ONE* 12, no. 7 (July 27, 2017):
e0181610, https://doi.org/10.1371/journal.pone.0181610; J. Martin Ramirez, "Hormones
and Aggression in Childhood and Adolescence," *Aggression and Violent Behavior* 8, no. 6
(November 2003): 621–44, https://doi.org/10.1016/S1359-1789(02)00102-7.

50. Frías-Lasserre, Villagra, and Guerrero-Bosagna, "Stress in the Educational System as
a Potential Source of Epigenetic Influences on Children's Development and Behavior."

51. Loi, Del Savio, and Stupka, "Social Epigenetics and Equality of Opportunity";
Simons, Beach, and Barr, "Differential Susceptibility to Context"; Tung and Gilad, "Social
Environmental Effects on Gene Regulation."

52. Polderman et al., "Meta-Analysis of the Heritability of Human Traits Based on 50
Years of Twin Studies."

53. Polderman et al., "Meta-Analysis of the Heritability of Human Traits Based on 50
Years of Twin Studies."

54. Loi, Del Savio, and Stupka, "Social Epigenetics and Equality of Opportunity";
David Rodenhiser and Mellissa Mann, "Epigenetics and Human Disease: Translating Basic
Biology into Clinical Applications," *CMAJ: Canadian Medical Association Journal* 174, no. 3
(January 31, 2006): 341–48, https://doi.org/10.1503/cmaj.050774; Rohde et al., "Genetics
and Epigenetics in Obesity."

Chapter 5. Architecture of the Human Brain

1. Joseph Carey, *Brain Facts: A Primer on the Brain and Nervous System*, 5th ed. (Washington, DC: Society for Neuroscience, 2006), https://www.brainline.org/sites/default/files/Brainfacts_Guide_Society_for_Neuroscience.pdf; K. Sukel, "The Synapse: A Primer," *DANA Foundation*, March 15, 2011, accessed February 26, 2019, http://www.dana.org/News/Details.aspx?id=43512.

2. Carey, *Brain Facts*; Sukel, "The Synapse"; Rand Swenson, *Review of Clinical and Functional Neuroscience* (Hanover, NH: Dartmouth Medical School, 2006), https://www.dartmouth.edu/~rswenson/NeuroSci/.

3. Carey, *Brain Facts*.

4. Ho Namkung, Sun-Hong Kim, and Akira Sawa, "The Insula: An Underestimated Brain Area in Clinical Neuroscience, Psychiatry, and Neurology," *Trends in Neurosciences* 40, no. 4 (April 2017): 200–207, https://doi.org/10.1016/j.tins.2017.02.002.

5. Carey, *Brain Facts*; Sukel, "The Synapse."

6. Carey, *Brain Facts*; Sukel, "The Synapse"; Swenson, *Review of Clinical and Functional Neuroscience*.

7. Frederico A. C. Azevedo et al., "Equal Numbers of Neuronal and Nonneuronal Cells Make the Human Brain an Isometrically Scaled-Up Primate Brain," *Journal of Comparative Neurology* 513, no. 5 (April 10, 2009): 532–41, https://doi.org/10.1002/cne.21974.

8. Swenson, *Review of Clinical and Functional Neuroscience*.

9. Carey, *Brain Facts*; Sukel, "The Synapse."

10. Carey, *Brain Facts*; Sukel, "The Synapse."

11. Valerie Ross, "Numbers: The Nervous System, from 268-MPH Signals to Trillions of Synapses," *Discover Magazine*, May 15, 2011, accessed February 26, 2019, http://discovermagazine.com/2011/mar/10-numbers-the-nervous-system.

12. Carl Huffman, "Alcmaeon," in *Stanford Encyclopedia of Philosophy*, ed. Edward N. Zalta (Stanford, CA: Metaphysics Research Lab, Stanford University, 2015), https://plato.stanford.edu/archives/spr2017/entries/alcmaeon/.

13. Stanley Finger, *Origins of Neuroscience: A History of Explorations into Brain Function*, 1st paperback ed. (Oxford, UK: Oxford University Press, 2001).

14. Katrin Amunts and Karl Zilles, "Architectonic Mapping of the Human Brain beyond Brodmann," *Neuron* 88, no. 6 (December 2015): 1,086–107, https://doi.org/10.1016/j.neuron.2015.12.001; Marios Loukas et al., "Korbinian Brodmann (1868–1918) and His Contributions to Mapping the Cerebral Cortex," *Neurosurgery* 68, no. 1 (January 1, 2011): 6–11, https://doi.org/10.1227/NEU.0b013e3181fc5cac.

15. Sarah Genon et al., "How to Characterize the Function of a Brain Region," *Trends in Cognitive Sciences* 22, no. 4 (April 2018): 350–64, https://doi.org/10.1016/j.tics.2018.01.010.

16. "The Discovery of the Neuron," *ScienceBlogs*, July 3, 2007, accessed February 26, 2019, https://scienceblogs.com/neurophilosophy/2007/07/03/the-discovery-of-the-neuron.

17. Carey, *Brain Facts*; Bryan Kolb and Robbin Gibb, "Brain Plasticity and Behavior in the Developing Brain," *Journal of the Canadian Academy of Child and Adolescent Psychiatry* 20, no. 4 (November 2011): 265–76.

18. Francisco R. M. Ferreira, Maria I. Nogueira, and Javier Defelipe, "The Influence of James and Darwin on Cajal and His Research into the Neuron Theory and Evolution of the Nervous System," *Frontiers in Neuroanatomy* 8 (2014): 1, https://doi.org/10.3389/fnana.2014.00001.

19. Lisa Feldman Barrett and Ajay Bhaskar Satpute, "Large-Scale Brain Networks in Affective and Social Neuroscience: Towards an Integrative Functional Architecture of the Brain," *Current Opinion in Neurobiology* 23, no. 3 (June 2013): 361–72, https://doi.org/10.1016/j.conb.2012.12.012.

20. Mark D'Esposito, Andrew S. Kayser, and Anthony J. W. Chen, "Functional MRI: Cognitive Neuroscience Applications," in *Functional Neuroradiology*, ed. Scott H. Faro et al. (Boston: Springer US, 2011), 687–706, https://doi.org/10.1007/978-1-4419-0345-7_34.

21. Christopher M. Filley and R. Douglas Fields, "White Matter and Cognition: Making the Connection," *Journal of Neurophysiology* 116, no. 5 (January 2016): 2,093–104, https://doi.org/10.1152/jn.00221.2016; Sukel, "The Synapse."

22. Filley and Fields, "White Matter and Cognition"; Sukel, "The Synapse."

23. Ivana Despotović, Bart Goossens, and Wilfried Philips, "MRI Segmentation of the Human Brain: Challenges, Methods, and Applications," *Computational and Mathematical Methods in Medicine* 2015 (2015), https://doi.org/10.1155/2015/450341.

24. Madhura Ingalhalikar et al., "Sex Differences in the Structural Connectome of the Human Brain," *Proceedings of the National Academy of Sciences of the United States of America* 111, no. 2 (January 14, 2014): 823–28, https://doi.org/10.1073/pnas.1316909110.

25. "Human Connectome Project: Mapping the Human Brain Connectivity," *University of Southern California, Mark and Mary Stevens Neuroimaging and Informatics Institute*, accessed February 26, 2019, http://www.humanconnectomeproject.org/.

26. D. C. Van Essen et al., "The Human Connectome Project: A Data Acquisition Perspective," *NeuroImage* 62, no. 4 (October 1, 2012): 2,222–31, https://doi.org/10.1016/j.neuroimage.2012.02.018.

Chapter 6. The Gendered Brain

1. Stephen Jay Gould, *The Mismeasure of Man*, rev. and expanded, with a new introduction (New York: W. W. Norton, 2008); Richard Lynn, "Sex Differences in Intelligence and Brain Size: A Paradox Resolved," *Personality and Individual Differences* 17, no. 2 (August 1994): 257–71, https://doi.org/10.1016/0191-8869(94)90030-2; Richard Lynn, "Sex Differences in Intelligence and Brain Size: A Developmental Theory," *Intelligence* 27, no. 1 (February 1999): 1–12, https://doi.org/10.1016/S0160-2896(99)00009-4.

2. Paul Irwing and Richard Lynn, "Sex Differences in Means and Variability on the Progressive Matrices in University Students: A Meta-Analysis," *British Journal of Psychology* 96, no. 4 (November 2005): 505–24, https://doi.org/10.1348/000712605X53542.

3. Lynn, "Sex Differences in Intelligence and Brain Size: A Paradox Resolved"; Lynn, "Sex Differences in Intelligence and Brain Size: A Developmental Theory"; Richard Lynn, Juri Allik, and Olev Must, "Sex Differences in Brain Size, Stature, and Intelligence in Children and Adolescents: Some Evidence from Estonia," *Personality and Individual Differences* 29, no. 3 (September 2000): 555–60, https://doi.org/10.1016/S0191-8869(99)00215-9.

4. R. C. Gur et al., "Sex Differences in Brain Gray and White Matter in Healthy Young Adults: Correlations with Cognitive Performance," *Journal of Neuroscience: The Official Journal of the Society for Neuroscience* 19, no. 10 (May 15, 1999): 4,065–72.

5. Martin Heil et al., "Adults' Sex Difference in a Dynamic Mental Rotation Task: Validating Infant Results," *Journal of Individual Differences* 39, no. 1 (January 2018): 48–52, https://doi.org/10.1027/1614-0001/a000248; Kirsten Jordan et al., "Women and Men

Exhibit Different Cortical Activation Patterns during Mental Rotation Tasks," *Neuropsychologia* 40, no. 13 (2002): 2,397–408.

6. Steven G. Vandenberg and Allan R. Kuse, "Mental Rotations: A Group Test of Three-Dimensional Spatial Visualization," *Perceptual and Motor Skills* 47, no. 2 (December 1978): 599–604, https://doi.org/10.2466/pms.1978.47.2.599.

7. Heil et al., "Adults' Sex Difference in a Dynamic Mental Rotation Task"; Mary Soares Masters and Barbara Sanders, "Is the Gender Difference in Mental Rotation Disappearing?" *Behavior Genetics* 23, no. 4 (July 1993): 337–41, https://doi.org/10.1007/BF01067434.

8. D. F. Halpern, "Sex Differences in Intelligence: Implications for Education," *American Psychologist* 52, no. 10 (October 1997): 1,091–102.

9. Ulrich S. Tran, Agnes A. Hofer, and Martin Voracek, "Sex Differences in General Knowledge: Meta-Analysis and New Data on the Contribution of School-Related Moderators among High-School Students," *PloS One* 9, no. 10 (2014): e110391, https://doi.org/10.1371/journal.pone.0110391.

10. Richard Lynn and Paul Irwing, "Sex Differences on the Progressive Matrices: A Meta-Analysis," *Intelligence* 32, no. 5 (September 2004): 481–98, https://doi.org/10.1016/j.intell.2004.06.008.

11. Cecile D. Ladouceur et al., "White Matter Development in Adolescence: The Influence of Puberty and Implications for Affective Disorders," *Developmental Cognitive Neuroscience* 2, no. 1 (January 2012): 36–54, https://doi.org/10.1016/j.dcn.2011.06.002.

12. Richard Lynn and Tatu Vanhanen, *IQ and the Wealth of Nations: Human Evolution, Behavior, and Intelligence* (Westport, CT: Praeger, 2002).

13. Gur et al., "Sex Differences in Brain Gray and White Matter in Healthy Young Adults"; Richard Lynn and Tatu Vanhanen, *Intelligence: A Unifying Construct for the Social Sciences* (London: Ulster Institute for Social Research, 2012).

14. Lara Menzies et al., "The Effects of Puberty on White Matter Development in Boys," *Developmental Cognitive Neuroscience* 11 (February 2015): 116–28, https://doi.org/10.1016/j.dcn.2014.10.002.

15. Gur et al., "Sex Differences in Brain Gray and White Matter in Healthy Young Adults."

16. Gur et al., "Sex Differences in Brain Gray and White Matter in Healthy Young Adults."

17. John S. Allen et al., "Sexual Dimorphism and Asymmetries in the Gray–White Composition of the Human Cerebrum," *NeuroImage* 18, no. 4 (April 2003): 880–94.

18. Michael Peters, "Does Brain Size Matter? A Reply to Rushton and Ankney," *Canadian Journal of Experimental Psychology* 49, no. 4 (1995): 570–76, https://doi.org/10.1037/1196-1961.49.4.570.

19. Jordan et al., "Women and Men Exhibit Different Cortical Activation Patterns during Mental Rotation Tasks"; E. Weiss et al., "Sex Differences in Brain Activation Pattern during a Visuospatial Cognitive Task: A Functional Magnetic Resonance Imaging Study in Healthy Volunteers," *Neuroscience Letters* 344, no. 3 (July 2003): 169–72, https://doi.org/10.1016/S0304-3940(03)00406-3.

20. Jordan et al., "Women and Men Exhibit Different Cortical Activation Patterns during Mental Rotation Tasks."

21. Weiss et al., "Sex Differences in Brain Activation Pattern during a Visuospatial Cognitive Task."

22. Margaret Semrud-Clikeman et al., "Gender Differences in Brain Activation on a Mental Rotation Task," *International Journal of Neuroscience* 122, no. 10 (January 26, 2012): 590–97, https://doi.org/10.3109/00207454.2012.693999.

23. Kaili Rimfeld et al., "Phenotypic and Genetic Evidence for a Unifactorial Structure of Spatial Abilities," *Proceedings of the National Academy of Sciences* 114, no. 10 (March 7, 2017): 2,777–82, https://doi.org/10.1073/pnas.1607883114.

24. Gur et al., "Sex Differences in Brain Gray and White Matter in Healthy Young Adults"; Melissa Hines, "Sex-Related Variation in Human Behavior and the Brain," *Trends in Cognitive Sciences* 14, no. 10 (October 2010): 448–56, https://doi.org/10.1016/j.tics.2010.07.005.

25. D. C. Van Essen et al., "The Human Connectome Project: A Data Acquisition Perspective," *NeuroImage* 62, no. 4 (October 1, 2012): 2,222–31, https://doi.org/10.1016/j.neuroimage.2012.02.018.

26. Madhura Ingalhalikar et al., "Sex Differences in the Structural Connectome of the Human Brain," *Proceedings of the National Academy of Sciences of the United States of America* 111, no. 2 (January 14, 2014): 823–28, https://doi.org/10.1073/pnas.1316909110.

27. Neda Jahanshad et al., "Sex Differences in the Human Connectome: 4-Tesla High Angular Resolution Diffusion Imaging (HARDI) Tractography in 234 Young Adult Twins," in *2011 IEEE International Symposium on Biomedical Imaging: From Nano to Macro* (2011): 939–43, https://doi.org/10.1109/ISBI.2011.5872558.

28. Jahanshad et al., "Sex Differences in the Human Connectome: 4-Tesla High Angular Resolution Diffusion Imaging (HARDI) Tractography in 234 Young Adult Twins"; Edmund T. Rolls, "The Functions of the Orbitofrontal Cortex," *Brain and Cognition* 55, no. 1 (June 2004): 11–29, https://doi.org/10.1016/S0278-2626(03)00277-X; Pierre Sacré et al., "Lucky Rhythms in Orbitofrontal Cortex Bias Gambling Decisions in Humans," *Scientific Reports* 6, no. 1 (December 2016), https://doi.org/10.1038/srep36206; Thomas A. Stalnaker, Nisha K. Cooch, and Geoffrey Schoenbaum, "What the Orbitofrontal Cortex Does Not Do," *Nature Neuroscience* 18, no. 5 (May 2015): 620–27, https://doi.org/10.1038/nn.3982.

29. Moheb Costandi, *The Human Brain: 50 Human Brain Ideas You Really Need to Know* (London: Quercus, 2013).

30. Ingalhalikar et al., "Sex Differences in the Structural Connectome of the Human Brain."

31. Ingalhalikar et al., "Sex Differences in the Structural Connectome of the Human Brain."

32. Ingalhalikar et al., "Sex Differences in the Structural Connectome of the Human Brain."

33. Ingalhalikar et al., "Sex Differences in the Structural Connectome of the Human Brain."

34. Ingalhalikar et al., "Sex Differences in the Structural Connectome of the Human Brain."

35. Cordelia Fine, "New Insights into Gendered Brain Wiring, or a Perfect Case Study in Neurosexism?" *Conversation*, December 3, 2013, accessed February 26, 2019, http://theconversation.com/new-insights-into-gendered-brain-wiring-or-a-perfect-case-study-in-neurosexism-21083.

36. Fine, "New Insights into Gendered Brain Wiring."

37. Fine, "New Insights into Gendered Brain Wiring."

38. Fine, "New Insights into Gendered Brain Wiring."

39. Fine, "New Insights into Gendered Brain Wiring."

40. Fine, "New Insights into Gendered Brain Wiring."

41. Fine, "New Insights into Gendered Brain Wiring."

42. Hines, "Sex-Related Variation in Human Behavior and the Brain."

43. Hines, "Sex-Related Variation in Human Behavior and the Brain"; Melissa Hines, "Gender Development and the Human Brain," *Annual Review of Neuroscience* 34 (2011): 69–88, https://doi.org/10.1146/annurev-neuro-061010-113654; C. E. Roselli, "Neurobiology of Gender Identity and Sexual Orientation," *Journal of Neuroendocrinology* 30, no. 7 (July 2018): e12562, https://doi.org/10.1111/jne.12562.

44. Hines, "Sex-Related Variation in Human Behavior and the Brain."

45. Hines, "Sex-Related Variation in Human Behavior and the Brain."

46. Hines, "Sex-Related Variation in Human Behavior and the Brain."

47. Fine, "New Insights into Gendered Brain Wiring."

48. Ramune Griksiene and Osvaldas Ruksenas, "Effects of Hormonal Contraceptives on Mental Rotation and Verbal Fluency," *Psychoneuroendocrinology* 36, no. 8 (September 2011): 1,239–48, https://doi.org/10.1016/j.psyneuen.2011.03.001.

49. Kirsten Jordan, "The Brain between Sex and Gender: Women and Men from a Neuroscientific Perspective," in *Sex and Gender in Biomedicine: Theories, Methodologies, Results,* ed. Ineke Klinge and Claudia Wiesemann (Göttingen: Universitätsverlag Göttingen, 2010), 79–100.

50. Jordan, "The Brain between Sex and Gender"; Dhruv Marwha, Meha Halari, and Lise Eliot, "Meta-Analysis Reveals a Lack of Sexual Dimorphism in Human Amygdala Volume," *NeuroImage* 147 (February 15, 2017): 282–94, https://doi.org/10.1016/j.neuroimage.2016.12.021; Stuart J. Ritchie et al., "Beyond a Bigger Brain: Multivariable Structural Brain Imaging and Intelligence," *Intelligence* 51 (August 2015): 47–56, https://doi.org/10.1016/j.intell.2015.05.001.

51. Jordan, "The Brain between Sex and Gender."

52. Stuart J. Ritchie et al., "Sex Differences in the Adult Human Brain: Evidence from 5,216 UK Biobank Participants," *Cerebral Cortex* 28, no. 8 (August 1, 2018): 2,959–75, https://doi.org/10.1093/cercor/bhy109.

53. Richard A. Kanaan et al., "Gender Differences in White Matter Microstructure," *PloS One* 7, no. 6 (2012): e38272, https://doi.org/10.1371/journal.pone.0038272.

54. Eileen Luders et al., "Gender Differences in Cortical Complexity," *Nature Neuroscience* 7, no. 8 (August 2004): 799–800, https://doi.org/10.1038/nn1277.

55. Virginia Fernández, Cristina Llinares-Benadero, and Víctor Borrell, "Cerebral Cortex Expansion and Folding: What Have We Learned?" *EMBO Journal* 35, no. 10 (May 17, 2016): 1,021–44, https://doi.org/10.15252/embj.201593701.

56. Jordan, "The Brain between Sex and Gender."

57. Jordan, "The Brain between Sex and Gender."

58. B. A. Ardekani, K. Figarsky, and J. J. Sidtis, "Sexual Dimorphism in the Human Corpus Callosum: An MRI Study Using the OASIS Brain Database," *Cerebral Cortex* 23, no. 10 (October 1, 2013): 2,514–20, https://doi.org/10.1093/cercor/bhs253.

59. Jordan, "The Brain between Sex and Gender."

60. Jordan, "The Brain between Sex and Gender."

61. T.-V. Nguyen et al., "Interactive Effects of Dehydroepiandrosterone and Testosterone on Cortical Thickness during Early Brain Development," *Journal of Neuroscience* 33, no. 26 (June 26, 2013): 10,840–48, https://doi.org/10.1523/JNEUROSCI.5747-12.2013; Hugo G. Schnack et al., "Changes in Thickness and Surface Area of the Human Cortex and Their Relationship with Intelligence," *Cerebral Cortex* 25, no. 6 (June 2015): 1,608–17, https://doi.org/10.1093/cercor/bht357.

62. Rachel M. Brouwer et al., "Genetic Associations between Intelligence and Cortical Thickness Emerge at the Start of Puberty," *Human Brain Mapping* 35, no. 8 (August 2014): 3,760–73, https://doi.org/10.1002/hbm.22435.

63. Schnack et al., "Changes in Thickness and Surface Area of the Human Cortex and Their Relationship with Intelligence."

64. Schnack et al., "Changes in Thickness and Surface Area of the Human Cortex and Their Relationship with Intelligence."

65. Ritchie et al., "Sex Differences in the Adult Human Brain."

66. P. Cédric, M. P. Koolschijn, Jiska S. Peper, and Eveline A. Crone, "The Influence of Sex Steroids on Structural Brain Maturation in Adolescence," *PloS One* 9, no. 1 (2014): e83929, https://doi.org/10.1371/journal.pone.0083929.

67. Cédric, Koolschijn, Peper, and Crone, "The Influence of Sex Steroids on Structural Brain Maturation in Adolescence."

68. Jessica A. Wojtalik et al., "Using Cognitive Neuroscience to Improve Mental Health Treatment: A Comprehensive Review," *Journal of the Society for Social Work and Research* 9, no. 2 (2018): 223–60, https://doi.org/10.1086/697566.

Chapter 7. The Racialized Brain

1. Ulric Neisser et al., "Intelligence: Knowns and Unknowns.," *American Psychologist* 51, no. 2 (1996): 77–101, https://doi.org/10.1037/0003-066X.51.2.77.

2. Neisser et al., "Intelligence"; Alan Reifman, "Revisiting the Bell Curve," *Psycoloquy* 11, no. 99 (2000): 19, http://academic.sun.ac.za/mathed/174/NatureNurture/Reifman.pdf.

3. Bärbel Inhelder and Jean Piaget, *The Growth of Logical Thinking: From Childhood to Adolescence*, trans. Anne Parsons and Stanley Milgram (New York: Basic Books, 1958), https://doi.org/10.1037/10034-000.

4. Philip E. Vernon, *Intelligence and Cultural Environment* (New York: Routledge, 2013), http://site.ebrary.com/id/10832478.

5. Arthur Robert Jensen, *The g Factor: The Science of Mental Ability*, Human Evolution, Behavior, and Intelligence (Westport, CT: Praeger, 1998); J. Philippe Rushton and C. Davison Ankney, "Brain Size and Cognitive Ability: Correlations with Age, Sex, Social Class, and Race," *Psychonomic Bulletin and Review* 3, no. 1 (March 1996): 21–36, https://doi.org/10.3758/BF03210739.

6. Leon J. Kamin, *The Science and Politics of I.Q.* (New York: Routledge, 2009); Richard E. Nisbett, *Intelligence and How to Get It: Why Schools and Cultures Count* (New York: W. W. Norton, 2010).

7. Stephen Jay Gould, *The Mismeasure of Man*, rev. and expanded, with a new introduction (New York: W. W. Norton, 2008).

8. Daniel Goleman, "An Emerging Theory on Blacks' I.Q. Scores," *New York Times*, April 10, 1988, accessed December 18, 2018, https://www.nytimes.com/1988/04/10/education/an-emerging-theory-on-blacks-iq-scores.html; Richard Lynn and Gerhard Meisenberg, "The Average IQ of Sub-Saharan Africans: Comments on Wicherts, Dolan, and van Der Maas," *Intelligence* 38, no. 1 (January 2010): 21–29, https://doi.org/10.1016/j.intell.2009.09.009; Richard Lynn and Tatu Vanhanen, *Intelligence: A Unifying Construct for the Social Sciences* (London: Ulster Institute for Social Research, 2012).

9. Lynn and Vanhanen, *Intelligence*.

10. Gregory Cochran, Jason Hardy, and Henry Harpending, "Natural History of Ashkenazi Intelligence," *Journal of Biosocial Science* 38, no. 5 (September 2006): 659–93, https://doi.org/10.1017/S0021932005027069; Richard Lynn, *The Chosen People: A Study of Jewish Intelligence and Achievement* (Whitefish, MT: Washington Summit, 2011); Richard Lynn and David Longley, "On the High Intelligence and Cognitive Achievements of Jews in Britain," *Intelligence* 34, no. 6 (November 2006): 541–47, https://doi.org/10.1016/j.intell.2006.03.011.

11. Gil Atzmon et al., "Abraham's Children in the Genome Era: Major Jewish Diaspora Populations Comprise Distinct Genetic Clusters with Shared Middle Eastern Ancestry," *American Journal of Human Genetics* 86, no. 6 (June 2010): 850–59, https://doi.org/10.1016/j.ajhg.2010.04.015.

12. Lynn and Longley, "On the High Intelligence and Cognitive Achievements of Jews in Britain"; Charles Murray, "Jewish Genius," *Commentary* (blog), April 1, 2007, accessed December 18, 2018, https://www.commentarymagazine.com/articles/jewish-genius/; Steven Pinker, "Jews, Genes, and Intelligence," accessed December 18, 2018, https://stevenpinker.com/presentations/steven-pinker-jews-genes-and-intelligence.

13. Gregory Cochran, Jason Hardy, and Henry Harpending, "Natural History of Ashkenazi Intelligence," *Journal of Biosocial Science* 38, no. 5 (September 2006): 659, https://doi.org/10.1017/S0021932005027069.

14. Richard Lynn and Tatu Vanhanen, *IQ and the Wealth of Nations: Human Evolution, Behavior, and Intelligence* (Westport, CT: Praeger, 2002); Lynn and Vanhanen, *Intelligence*.

15. Arthur R. Jensen, "The Nature of the Black–White Difference on Various Psychometric Tests: Spearman's Hypothesis," *Behavioral and Brain Sciences* 8, no. 2 (July 1985): 193, https://doi.org/10.1017/S0140525X00020392; Jensen, *The g Factor*; Satoshi Kanazawa, "The Evolution of General Intelligence," *Personality and Individual Differences* 53, no. 2 (July 2012): 90–93, https://doi.org/10.1016/j.paid.2011.05.015; Lynn and Vanhanen, *IQ and the Wealth of Nations*.

16. Lynn and Meisenberg, "The Average IQ of Sub-Saharan Africans"; Lynn and Vanhanen, *IQ and the Wealth of Nations*; J. Philippe Rushton, *Race, Evolution, and Behavior: A Life History Perspective*, 3rd ed. (Puerto Ayora: Charles Darwin Research Institute, 2000).

17. Lynn and Meisenberg, "The Average IQ of Sub-Saharan Africans."

18. Jelte M. Wicherts, Conor V. Dolan, and Han L. J. van der Maas, "A Systematic Literature Review of the Average IQ of Sub-Saharan Africans," *Intelligence* 38, no. 1 (January 2010): 1–20, https://doi.org/10.1016/j.intell.2009.05.002.

19. Wicherts, Dolan, and van der Maas, "A Systematic Literature Review of the Average IQ of Sub-Saharan Africans."

20. Tamara C. Daley et al., "IQ on the Rise: The Flynn Effect in Rural Kenyan Children," *Psychological Science* 14, no. 3 (May 2003): 215–19, https://doi.org/10.1111/1467-9280.02434; Wicherts, Dolan, and van der Maas, "A Systematic Literature Review of the Average IQ of Sub-Saharan Africans."

21. William T. Dickens and James R. Flynn, "Black Americans Reduce the Racial IQ Gap: Evidence from Standardization Samples," *Psychological Science* 17, no. 10 (October 2006): 913–20, https://doi.org/10.1111/j.1467-9280.2006.01802.x.

22. Eric Turkheimer, Kathryn Paige Harden, and Richard E. Nisbett, "There's Still No Good Reason to Believe Black–White IQ Differences Are Due to Genes," *Vox*, June 15, 2017, accessed December 18, 2018, https://www.vox.com/the-big-idea/2017/6/15/15797120/race-black-white-iq-response-critics.

23. Roland G. Fryer and Steven D. Levitt, "Understanding the Black–White Test Score Gap in the First Two Years of School," *Review of Economics and Statistics* 86, no. 2 (May 2004): 447–64, https://doi.org/10.1162/003465304323031049.

24. Roland G. Fryer and Steven D. Levitt, "Testing for Racial Differences in the Mental Ability of Young Children," *American Economic Review* 103, no. 2 (April 2013): 981–1,005, https://doi.org/10.1257/aer.103.2.981; W. Jean Yeung and Dalton Conley, "Black–White Achievement Gap and Family Wealth," *Child Development* 79, no. 2 (March 2008): 303–24, https://doi.org/10.1111/j.1467-8624.2007.01127.x.

25. David F. Marks, "IQ Variations across Time, Race, and Nationality: An Artifact of Differences in Literacy Skills," *Psychological Reports* 106, no. 3 (June 2010): 643–64, https://doi.org/10.2466/PR0.106.3.643-664.

26. Michael E. Levin, *Why Race Matters: Race Differences and What They Mean*, Human Evolution, Behavior, and Intelligence (Westport, CT: Praeger, 1997); Rushton, *Race, Evolution, and Behavior*; Anthony Walsh and Ilhong Yun, "Evoked Culture and Evoked Nature: The Promise of Gene-Culture Co-Evolution Theory for Sociology," *Frontiers in Sociology* 1 (July 19, 2016), https://doi.org/10.3389/fsoc.2016.00008.

27. Jensen, *The g Factor*.

28. Robert Plomin and Ian J. Deary, "Genetics and Intelligence Differences: Five Special Findings," *Molecular Psychiatry* 20, no. 1 (February 2015): 98–108, https://doi.org/10.1038/mp.2014.105.

29. Ken Richardson and Sarah H. Norgate, "Does IQ Really Predict Job Performance?" *Applied Developmental Science* 19, no. 3 (July 3, 2015): 153–69, https://doi.org/10.1080/10888691.2014.983635.

30. Jensen, "The Nature of the Black–White Difference on Various Psychometric Tests."

31. John Massey and Dale Vogt, "Heritability and Its Use in Animal Breeding" *University of Missouri Extension*, 1993, accessed December 18, 2018, http://hdl.handle.net/10355/3614.

32. Massey and Vogt, "Heritability and Its Use in Animal Breeding."

33. Genetics Home Reference, "What Is Heritability?" *National Institutes of Health U.S. National Library of Medicine*, accessed December 18, 2018, https://ghr.nlm.nih.gov/primer/inheritance/heritability; Leonard Lieberman, "How 'Caucasoids' Got Such Big Crania and Why They Shrank: From Morton to Rushton," *Current Anthropology* 42, no. 1 (February 2001): 69–95; Scott MacEachern, "Africanist Archaeology and Ancient IQ: Racial Science and Cultural Evolution in the Twenty-First Century," *World Archaeology* 38, no. 1 (March 2006): 72–92, https://doi.org/10.1080/00438240500509918.

34. Eric Turkheimer et al., "Socioeconomic Status Modifies Heritability of IQ in Young Children," *Psychological Science* 14, no. 6 (November 2003): 623–28, https://doi.org/10.1046/j.0956-7976.2003.psci_1475.x.

35. Arthur R. Jensen, "Level I/Level II: Factors or Categories?" *Journal of Educational Psychology* 74, no. 6 (1982): 868–73, https://doi.org/10.1037/0022-0663.74.6.868.

36. Jensen, "Level I/Level II."

37. Sarah H. Broman, "Prenatal Risk Factors for Mental Retardation in Young Children," *Public Health Reports* 102, no. 4 suppl (July 1987): 55–57; Lee J. Cronbach, "Measured Mental Abilities: Lingering Questions and Loose Ends," in *Human Diversity: Its Causes and Social Significance*, ed. B. Davis and P. Flaherty (Cambridge, MA: Ballinger, 1976), 130–47; Lee J. Cronbach, "Five Decades of Public Controversy over Mental Testing," *American Psychologist* 30, no. 1 (1975): 1–14, https://doi.org/10.1037/0003-066X.30.1.1; Jerrold J. Hein-

del and Laura N. Vandenberg, "Developmental Origins of Health and Disease: A Paradigm for Understanding Disease Cause and Prevention," *Current Opinion in Pediatrics* 27, no. 2 (April 2015): 248–53, https://doi.org/10.1097/MOP.0000000000000191.

38. Alan Reifman, "Revisiting the Bell Curve," *Psycoloquy* 11, no. 99 (2000): 19, http://aca demic.sun.ac.za/mathed/174/NatureNurture/Reifman.pdf.

39. Donald C. Johanson and Kate Wong, *Lucy's Legacy: The Quest for Human Origins*, 1st paperback ed. (New York: Three Rivers Press, 2010).

40. Johanson and Wong, *Lucy's Legacy*; H. Pontzer, "Overview of Hominin Evolution," *Nature Education Knowledge* 3, no. 10 (2012): 8, https://www.nature.com/scitable/knowl edge/library/overview-of-hominin-evolution-89010983; Chris Stringer, *The Origin of Our Species* (London: Allen Lane, 2011).

41. Robin Anne Smith, "How Did Human Brains Get to Be So Big?" *Scientific American Blog Network*, 2012, accessed December 18, 2018, https://blogs.scientificamerican.com/ guest-blog/how-did-human-brains-get-to-be-so-big/.

42. M. W. Berger, "A New Take on the Nineteenth-Century Skull Collection of Samuel Morton," *Penn Today*, October 4, 2018, accessed December 18, 2018, https://penntoday .upenn.edu/news/new-take-on-infamous-Morton-skulls; Gould, *The Mismeasure of Man*; Audrey Smedley and Brian D. Smedley, *Race in North America: Origin and Evolution of a Worldview*, 4th ed. (Boulder, CO: Westview Press, 2012).

43. Gould, *The Mismeasure of Man*; Smedley and Smedley, *Race in North America*.

44. Gould, *The Mismeasure of Man*; Smedley and Smedley, *Race in North America*.

45. Gould, *The Mismeasure of Man*.

46. John S. Michael, "A New Look at Morton's Craniological Research," *Current Anthropology* 29, no. 2 (April 1988): 349–54, https://doi.org/10.1086/203646.

47. Jason E. Lewis et al., "The Mismeasure of Science: Stephen Jay Gould versus Samuel George Morton on Skulls and Bias," *PLOS Biology* 9, no. 6 (June 7, 2011): e1001071, https://doi.org/10.1371/journal.pbio.1001071.

48. Lewis et al., "The Mismeasure of Science."

49. John S. Michael, "Personal Commentary on Morton and Gould. Part 1–Part 4," *Myths about Race That Racists and Antiracists Still Believe* (blog), 2013, accessed December 18, 2018, http://michael1988.com/?page_id=424.

50. Luke Holman et al., "Evidence of Experimental Bias in the Life Sciences: Why We Need Blind Data Recording," *PLOS Biology* 13, no. 7 (July 8, 2015): e1002190, https://doi .org/10.1371/journal.pbio.1002190.

51. Rushton, *Race, Evolution, and Behavior*.

52. Kanazawa, "The Evolution of General Intelligence"; Levin, *Why Race Matters*; Lynn and Vanhanen, *IQ and the Wealth of Nations*; Rushton, *Race, Evolution, and Behavior*.

53. Rushton, *Race, Evolution, and Behavior*.

54. Rushton, *Race, Evolution, and Behavior*.

55. Rushton, *Race, Evolution, and Behavior*.

56. C. Loring Brace, "Racialism and Racist Agendas: *Race, Evolution, and Behavior: A Life History Perspective*, J. Philippe Rushton," *American Anthropologist* 98, no. 1 (March 1996): 176–77, https://doi.org/10.1525/aa.1996.98.1.02a00250.

57. Lieberman, "How 'Caucasoids' Got Such Big Crania and Why They Shrank."

58. Rushton, *Race, Evolution, and Behavior*.

59. Lieberman, "How 'Caucasoids' Got Such Big Crania and Why They Shrank."

60. Lieberman, "How 'Caucasoids' Got Such Big Crania and Why They Shrank."

61. Rushton, *Race, Evolution, and Behavior*.

62. Rushton and Ankney, "Brain Size and Cognitive Ability."

63. Mark A. Klebanoff, "The Collaborative Perinatal Project: A 50-Year Retrospective," *Pediatric and Perinatal Epidemiology* 23, no. 1 (January 2009): 2–8, https://doi.org/10.1111/j.1365-3016.2008.00984.x.

64. Rushton and Ankney, "Brain Size and Cognitive Ability."

65. Broman, "Prenatal Risk Factors for Mental Retardation in Young Children."

66. A. H. Hemachandra, "Racial Disparities in the Association between Birth Weight in the Term Infant and Blood Pressure at Age 7 Years: Results from the Collaborative Perinatal Project," *Journal of the American Society of Nephrology* 17, no. 9 (August 9, 2006): 2,576–81, https://doi.org/10.1681/ASN.2005090898.

67. Adrienne L. Tierney and Charles A. Nelson, "Brain Development and the Role of Experience in the Early Years," *Zero to Three* 30, no. 2 (November 1, 2009): 9–13.

68. Rushton and Ankney, "Brain Size and Cognitive Ability."

69. J. Philippe Rushton, "Cranial Capacity Related to Sex, Rank, and Race in a Stratified Random Sample of 6,325 U.S. Military Personnel," *Intelligence* 16, no. 3–4 (July 1992): 401–13, https://doi.org/10.1016/0160-2896(92)90017-L.

70. Michael Peters, "Does Brain Size Matter? A Reply to Rushton and Ankney," *Canadian Journal of Experimental Psychology* 49, no. 4 (1995): 570–76, https://doi.org/10.1037/1196-1961.49.4.570.

71. Harry J. Jerison, "Brain to Body Ratios and the Evolution of Intelligence," *Science* 121, no. 3,144 (April 1, 1955): 447–49.

72. Jerison, "Brain to Body Ratios and the Evolution of Intelligence."

73. Heindel and Vandenberg, "Developmental Origins of Health and Disease"; Pathik D. Wadhwa et al., "Developmental Origins of Health and Disease: Brief History of the Approach and Current Focus on Epigenetic Mechanisms," *Seminars in Reproductive Medicine* 27, no. 5 (September 2009): 358–68, https://doi.org/10.1055/s-0029-1237424.

74. Rushton and Ankney, "Brain Size and Cognitive Ability."

75. Rushton, "Cranial Capacity Related to Sex, Rank, and Race in a Stratified Random Sample of 6,325 U.S. Military Personnel."

76. I. E. Odokuma et al., "Craniometric Patterns of Three Nigerian Ethnic Groups," *African Journal of Biotechnology* 9, no. 10 (March 8, 2010): 1,510–13, https://doi.org/10.5897/AJB09.1773.

77. Odokuma et al., "Craniometric Patterns of Three Nigerian Ethnic Groups"; Rushton, *Race, Evolution, and Behavior*.

78. S. J. Mack et al., "Evolution of Pacific/Asian Populations Inferred from HLA Class II Allele Frequency Distributions," *Tissue Antigens* 55, no. 5 (May 2000): 383–400; Yuchen Wang et al., "Genetic Structure, Divergence, and Admixture of Han Chinese, Japanese, and Korean Populations," *Hereditas* 155, no. 1 (December 2018), https://doi.org/10.1186/s41065-018-0057-5.

79. Harry J. Jerison, "The Evolution of Diversity in Brain Size," in *Development and Evolution in Brain Size: Behavioral Implications*, ed. M. E. Hahn, B. C. Jensen, and B. C. Dudek (New York: Academic Press, 1979), 30–60.

80. Stuart J. Ritchie et al., "Beyond a Bigger Brain: Multivariable Structural Brain Imaging and Intelligence," *Intelligence* 51 (August 2015): 47–56, https://doi.org/10.1016/j.intell.2015.05.001.

81. Jerison, "The Evolution of Diversity in Brain Size."

82. Phillip V. Tobias, "Brain-Size, Grey Matter, and Race—Fact or Fiction?" *American Journal of Physical Anthropology* 32, no. 1 (January 1970): 3–25, https://doi.org/10.1002/ajpa.1330320103.

83. Tobias, "Brain-Size, Grey Matter, and Race."

84. Tobias, "Brain-Size, Grey Matter, and Race."

85. Zack Z. Cernovsky, "J. P. Rushton on Negroids and Caucasoids: Statistical Concepts and Disconfirmatory Evidence," *International Journal of Dynamic Assessment and Instruction* 2 (1992): 55–67; Lieberman, "How 'Caucasoids' Got Such Big Crania and Why They Shrank."

86. Cernovsky, "J. P. Rushton on Negroids and Caucasoids"; Peters, "Does Brain Size Matter?"

87. Lieberman, "How 'Caucasoids' Got Such Big Crania and Why They Shrank."

88. Peters, "Does Brain Size Matter?"

89. Peters, "Does Brain Size Matter?"

90. S. A. Tishkoff et al., "The Genetic Structure and History of Africans and African Americans," *Science* 324, no. 5,930 (May 22, 2009): 1,035–44, https://doi.org/10.1126/science.1172257.

91. Rushton, *Race, Evolution, and Behavior.*

92. Rushton, *Race, Evolution, and Behavior.*

93. Lieberman, "How 'Caucasoids' Got Such Big Crania and Why They Shrank."

94. K. S. Lashley, "Persistent Problems in the Evolution of Mind," *Quarterly Review of Biology* 24, no. 1 (March 1949): 28–42, https://doi.org/10.1086/396806.

95. Rushton, *Race, Evolution, and Behavior.*

96. Lieberman, "How 'Caucasoids' Got Such Big Crania and Why They Shrank."

97. Lieberman, "How 'Caucasoids' Got Such Big Crania and Why They Shrank"; Fredric Weizmann et al., "Differential K Theory and Racial Hierarchies," *Canadian Psychology* 31, no. 1 (1990): 1–13, https://doi.org/10.1037/h0078934.

98. Lieberman, "How 'Caucasoids' Got Such Big Crania and Why They Shrank."

99. Lieberman, "How 'Caucasoids' Got Such Big Crania and Why They Shrank."

100. Lieberman, "How 'Caucasoids' Got Such Big Crania and Why They Shrank."

101. Vincent Sarich, "Race and Language in Prehistory," Eighth Annual Meeting of the Human Behavior and Evolution Society, Evanston, IL, 1995.

102. Gould, *The Mismeasure of Man*; Earl B. Hunt, *Human Intelligence* (Cambridge and New York: Cambridge University Press, 2011); Lieberman, "How 'Caucasoids' Got Such Big Crania and Why They Shrank"; Shelley L. Smith, "Single (Sub)Species Then and Now: An Examination of the Nonracial Perspective of C. Loring Brace," *American Journal of Physical Anthropology* 165 (February 2018): 104–25, https://doi.org/10.1002/ajpa.23385.

103. Hunt, *Human Intelligence.*

104. Lieberman, "How 'Caucasoids' Got Such Big Crania and Why They Shrank," 74.

105. Lieberman, "How 'Caucasoids' Got Such Big Crania and Why They Shrank."

106. Kenneth L. Beals et al., "Brain Size, Cranial Morphology, Climate, and Time Machines [and Comments and Reply]," *Current Anthropology* 25, no. 3 (June 1984): 301–30,

https://doi.org/10.1086/203138; Michel A. Hofman, "Evolution of the Human Brain: When Bigger Is Better," *Frontiers in Neuroanatomy* 8 (March 27, 2014), https://doi.org/10.3389/fnana.2014.00015; Kamran Safi, Marc A. Seid, and Dina K. N. Dechmann, "Bigger Is Not Always Better: When Brains Get Smaller," *Biology Letters* 1, no. 3 (September 22, 2005): 283–86, https://doi.org/10.1098/rsbl.2005.0333.

107. Jean-Louis Santini, "Are Brains Shrinking to Make Us Smarter?" *Phys.org*, February 6, 2011, accessed December 18, 2018, https://phys.org/news/2011-02-brains-smarter.html#jCp.

108. Nneka Isamah et al., "Variability in Frontotemporal Brain Structure: The Importance of Recruitment of African Americans in Neuroscience Research," *PloS One* 5, no. 10 (October 26, 2010): e13642, https://doi.org/10.1371/journal.pone.0013642.

109. Sarich, "Race and Language in Prehistory."

110. Hofman, "Evolution of the Human Brain."

111. Hofman, "Evolution of the Human Brain."

112. Richard Milner, *The Encyclopedia of Evolution: Humanity's Search for Its Origins*, 1st Owl book ed. (New York: H. Holt and Co., 1993); Stephen Molnar, *Human Variation: Races, Types, and Ethnic Groups*, 6th ed. (Upper Saddle River, NJ: Pearson Prentice Hall, 2006); Stringer, *The Origin of Our Species*.

113. P. K. Reardon et al., "Normative Brain Size Variation and Brain Shape Diversity in Humans," *Science* 360, no. 6,394 (June 15, 2018): 1,222–27, https://doi.org/10.1126/science.aar2578; "Bigger Human Brain Prioritizes Thinking Hub—at a Cost: Information Integration Trumps Emotional, Sensory, Motor Functions, Study Finds," *Science-Daily*, May 31, 2018, accessed December 18, 2018, https://www.sciencedaily.com/releases/2018/05/180531171604.htm.

Chapter 8. Brain Traits: The View from MRIs

1. Tal Geva, "Magnetic Resonance Imaging: Historical Perspective," *Journal of Cardiovascular Magnetic Resonance* 8, no. 4 (August 1, 2006): 573–80, https://doi.org/10.1080/10976640600755302.

2. John C. Mazziotta, "Imaging: Window on the Brain," *Archives of Neurology* 57, no. 10 (October 1, 2000): 1,413–21, https://doi.org/10.1001/archneur.57.10.1413.

3. Christof Koch, "Does Size Matter—for Brains?" *Scientific American*, January 1, 2016, accessed December 18, 2018, https://www.scientificamerican.com/article/does-brain-size-matter1/.

4. R. Van de Walle, I. Lemahieu, and E. Achten, "Magnetic Resonance Imaging and the Reduction of Motion Artifacts: Review of the Principles," *Technology and Health Care: Official Journal of the European Society for Engineering and Medicine* 5, no. 6 (December 1997): 419–35.

5. Ivana Despotović, Bart Goossens, and Wilfried Philips, "MRI Segmentation of the Human Brain: Challenges, Methods, and Applications," *Computational and Mathematical Methods in Medicine* 2015 (2015): 450341, https://doi.org/10.1155/2015/450341.

6. Michael Peters, "Does Brain Size Matter? A Reply to Rushton and Ankney," *Canadian Journal of Experimental Psychology* 49, no. 4 (1995): 570–76, https://doi.org/10.1037/1196-1961.49.4.570.

7. Koch, "Does Size Matter—for Brains?"

8. Lee Willerman et al., "In Vivo Brain Size and Intelligence," *Intelligence* 15, no. 2 (April 1991): 223–28, https://doi.org/10.1016/0160-2896(91)90031-8.

9. Koch, "Does Size Matter—for Brains?"

10. Jakob Pietschnig et al., "Meta-Analysis of Associations between Human Brain Volume and Intelligence Differences: How Strong Are They and What Do They Mean?" *Neuroscience and Biobehavioral Reviews* 57 (October 2015): 411–32, https://doi.org/10.1016/j.neubiorev.2015.09.017.

11. Gilles E. Gignac and Timothy C. Bates, "Brain Volume and Intelligence: The Moderating Role of Intelligence Measurement Quality," *Intelligence* 64 (September 2017): 18–29, https://doi.org/10.1016/j.intell.2017.06.004.

12. Hugo G. Schnack et al., "Changes in Thickness and Surface Area of the Human Cortex and Their Relationship with Intelligence," *Cerebral Cortex* 25, no. 6 (June 2015): 1,608–17, https://doi.org/10.1093/cercor/bht357; P. Shaw et al., "Intellectual Ability and Cortical Development in Children and Adolescents," *Nature* 440, no. 7,084 (March 30, 2006): 676–79, https://doi.org/10.1038/nature04513; Peng Yuan and Naftali Raz, "Prefrontal Cortex and Executive Functions in Healthy Adults: A Meta-Analysis of Structural Neuroimaging Studies," *Neuroscience and Biobehavioral Reviews* 42 (May 2014): 180–92, https://doi.org/10.1016/j.neubiorev.2014.02.005.

13. Koch, "Does Size Matter—for Brains?"

14. Stuart J. Ritchie et al., "Beyond a Bigger Brain: Multivariable Structural Brain Imaging and Intelligence," *Intelligence* 51 (August 2015): 47–56, https://doi.org/10.1016/j.intell.2015.05.001.

15. David Huepe and Natalia Salas, "Fluid Intelligence, Social Cognition, and Perspective Changing Abilities as Pointers of Psychosocial Adaptation," *Frontiers in Human Neuroscience* 7 (2013): 287, https://doi.org/10.3389/fnhum.2013.00287; Ritchie et al., "Beyond a Bigger Brain."

16. Shaw et al., "Intellectual Ability and Cortical Development in Children and Adolescents."

17. Oksana Yu Naumova et al., "Gene Expression in the Human Brain: The Current State of the Study of Specificity and Spatiotemporal Dynamics," *Child Development* 84, no. 1 (February 2013): 76–88, https://doi.org/10.1111/cdev.12014.

18. A. P. Shimamura, "Humans," in *Encyclopedia of Neuroscience*, ed. Larry R. Squire (Amsterdam: Elsevier, 2009), 29–34, https://doi.org/10.1016/B978-008045046-9.00763-4.

19. Richard J. Haier et al., "The Neuroanatomy of General Intelligence: Sex Matters," *NeuroImage* 25, no. 1 (March 2005): 320–27, https://doi.org/10.1016/j.neuroimage.2004.11.019.

20. Haier et al., "The Neuroanatomy of General Intelligence."

21. R. C. Gur et al., "Sex Differences in Brain Gray and White Matter in Healthy Young Adults: Correlations with Cognitive Performance," *Journal of Neuroscience: The Official Journal of the Society for Neuroscience* 19, no. 10 (May 15, 1999): 4,065–72; Melissa Hines, "Sex-Related Variation in Human Behavior and the Brain," *Trends in Cognitive Sciences* 14, no. 10 (October 2010): 448–56, https://doi.org/10.1016/j.tics.2010.07.005.

22. Jordan Bai et al., "Population Differences in Brain Morphology and Microstructure among Chinese, Malay, and Indian Neonates," *PLoS ONE* 7, no. 10 (October 24, 2012): e47816, https://doi.org/10.1371/journal.pone.0047816.

23. Bai et al., "Population Differences in Brain Morphology and Microstructure among Chinese, Malay, and Indian Neonates."

24. Garrett E. Alexander, "Biology of Parkinson's Disease: Pathogenesis and Pathophysiology of a Multisystem Neurodegenerative Disorder," *Dialogues in Clinical Neuroscience* 6, no. 3 (September 2004): 259–80.

25. Nneka Isamah et al., "Variability in Frontotemporal Brain Structure: The Importance of Recruitment of African Americans in Neuroscience Research," *PloS One* 5, no. 10 (October 26, 2010): e13642, https://doi.org/10.1371/journal.pone.0013642.

26. Isamah et al., "Variability in Frontotemporal Brain Structure."

27. Isamah et al., "Variability in Frontotemporal Brain Structure."

28. Isamah et al., "Variability in Frontotemporal Brain Structure."

29. Isamah et al., "Variability in Frontotemporal Brain Structure."

30. Thomas A. Stalnaker, Nisha K. Cooch, and Geoffrey Schoenbaum, "What the Orbitofrontal Cortex Does Not Do," *Nature Neuroscience* 18, no. 5 (May 2015): 620–27, https://doi.org/10.1038/nn.3982.

31. Pierre Sacré et al., "Lucky Rhythms in Orbitofrontal Cortex Bias Gambling Decisions in Humans," *Scientific Reports* 6, no. 1 (December 2016), https://doi.org/10.1038/srep36206.

32. Isamah et al., "Variability in Frontotemporal Brain Structure."

33. Isamah et al., "Variability in Frontotemporal Brain Structure."

34. Martha J. Farah, "The Neuroscience of Socioeconomic Status: Correlates, Causes, and Consequences," *Neuron* 96, no. 1 (September 2017): 56–71, https://doi.org/10.1016/j.neuron.2017.08.034.

35. Isamah et al., "Variability in Frontotemporal Brain Structure."

36. Bai et al., "Population Differences in Brain Morphology and Microstructure among Chinese, Malay, and Indian Neonates."

37. Bai et al., "Population Differences in Brain Morphology and Microstructure among Chinese, Malay, and Indian Neonates."

38. Osvaldo Cairó, "External Measures of Cognition," *Frontiers in Human Neuroscience* 5 (2011), https://doi.org/10.3389/fnhum.2011.00108.

39. Michael McDaniel, "Big-Brained People Are Smarter: A Meta-Analysis of the Relationship between in Vivo Brain Volume and Intelligence," *Intelligence* 33, no. 4 (July 2005): 337–46, https://doi.org/10.1016/j.intell.2004.11.005.

40. McDaniel, "Big-Brained People Are Smarter."

41. McDaniel, "Big-Brained People Are Smarter."

42. McDaniel, "Big-Brained People Are Smarter."

43. Committee on Pesticides in the Diets of Infants and Children and Board on Agriculture and Board on Environmental Studies and Toxicology, *Pesticides in the Diets of Infants and Children* (Washington, DC: National Academies Press, 1993), https://doi.org/10.17226/2126; D. F. Huelke, "An Overview of Anatomical Considerations of Infants and Children in the Adult World of Automobile Safety Design," *Annual Proc Association for the Advancement of Automotive Medicine* 42 (1998): 93–113.

44. Tracey McLellan et al., "Synaptic Pruning," in *Encyclopedia of Child Behavior and Development*, ed. Sam Goldstein and Jack A. Naglieri (Boston: Springer US, 2011), 1,464–65, https://doi.org/10.1007/978-0-387-79061-9_2856.

45. McLellan et al., "Synaptic Pruning"; Joan Stiles and Terry L. Jernigan, "The Basics of Brain Development," *Neuropsychology Review* 20, no. 4 (December 2010): 327–48, https://doi.org/10.1007/s11065-010-9148-4; Gregory Z. Tau and Bradley S. Peterson, "Normal Development of Brain Circuits," *Neuropsychopharmacology: Official Publication of the American College of Neuropsychopharmacology* 35, no. 1 (January 2010): 147–68, https://doi.org/10.1038/npp.2009.115.

46. K. Sukel, "The Synapse: A Primer," *DANA Foundation*, March 15, 2011, accessed December 18, 2018, http://www.dana.org/News/Details.aspx?id=43512.

47. J. P. Shonkoff and D. A. Phillips, eds., with the National Research Council and Institute of Medicine Committee on Integrating the Science of Early Childhood Development, *From Neurons to Neighborhoods: The Science of Early Childhood Development* (Washington, DC: National Academies Press, 2000), 8.

48. Suzana Herculano-Houzel, "The Human Brain in Numbers: A Linearly Scaled-Up Primate Brain," *Frontiers in Human Neuroscience* 3 (2009), https://doi.org/10.3389/neuro .09.031.2009.

49. Nicholas Lange et al., "Associations between IQ, Total and Regional Brain Volumes, and Demography in a Large Normative Sample of Healthy Children and Adolescents," *Developmental Neuropsychology* 35, no. 3 (May 5, 2010): 296–317, https://doi.org/10.1080/ 87565641003696833.

50. Chun Chieh Fan et al., "Modeling the 3-D Geometry of the Cortical Surface with Genetic Ancestry," *Current Biology* 25, no. 15 (August 2015): 1,988–92, https://doi.org/ 10.1016/j.cub.2015.06.006.

51. Fan et al., "Modeling the 3-D Geometry of the Cortical Surface with Genetic Ancestry."

52. Fan et al., "Modeling the 3-D Geometry of the Cortical Surface with Genetic Ancestry."

53. "Neurons That Fire Together Don't Always Wire Together: Neuroscientists Uncover a New Rule of Connectivity of Neurons in the Neocortex," *ScienceDaily*, November 8, 2018, accessed December 18, 2018, https://www.sciencedaily.com/releases/2018/11/ 181108130537.htm.

54. R. Douglas Fields, "Change in the Brain's White Matter," *Science* 330, no. 6,005 (November 5, 2010): 768–69, https://doi.org/10.1126/science.1199139; Robert J. Zatorre, R. Douglas Fields, and Heidi Johansen-Berg, "Plasticity in Gray and White: Neuroimaging Changes in Brain Structure during Learning," *Nature Neuroscience* 15, no. 4 (April 2012): 528–36, https://doi.org/10.1038/nn.3045.

55. Richard E. Brown, "Hebb and Cattell: The Genesis of the Theory of Fluid and Crystallized Intelligence," *Frontiers in Human Neuroscience* 10 (2016): 606, https://doi. org/10.3389/fnhum.2016.00606.

56. Brown, "Hebb and Cattell"; David Suzuki, "Donald Olding Hebb," *Canadian Encyclopedia*, August 1, 2014, accessed December 18, 2018, https://www.thecanadianencyclopedia.ca/en/article/donald-olding-hebb.

57. Mean-Hwan Kim et al., "Segregated Subnetworks of Intracortical Projection Neurons in Primary Visual Cortex," *Neuron* 100, no. 6 (December 2018), https://doi.org/10.1016/ j.neuron.2018.10.023.

58. Kim et al., "Segregated Subnetworks of Intracortical Projection Neurons in Primary Visual Cortex."

59. Cassandra Sampaio-Baptista and Heidi Johansen-Berg, "White Matter Plasticity in the Adult Brain," *Neuron* 96, no. 6 (December 2017): 1,239–51, https://doi.org/10.1016/j. neuron.2017.11.026.

60. Sara L. Bengtsson et al., "Extensive Piano Practicing Has Regionally Specific Effects on White Matter Development," *Nature Neuroscience* 8, no. 9 (September 2005): 1,148–50, https://doi.org/10.1038/nn1516.

61. Jan Scholz et al., "Training Induces Changes in White-Matter Architecture," *Nature Neuroscience* 12, no. 11 (November 2009): 1,370–71, https://doi.org/10.1038/nn.2412.

62. Gianna C. Cannonieri et al., "Practice and Perfect: Length of Training and Structural Brain Changes in Experienced Typists," *Neuroreport* 18, no. 10 (July 2, 2007): 1,063–66, https://doi.org/10.1097/WNR.0b013e3281a030e5; Sampaio-Baptista and Johansen-Berg, "White Matter Plasticity in the Adult Brain."

63. Fields, "Change in the Brain's White Matter"; Jürgen Hänggi et al., "Structural Neuroplasticity in the Sensorimotor Network of Professional Female Ballet Dancers," *Human Brain Mapping* 31, no. 8 (August 2010): 1,196–206, https://doi.org/10.1002/hbm.20928; Sampaio-Baptista and Johansen-Berg, "White Matter Plasticity in the Adult Brain."

64. Sampaio-Baptista and Johansen-Berg, "White Matter Plasticity in the Adult Brain."

65. Sampaio-Baptista and Johansen-Berg, "White Matter Plasticity in the Adult Brain."

66. Scholz et al., "Training Induces Changes in White-Matter Architecture."

67. M. Neiss and D. C. Rowe, "Parental Education and Child's Verbal IQ in Adoptive and Biological Families in the National Longitudinal Study of Adolescent Health," *Behavior Genetics* 30, no. 6 (November 2000): 487–95.

68. Lange et al., "Associations between IQ, Total and Regional Brain Volumes, and Demography in a Large Normative Sample of Healthy Children and Adolescents."

69. Lange et al., "Associations between IQ, Total and Regional Brain Volumes, and Demography in a Large Normative Sample of Healthy Children and Adolescents."

70. Lange et al., "Associations between IQ, Total and Regional Brain Volumes, and Demography in a Large Normative Sample of Healthy Children and Adolescents."

71. Rhoshel K. Lenroot and Jay N. Giedd, "Brain Development in Children and Adolescents: Insights from Anatomical Magnetic Resonance Imaging," *Neuroscience and Biobehavioral Reviews* 30, no. 6 (2006): 718–29, https://doi.org/10.1016/j.neubiorev.2006.06.001; Shimamura, "Humans."

72. Lange et al., "Associations between IQ, Total and Regional Brain Volumes, and Demography in a Large Normative Sample of Healthy Children and Adolescents."

73. Lange et al., "Associations between IQ, Total and Regional Brain Volumes, and Demography in a Large Normative Sample of Healthy Children and Adolescents."

74. Lange et al., "Associations between IQ, Total and Regional Brain Volumes, and Demography in a Large Normative Sample of Healthy Children and Adolescents."

75. Kimberly G. Noble et al., "Family Income, Parental Education, and Brain Structure in Children and Adolescents," *Nature Neuroscience* 18, no. 5 (May 2015): 773–78, https://doi.org/10.1038/nn.3983.

76. Noble et al., "Family Income, Parental Education, and Brain Structure in Children and Adolescents."

77. Noble et al., "Family Income, Parental Education, and Brain Structure in Children and Adolescents."

78. Allyson P. Mackey et al., "Neuroanatomical Correlates of the Income-Achievement Gap," *Psychological Science* 26, no. 6 (June 2015): 925–33, https://doi.org/10.1177/0956797615572233.

79. Mackey et al., "Neuroanatomical Correlates of the Income-Achievement Gap."

80. Mackey et al., "Neuroanatomical Correlates of the Income-Achievement Gap."

81. Brouwer et al., "Genetic Associations between Intelligence and Cortical Thickness Emerge at the Start of Puberty."

82. Lyndsey Layton, "New Brain Science Shows Poor Kids Have Smaller Brains Than Affluent Kids," *Washington Post*, April 15, 2015, accessed December 18, 2018, https://

www.washingtonpost.com/local/education/new-brain-science-shows-poor-kids-have-small
er-brains-than-affluent-kids/2015/04/15/3b679858-e2bc-11e4-b510-962fcfabc310_story
.html?noredirect=on&utm_term=.891cb8b09e20.

83. Christopher Bergland, "Socioeconomic Factors Impact a Child's Brain Structure," *Athlete's Way*, March 31, 2015, accessed January 5, 2019, https://www.psychologytoday.com/us/blog/the-athletes-way/201503/socioeconomic-factors-impact-childs-brain-structure.

84. Noble et al., "Family Income, Parental Education, and Brain Structure in Children and Adolescents."

85. Amy S. Finn et al., "Functional Brain Organization of Working Memory in Adolescents Varies in Relation to Family Income and Academic Achievement," *Developmental Science* 20, no. 5 (September 2017): e12450, https://doi.org/10.1111/desc.12450.

86. Finn et al., "Functional Brain Organization of Working Memory in Adolescents Varies in Relation to Family Income and Academic Achievement."

87. Farah, "The Neuroscience of Socioeconomic Status."

88. Noble et al., "Family Income, Parental Education, and Brain Structure in Children and Adolescents."

89. Nandini Mundkur, "Neuroplasticity in Children," *Indian Journal of Pediatrics* 72, no. 10 (October 2005): 855–57, https://doi.org/10.1007/BF02731115.

90. Jason L. Stein et al., "Identification of Common Variants Associated with Human Hippocampal and Intracranial Volumes," *Nature Genetics* 44 (April 15, 2012): 552.

Chapter 9. The Search for IQ-Linked Genes

1. Robert Plomin and Sophie von Stumm, "The New Genetics of Intelligence," *Nature Reviews Genetics* 19, no. 3 (January 8, 2018): 148–59, https://doi.org/10.1038/nrg.2017.104.

2. Anthony J. F. Griffiths, *An Introduction to Genetic Analysis*, 7th ed. (New York: W. H. Freeman, 2000).

3. Genetics Home Reference, "What Is Heritability?" *National Institutes of Health U.S. National Library of Medicine*, accessed December 18, 2018, https://ghr.nlm.nih.gov/primer/inheritance/heritability; Leonard Lieberman, "How 'Caucasoids' Got Such Big Crania and Why They Shrank: From Morton to Rushton," *Current Anthropology* 42, no. 1 (February 2001): 69–95.

4. Chao-Qiang Lai, "How Much of Human Height Is Genetic and How Much Is Due to Nutrition?" *Scientific American*, December 11, 2006, accessed December 18, 2018, https://www.scientificamerican.com/article/how-much-of-human-height/.

5. Lai, "How Much of Human Height Is Genetic and How Much Is Due to Nutrition?"

6. Lai, "How Much of Human Height Is Genetic and How Much Is Due to Nutrition?"

7. Aline Jelenkovic et al., "Genetic and Environmental Influences on Height from Infancy to Early Adulthood: An Individual-Based Pooled Analysis of 45 Twin Cohorts," *Scientific Reports* 6 (June 23, 2016): 28,496, https://doi.org/10.1038/srep28496.

8. David S. Moore and David Shenk, "The Heritability Fallacy," *Wiley Interdisciplinary Reviews. Cognitive Science* 8, no. 1–2 (January 2017), https://doi.org/10.1002/wcs.1400.

9. Tinca J. C. Polderman et al., "Meta-Analysis of the Heritability of Human Traits Based on 50 Years of Twin Studies," *Nature Genetics* 47, no. 7 (July 2015): 702–9, https://doi.org/10.1038/ng.3285.

10. Genetics Home Reference, "What Is Heritability?"

11. Polderman et al., "Meta-Analysis of the Heritability of Human Traits Based on 50 Years of Twin Studies."

12. Polderman et al., "Meta-Analysis of the Heritability of Human Traits Based on 50 Years of Twin Studies."

13. Polderman et al., "Meta-Analysis of the Heritability of Human Traits Based on 50 Years of Twin Studies."

14. S. Scarr-Salapatek, "Race, Social Class, and IQ," *Science* 174, no. 4,016 (December 24, 1971): 1,285–95.

15. Eric Turkheimer et al., "Socioeconomic Status Modifies Heritability of IQ in Young Children," *Psychological Science* 14, no. 6 (November 2003): 623–28, https://doi.org/10.1046/j.0956-7976.2003.psci_1475.x.

16. Turkheimer et al., "Socioeconomic Status Modifies Heritability of IQ in Young Children."

17. Turkheimer et al., "Socioeconomic Status Modifies Heritability of IQ in Young Children."

18. Bruno Sauce and Louis D. Matzel, "The Paradox of Intelligence: Heritability and Malleability Coexist in Hidden Gene–Environment Interplay," *Psychological Bulletin* 144, no. 1 (January 2018): 26–47, https://doi.org/10.1037/bul0000131.

19. W. D. Hill et al., "A Combined Analysis of Genetically Correlated Traits Identifies 187 Loci and a Role for Neurogenesis and Myelination in Intelligence," *Molecular Psychiatry* 24 (January 11, 2018): 169–81, https://doi.org/10.1038/s41380-017-0001-5; Sauce and Matzel, "The Paradox of Intelligence"; Nicholas Wade, "First Gene to Be Linked with High Intelligence Is Reported Found," *New York Times*, May 14, 1998, accessed December 18, 2018, https://www.nytimes.com/1998/05/14/us/first-gene-to-be-linked-with-high-intelligence-is-reported-found.html.

20. Min Zhao, Lei Kong, and Hong Qu, "A Systems Biology Approach to Identify Intelligence Quotient Score-Related Genomic Regions and Pathways Relevant to Potential Therapeutic Treatments," *Scientific Reports* 4, no. 1 (May 2015), https://doi.org/10.1038/srep04176.

21. Christopher F. Chabris et al., "Most Reported Genetic Associations with General Intelligence Are Probably False Positives," *Psychological Science* 23, no. 11 (November 2012): 1,314–23, https://doi.org/10.1177/0956797611435528; Hill et al., "A Combined Analysis of Genetically Correlated Traits Identifies 187 Loci and a Role for Neurogenesis and Myelination in Intelligence"; Sauce and Matzel, "The Paradox of Intelligence."

22. Genetics Home Reference, "What Are Single Nucleotide Polymorphisms (SNPs)?" *National Institutes of Health U.S. National Library of Medicine*, accessed December 18, 2018, https://ghr.nlm.nih.gov/primer/genomicresearch/snp.

23. Genetics Home Reference, "What Are Single Nucleotide Polymorphisms (SNPs)?"

24. Genetics Home Reference, "What Are Single Nucleotide Polymorphisms (SNPs)?"

25. Colin G. DeYoung et al., "Testing Predictions from Personality Neuroscience. Brain Structure and the Big Five," *Psychological Science* 21, no. 6 (June 2010): 820–28, https://doi.org/10.1177/0956797610370159; R. A. Power and M. Pluess, "Heritability Estimates of the Big Five Personality Traits Based on Common Genetic Variants," *Translational Psychiatry* 5 (July 14, 2015): e604, https://doi.org/10.1038/tp.2015.96.

26. Genetics Home Reference, "What Are Genome-Wide Association Studies?" *National Institutes of Health U.S. National Library of Medicine*, accessed December 18, 2018, https://ghr.nlm.nih.gov/primer/genomicresearch/gwastudies.

27. Paul M. Thompson et al., "Genetics of the Connectome and the ENIGMA Project," in *Micro-, Meso-, and Macro-Connectomics of the Brain*, ed. Henry Kennedy, David C. Van

Essen, and Yves Christen (Cham: Springer, 2016), 147–64, http://www.ncbi.nlm.nih.gov/books/NBK435765/.

28. T. J. Jorgensen et al., "Hypothesis-Driven Candidate Gene Association Studies: Practical Design and Analytical Considerations," *American Journal of Epidemiology* 170, no. 8 (October 15, 2009): 986–93, https://doi.org/10.1093/aje/kwp242; Genetics Home Reference, "What Are Single Nucleotide Polymorphisms (SNPs)?"; Ken Richardson, "GWAS and Cognitive Abilities: Why Correlations Are Inevitable and Meaningless: GWA Studies to Identify Genetic Factors for Educational Achievements Largely Ignore Underlying Social Structures and Dynamics," *EMBO Reports* 18, no. 8 (August 2017): 1,279–83, https://doi.org/10.15252/embr.201744140.

29. G. Davies et al., "Genome-Wide Association Studies Establish That Human Intelligence Is Highly Heritable and Polygenic," *Molecular Psychiatry* 16, no. 10 (October 2011): 996–1,005, https://doi.org/10.1038/mp.2011.85.

30. Hefziba Lifshitz et al., "Crystallized and Fluid Intelligence of University Students with Intellectual Disability Who Are Fully Integrated versus Those Who Studied in Adapted Enrichment Courses," *PloS One* 13, no. 4 (2018): e0193351, https://doi.org/10.1371/journal.pone.0193351.

31. Jason Stein et al., "Identification of Common Variants Associated with Human Hippocampal and Intracranial Volumes," *Nature Genetics* 44, no. 5 (May 2012): 552–61, https://doi.org/10.1038/ng.2250.

32. Thompson et al., "Genetics of the Connectome and the ENIGMA Project."

33. Andy Coghlan, "Best Evidence Yet That a Single Gene Can Affect IQ," *New Scientist* 214, no. 2,861 (April 2012): 14, https://doi.org/10.1016/S0262-4079(12)60994-4; Stein et al., "Identification of Common Variants Associated with Human Hippocampal and Intracranial Volumes."

34. Stein et al., "Identification of Common Variants Associated with Human Hippocampal and Intracranial Volumes."

35. Razib Khan, "Common Variant for 'IQ Gene'? Gene Expression," *Discovermagazine.com* (blog), April 15, 2012, accessed December 18, 2018, http://blogs.discovermagazine.com/gnxp/2012/04/common-variant-for-iq-gene/#.XCL_2C57ksk.

36. Khan, "Common Variant for 'IQ Gene'? Gene Expression."

37. Chabris et al., "Most Reported Genetic Associations with General Intelligence Are Probably False Positives."

38. Eva Krapohl et al., "The High Heritability of Educational Achievement Reflects Many Genetically Influenced Traits, Not Just Intelligence," *Proceedings of the National Academy of Sciences* 111, no. 42 (October 21, 2014): 15,273–78, https://doi.org/10.1073/pnas.1408777111.

39. Joey W. Trampush et al., "Independent Evidence for an Association between General Cognitive Ability and a Genetic Locus for Educational Attainment: Genetic Overlap between Cognition and Education," *American Journal of Medical Genetics Part B: Neuropsychiatric Genetics* 168, no. 5 (July 2015): 363–73, https://doi.org/10.1002/ajmg.b.32319.

40. Trampush et al., "Independent Evidence for an Association between General Cognitive Ability and a Genetic Locus for Educational Attainment."

41. Cornelius A. Rietveld et al., "GWAS of 126,559 Individuals Identifies Genetic Variants Associated with Educational Attainment," *Science* 340, no. 6,139 (June 21, 2013): 1,467–71, https://doi.org/10.1126/science.1235488; Cornelius A. Rietveld et al., "Common

Genetic Variants Associated with Cognitive Performance Identified Using the Proxy-Phenotype Method," *Proceedings of the National Academy of Sciences* 111, no. 38 (September 23, 2014): 13,790–94, https://doi.org/10.1073/pnas.1404623111.

42. Rietveld et al., "Common Genetic Variants Associated with Cognitive Performance Identified Using the Proxy-Phenotype Method."

43. Ewen Callaway, "'Smart Genes' Prove Elusive," *Nature*, September 8, 2014, https://doi.org/10.1038/nature.2014.15858.

44. Rietveld et al., "Common Genetic Variants Associated with Cognitive Performance Identified Using the Proxy-Phenotype Method."

45. Rietveld et al., "Common Genetic Variants Associated with Cognitive Performance Identified Using the Proxy-Phenotype Method."

46. Callaway, "'Smart Genes' Prove Elusive."

47. Sauce and Matzel, "The Paradox of Intelligence."

48. Callaway, "'Smart Genes' Prove Elusive."

49. Chabris et al., "Most Reported Genetic Associations with General Intelligence Are Probably False Positives."

50. Chabris et al., "Most Reported Genetic Associations with General Intelligence Are Probably False Positives."

51. Chabris et al., "Most Reported Genetic Associations with General Intelligence Are Probably False Positives."

52. B. Pourcain et al., "Childhood Intelligence Is Heritable, Highly Polygenic, and Associated with FNBP1L," *Molecular Psychiatry* 19, no. 2 (February 2014): 253–58, https://doi.org/10.1038/mp.2012.184.

53. Pourcain et al., "Childhood Intelligence Is Heritable, Highly Polygenic, and Associated with FNBP1L."

54. Pourcain et al., "Childhood Intelligence Is Heritable, Highly Polygenic, and Associated with FNBP1L."

55. Suzanne Sniekers et al., "Genome-Wide Association Meta-Analysis of 78,308 Individuals Identifies New Loci and Genes Influencing Human Intelligence," *Nature Genetics* 49, no. 7 (July 2017): 1,107–12, https://doi.org/10.1038/ng.3869.

56. Laura Sanders, "40 More 'Intelligence' Genes Found," *Science News*, May 22, 2017, accessed December 18, 2018, https://www.sciencenews.org/article/40-more-intelligence-genes-found; Sniekers et al., "Genome-Wide Association Meta-Analysis of 78,308 Individuals Identifies New Loci and Genes Influencing Human Intelligence."

57. Sniekers et al., "Genome-Wide Association Meta-Analysis of 78,308 Individuals Identifies New Loci and Genes Influencing Human Intelligence."

58. Eva Krapohl et al., "Widespread Covariation of Early Environmental Exposures and Trait-Associated Polygenic Variation," *Proceedings of the National Academy of Sciences* 114, no. 44 (October 31, 2017): 11,727–32, https://doi.org/10.1073/pnas.1707178114.

59. Davide Piffer, "Estimating the Genotypic Intelligence of Populations and Assessing the Impact of Socioeconomic Factors and Migrations," *Winnower*, 2015, accessed December 18, 2018, https://doi.org/10.15200/winn.142299.93508.

60. Piffer, "Estimating the Genotypic Intelligence of Populations and Assessing the Impact of Socioeconomic Factors and Migrations."

61. Richard Lynn and Tatu Vanhanen, *Intelligence: A Unifying Construct for the Social Sciences* (London: Ulster Institute for Social Research, 2012); Piffer, "Estimating the Genotypic Intelligence of Populations and Assessing the Impact of Socioeconomic Factors and Migrations."

62. Piffer, "Estimating the Genotypic Intelligence of Populations and Assessing the Impact of Socioeconomic Factors and Migrations."

63. Davide Piffer, "A Review of Intelligence GWAS Hits: Their Relationship to Country IQ and the Issue of Spatial Autocorrelation," *Intelligence* 53 (November 2015): 43–50, https://doi.org/10.1016/j.intell.2015.08.008.

64. James Thompson, "Comments on Piffer from Prof Posthuma," *Unz Review*, June 7, 2017, accessed December 18, 2018, http://www.unz.com/jthompson/comments-on-piffer -from-prof-posthuma/.

65. Thompson, "Comments on Piffer from Prof Posthuma."

66. Plomin and von Stumm, "The New Genetics of Intelligence."

67. I. Sample, "Scientists Identify 40 Genes That Shed New Light on Biology of Intelligence," *Guardian*, May 22, 2017, accessed December 18, 2018, https://www.theguardian .com/science/2017/may/22/scientists-uncover-40-genes-iq-einstein-genius.

68. Davide Piffer, "Evidence for Recent Polygenic Selection on Educational Attainment and Intelligence Inferred from GWAS Hits: A Replication of Previous Findings Using Recent Data," *Preprints*, accessed December 18, 2018, https://www.preprints.org/manu script/201706.0039/v1.

69. Krapohl et al., "The High Heritability of Educational Achievement Reflects Many Genetically Influenced Traits, Not Just Intelligence."

70. Krapohl et al., "The High Heritability of Educational Achievement Reflects Many Genetically Influenced Traits, Not Just Intelligence."

71. Sarah C. P. Williams, "Genes Don't Just Influence Your IQ—They Determine How Well You Do in School," *Science*, October 6, 2014, accessed December 18, 2018, https:// www.sciencemag.org/news/2014/10/genes-dont-just-influence-your-iq-they-determine -how-well-you-do-school.

72. Williams, "Genes Don't Just Influence Your IQ."

73. Ken B. Hanscombe et al., "Socioeconomic Status (SES) and Children's Intelligence (IQ): In a UK-Representative Sample SES Moderates the Environmental, Not Genetic, Effect on IQ," ed. James G. Scott, *PLoS ONE* 7, no. 2 (February 1, 2012): e30320, https:// doi.org/10.1371/journal.pone.0030320; Turkheimer et al., "Socioeconomic Status Modifies Heritability of IQ in Young Children."

74. Krapohl et al., "The High Heritability of Educational Achievement Reflects Many Genetically Influenced Traits, Not Just Intelligence"; Frank Pajares, "Self-Efficacy Beliefs in Academic Settings," *Review of Educational Research* 66, no. 4 (December 1996): 543–78, https://doi.org/10.3102/00346543066004543.

75. Pajares, "Self-Efficacy Beliefs in Academic Settings."

76. Robert Plomin and Ian J. Deary, "Genetics and Intelligence Differences: Five Special Findings," *Molecular Psychiatry* 20, no. 1 (February 2015): 98–108, https://doi.org/10.1038/ mp.2014.105.

77. Eva Krapohl and Robert Plomin, "Genetic Link between Family Socioeconomic Status and Children's Educational Achievement Estimated from Genome-Wide SNPs," *Molecular Psychiatry* 21, no. 3 (March 2015):437–43, http://doi:10.1038/mp.2015.2.

78. Krapohl and Plomin, "Genetic Link between Family Socioeconomic Status and Children's Educational Achievement Estimated from Genome-Wide SNPs."

79. Maciej Trzaskowski et al., "Genetic Influence on Family Socioeconomic Status and Children's Intelligence," *Intelligence* 42 (January 2014): 83–88, https://doi.org/10.1016/j.intell.2013.11.002.

80. Trzaskowski et al., "Genetic Influence on Family Socioeconomic Status and Children's Intelligence."

81. Trzaskowski et al., "Genetic Influence on Family Socioeconomic Status and Children's Intelligence."

82. Hanscombe et al., "Socioeconomic Status (SES) and Children's Intelligence (IQ)."

83. Hanscombe et al., "Socioeconomic Status (SES) and Children's Intelligence (IQ)."

84. Hanscombe et al., "Socioeconomic Status (SES) and Children's Intelligence (IQ)."

85. Hanscombe et al., "Socioeconomic Status (SES) and Children's Intelligence (IQ)."

86. Hanscombe et al., "Socioeconomic Status (SES) and Children's Intelligence (IQ)"; Trzaskowski et al., "Genetic Influence on Family Socioeconomic Status and Children's Intelligence"; Krapohl et al., "The High Heritability of Educational Achievement Reflects Many Genetically Influenced Traits, Not Just Intelligence."

87. Hanscombe et al., "Socioeconomic Status (SES) and Children's Intelligence (IQ)"; Krapohl et al., "The High Heritability of Educational Achievement Reflects Many Genetically Influenced Traits, Not Just Intelligence"; Trzaskowski et al., "Genetic Influence on Family Socioeconomic Status and Children's Intelligence."

88. Natalie H. Brito, Luciane R. Piccolo, and Kimberly G. Noble, "Associations between Cortical Thickness and Neurocognitive Skills during Childhood Vary by Family Socioeconomic Factors," *Brain and Cognition* 116 (August 2017): 54–62, https://doi.org/10.1016/j.bandc.2017.03.007.

89. Kimberly Noble et al., "Family Income, Parental Education, and Brain Structure in Children and Adolescents," *Nature Neuroscience* 18, no. 5 (May 2015): 773–78, https://doi.org/10.1038/nn.3983.

90. Brito, Piccolo, and Noble, "Associations between Cortical Thickness and Neurocognitive Skills during Childhood Vary by Family Socioeconomic Factors."

91. Noble et al., "Family Income, Parental Education, and Brain Structure in Children and Adolescents."

92. Noble et al., "Family Income, Parental Education, and Brain Structure in Children and Adolescents."

93. Noble et al., "Family Income, Parental Education, and Brain Structure in Children and Adolescents."

94. Rommy von Bernhardi, Laura Eugenín-von Bernhardi, and Jaime Eugenín, "What Is Neural Plasticity?" *Advances in Experimental Medicine and Biology* 1,015 (2017): 1–15, https://doi.org/10.1007/978-3-319-62817-2_1; Ole P. Ottersen, "How Hardwired Is the Brain? Technological Advances Provide New Insight into Brain Malleability and Neurotransmission," *Nutrition Reviews* 68 suppl 2 (December 2010): S60-64, https://doi.org/10.1111/j.1753-4887.2010.00350.x.

95. Sunyoung Pak, "The Growth Status of North Korean Refugee Children and Adolescents from Six to 19 Years of Age," *Economics and Human Biology* 8, no. 3 (December 2010): 385–95, https://doi.org/10.1016/j.ehb.2010.05.006.

96. Barbara Demick, "A Small Problem Growing," *Los Angeles Times*, February 12, 2004, accessed December 18, 2018, http://articles.latimes.com/2004/feb/12/world/fg-short12/2.

97. Claire Bindschaedler et al., "Growing up with Bilateral Hippocampal Atrophy: From Childhood to Teenage," *Cortex* 47, no. 8 (September 2011): 931–44, https://doi.org/10.1016/j.cortex.2010.09.005.

98. Bindschaedler et al., "Growing up with Bilateral Hippocampal Atrophy."

99. Neuroskeptic, "The Limits of Neuroplasticity," *Discover Magazine* (blog), November 13, 2010, accessed December 18, 2018, http://blogs.discovermagazine.com/neuroskeptic/2010/11/13/the-limits-of-neuroplasticity/#.XC8CTi57ksk.

100. Bindschaedler et al., "Growing up with Bilateral Hippocampal Atrophy."

101. E. A. Maguire, R. S. Frackowiak, and C. D. Frith, "Recalling Routes around London: Activation of the Right Hippocampus in Taxi Drivers," *Journal of Neuroscience: The Official Journal of the Society for Neuroscience* 17, no. 18 (September 15, 1997): 7,103–10.

102. Maguire, Frackowiak, and Frith, "Recalling Routes around London.".

103. E. A. Maguire et al., "Navigation-Related Structural Change in the Hippocampi of Taxi Drivers," *Proceedings of the National Academy of Sciences* 97, no. 8 (April 11, 2000): 4,398–403, https://doi.org/10.1073/pnas.070039597.

104. Maguire et al., "Navigation-Related Structural Change in the Hippocampi of Taxi Drivers."

105. Maguire et al., "Navigation-Related Structural Change in the Hippocampi of Taxi Drivers."

106. Maguire et al., "Navigation-Related Structural Change in the Hippocampi of Taxi Drivers."

107. Maguire et al., "Navigation-Related Structural Change in the Hippocampi of Taxi Drivers."

108. Katherine Woollett and Eleanor A. Maguire, "Acquiring 'the Knowledge' of London's Layout Drives Structural Brain Changes," *Current Biology* 21, no. 24 (December 2011): 2109–14, https://doi.org/10.1016/j.cub.2011.11.018.

109. Stein et al., "Identification of Common Variants Associated with Human Hippocampal and Intracranial Volumes."

110. Jamie L. Hanson et al., "Association between Income and the Hippocampus," *PloS One* 6, no. 5 (May 4, 2011): e18712, https://doi.org/10.1371/journal.pone.0018712.

111. Hanson et al., "Association between Income and the Hippocampus"; Hengyi Rao et al., "Early Parental Care Is Important for Hippocampal Maturation: Evidence from Brain Morphology in Humans," *NeuroImage* 49, no. 1 (January 1, 2010): 1,144–50, https://doi.org/10.1016/j.neuroimage.2009.07.003.

112. Ottersen, "How Hardwired Is the Brain?"

113. Elizabeth Huber et al., "Rapid and Widespread White Matter Plasticity during an Intensive Reading Intervention," *Nature Communications* 9, no. 1 (December 2018), https://doi.org/10.1038/s41467-018-04627-5.

114. Huber et al., "Rapid and Widespread White Matter Plasticity during an Intensive Reading Intervention."

115. Huber et al., "Rapid and Widespread White Matter Plasticity during an Intensive Reading Intervention."

116. "How Instruction Changes Brain Circuitry with Struggling Readers," *ScienceDaily*, June 14, 2018, accessed November 11, 2018, www.sciencedaily.com/releases/2018/06/180614213556.htm.

117. Bilian Jin et al., "Linking DNA Methyltransferases to Epigenetic Marks and Nucleosome Structure Genome-Wide in Human Tumor Cells," *Cell Reports* 2, no. 5 (November 2012): 1,411–24, https://doi.org/10.1016/j.celrep.2012.10.017.

118. A. Córdova-Palomera et al., "Polymorphic Variation in the Epigenetic Gene DNMT3B Modulates the Environmental Impact on Cognitive Ability: A Twin Study," *European Psychiatry: Journal of the Association of European Psychiatrists* 30, no. 2 (February 2015): 303–8, https://doi.org/10.1016/j.eurpsy.2014.11.011; Paul Haggarty et al., "Human Intelligence and Polymorphisms in the DNA Methyltransferase Genes Involved in Epigenetic Marking," *PLoS ONE* 5, no. 6 (June 25, 2010): e11329, https://doi.org/10.1371/journal.pone.0011329.

119. Córdova-Palomera et al., "Polymorphic Variation in the Epigenetic Gene DNMT3B Modulates the Environmental Impact on Cognitive Ability"; Haggarty et al., "Human Intelligence and Polymorphisms in the DNA Methyltransferase Genes Involved in Epigenetic Marking."

120. Córdova-Palomera et al., "Polymorphic Variation in the Epigenetic Gene DNMT3B Modulates the Environmental Impact on Cognitive Ability."

121. Córdova-Palomera et al., "Polymorphic Variation in the Epigenetic Gene DNMT3B Modulates the Environmental Impact on Cognitive Ability."

122. Haggarty et al., "Human Intelligence and Polymorphisms in the DNA Methyltransferase Genes Involved in Epigenetic Marking."

123. Haggarty et al., "Human Intelligence and Polymorphisms in the DNA Methyltransferase Genes Involved in Epigenetic Marking."

124. Rodney R. Dietert and Amrie C. Grammer, "Immune Disorders, Epigenetics, and the Developmental Origins of Health and Disease," in *The Epigenome and Developmental Origins of Health and Disease*, ed. Cheryl S. Rosenfeld (Amsterdam: Elsevier, 2016), 211–34, https://doi.org/10.1016/B978-0-12-801383-0.00012-8.

125. Santiago Sandoval-Motta et al., "The Human Microbiome and the Missing Heritability Problem," *Frontiers in Genetics* 8 (2017): 80, https://doi.org/10.3389/fgene.2017.00080.

Chapter 10. Smart Genes, Big Brains, and Civilization

1. L. Loewe and W. G. Hill, "The Population Genetics of Mutations: Good, Bad, and Indifferent," *Philosophical Transactions of the Royal Society B: Biological Sciences* 365, no. 1,544 (April 27, 2010): 1,153–67, https://doi.org/10.1098/rstb.2009.0317.

2. Avshalom Caspi et al., "Moderation of Breastfeeding Effects on the IQ by Genetic Variation in Fatty Acid Metabolism," *Proceedings of the National Academy of Sciences* 104, no. 47 (November 20, 2007): 18,860–65, https://doi.org/10.1073/pnas.0704292104; Colin D. Steer et al., "FADS2 Polymorphisms Modify the Effect of Breastfeeding on Child IQ," *PLoS ONE* 5, no. 7 (July 13, 2010): e11570, https://doi.org/10.1371/journal.pone.0011570.

3. Cornelius A. Rietveld et al., "Common Genetic Variants Associated with Cognitive Performance Identified Using the Proxy-Phenotype Method," *Proceedings of the National Academy of Sciences* 111, no. 38 (September 23, 2014): 13,790–94, https://doi.org/10.1073/pnas.1404623111.

4. Christopher F. Chabris et al., "Most Reported Genetic Associations with General Intelligence Are Probably False Positives," *Psychological Science* 23, no. 11 (November 2012): 1,314–23, https://doi.org/10.1177/0956797611435528.

5. W. D. Hill et al., "A Combined Analysis of Genetically Correlated Traits Identifies 187 Loci and a Role for Neurogenesis and Myelination in Intelligence," *Molecular Psychiatry* 24 (January 11, 2018): 169–81, https://doi.org/10.1038/s41380-017-0001-5.

6. Genetics Home Reference, "APSM Gene," *National Institutes of Health U.S. National Library of Medicine*, accessed November 13, 2018, https://ghr.nlm.nih.gov/gene/ASPM.

7. Nitzan Mekel-Bobrov et al., "The Ongoing Adaptive Evolution of ASPM and Microcephalin Is Not Explained by Increased Intelligence," *Human Molecular Genetics* 16, no. 6 (March 15, 2007): 600–608, https://doi.org/10.1093/hmg/ddl487.

8. Michael Balter, "Evolution. Are Human Brains Still Evolving? Brain Genes Show Signs of Selection," *Science* 309, no. 5,741 (September 9, 2005): 1,662–63, https://doi.org/10.1126/science.309.5741.1662; P. D. Evans, "Microcephalin, a Gene Regulating Brain Size, Continues to Evolve Adaptively in Humans," *Science* 309, no. 5,741 (September 9, 2005): 1,717–20, https://doi.org/10.1126/science.1113722.

9. Balter, "Evolution. Are Human Brains Still Evolving?"; Michael Balter, "Brain Man Makes Waves with Claims of Recent Human Evolution," *Science* 314, no. 5,807 (December 22, 2006): 1,871–73, https://doi.org/10.1126/science.314.5807.1871.

10. Alondra Yvette Oubré, "The Adaptionist Yardstick: Rethinking the Social Implications of Sarich's and Miele's Fast-Track Micro-Evolution," *Skeptic*, February 18, 2005, accessed December 18, 2018, https://www.skeptic.com/eskeptic/05-02-18/; Vincent Sarich, "Race," paper presented the eighth annual meeting of the Human Behavior and Evolution Society (HBES), Northwestern University, Evanston, Illinois, 1995.

11. Balter, "Evolution. Are Human Brains Still Evolving?"; Evans, "Microcephalin, a Gene Regulating Brain Size, Continues to Evolve Adaptively in Humans."

12. Evans, "Microcephalin, a Gene Regulating Brain Size, Continues to Evolve Adaptively in Humans."

13. Evans, "Microcephalin, a Gene Regulating Brain Size, Continues to Evolve Adaptively in Humans."

14. Balter, "Evolution. Are Human Brains Still Evolving?"; Balter, "Brain Man Makes Waves with Claims of Recent Human Evolution."

15. J. P. Rushton, P. A. Vernon, and T. A. Bons, "No Evidence That Polymorphisms of Brain Regulator Genes Microcephalin and ASPM Are Associated with General Mental Ability, Head Circumference, or Altruism," *Biology Letters* 3, no. 2 (April 22, 2007): 157–60, https://doi.org/10.1098/rsbl.2006.0586.

16. Antonio Regalado, "Scientist's Study of Brain Genes Sparks a Backlash; Dr. Lahn Connects Evolution in Some Groups to IQ; Debate on Race and DNA; 'Speculating Is Dangerous,'" *Wall Street Journal*, June 16, 2006, A1, A12.

17. Alondra Yvette Oubré, *Race, Genes, and Ability: Rethinking Ethnic Differences*, vol. 2 (Woodland Hills, CA: BTI Press, 2011); Regalado, "Scientist's Study of Brain Genes Sparks a Backlash."

18. Evans, "Microcephalin, a Gene Regulating Brain Size, Continues to Evolve Adaptively in Humans."

19. Roger P. Woods et al., "Normal Variants of Microcephalin and ASPM Do Not Account for Brain Size Variability," *Human Molecular Genetics* 15, no. 12 (June 15, 2006): 2,025–29, https://doi.org/10.1093/hmg/ddl126.

20. S. Mitaki et al., "Impact of Five SNPs in Dopamine-Related Genes on Executive Function," *Acta Neurologica Scandinavica* 127, no. 1 (January 2013): 70–76, https://doi.org/10.1111/j.1600-0404.2012.01673.x.

21. D. Dediu and D. R. Ladd, "Linguistic Tone Is Related to the Population Frequency of the Adaptive Haplogroups of Two Brain Size Genes, ASPM and Microcephalin," *Proceedings of the National Academy of Sciences* 104, no. 26 (June 26, 2007): 10,944–49, https://doi.org/10.1073/pnas.0610848104.

22. Patrick C. M. Wong, Bharath Chandrasekaran, and Jing Zheng, "The Derived Allele of ASPM Is Associated with Lexical Tone Perception," *PLoS ONE* 7, no. 4 (April 17, 2012): e34243, https://doi.org/10.1371/journal.pone.0034243.

23. Wong, Chandrasekaran, and Zheng, "The Derived Allele of ASPM Is Associated with Lexical Tone Perception."

24. Dediu and Ladd, "Linguistic Tone Is Related to the Population Frequency of the Adaptive Haplogroups of Two Brain Size Genes, ASPM and Microcephalin"; Wong, Chandrasekaran, and Zheng, "The Derived Allele of ASPM Is Associated with Lexical Tone Perception."

25. Wong, Chandrasekaran, and Zheng, "The Derived Allele of ASPM Is Associated with Lexical Tone Perception."

26. Joshua J. Buchman, Omer Durak, and Li-Huei Tsai, "ASPM Regulates Wnt Signaling Pathway Activity in the Developing Brain," *Genes and Development* 25, no. 18 (September 15, 2011): 1,909–14, https://doi.org/10.1101/gad.16830211.

27. Mekel-Bobrov et al., "The Ongoing Adaptive Evolution of ASPM and Microcephalin Is Not Explained by Increased Intelligence."

28. Michael A. Woodley et al., "The Relationship between Microcephalin, ASPM, and Intelligence: A Reconsideration," *Intelligence* 44 (May 2014): 51–63, https://doi.org/10.1016/j.intell.2014.02.011.

29. Woodley et al., "The Relationship between Microcephalin, ASPM, and Intelligence: A Reconsideration."

30. Ben van der Merwe, "It Might Be a Pseudoscience, but Students Take the Threat of Eugenics Seriously," *NewStatesmanAmerica*, February 19, 2018, accessed December 18, 2018, https://www.newstatesman.com/politics/education/2018/02/it-might-be-pseudo-science-students-take-threat-eugenics-seriously.

31. Woodley et al., "The Relationship between Microcephalin, ASPM, and Intelligence."

32. S. Pinker, "The Cognitive Niche: Coevolution of Intelligence, Sociality, and Language," *Proceedings of the National Academy of Sciences* 107, no. 2 suppl. (May 11, 2010): 8,993–99, https://doi.org/10.1073/pnas.0914630107; Woodley et al., "The Relationship between Microcephalin, ASPM, and Intelligence."

33. C. Eppig, C. L. Fincher, and R. Thornhill, "Parasite Prevalence and the Worldwide Distribution of Cognitive Ability," *Proceedings of the Royal Society B: Biological Sciences* 277, no. 1,701 (December 22, 2010): 3,801–8, https://doi.org/10.1098/rspb.2010.0973.

34. Hill et al., "A Combined Analysis of Genetically Correlated Traits Identifies 187 Loci and a Role for Neurogenesis and Myelination in Intelligence"; Sample, "Scientists Identify 40 Genes That Shed New Light on Biology of Intelligence."

35. Andrew A. Crawford et al., "Systematic Review and Meta-Analysis of Serotonin Transporter Genotype and Discontinuation from Antidepressant Treatment," *European Neuropsychopharmacology* 23, no. 10 (October 2013): 1,143–50, https://doi.org/10.1016/j.euroneuro.2012.12.001.

36. Regalado, "Scientist's Study of Brain Genes Sparks a Backlash."

37. Razib Khan, "Common Variant for 'IQ Gene'?" *Discover Magazine* (blog), April 15, 2012, accessed December 18, 2018, http://blogs.discovermagazine.com/gnxp/2012/04/common-variant-for-iq-gene/.

38. "Intelligence Gene Identified," *ScienceDaily*, April 27, 2006, accessed December 18, 2018, https://www.sciencedaily.com/releases/2006/04/060427161424.htm; Nicholas Wade, "First Gene to Be Linked with High Intelligence Is Reported Found," *New York Times*, May 14, 1998, accessed December 18, 2018, https://www.nytimes.com/1998/05/14/us/first-gene-to-be-linked-with-high-intelligence-is-reported-found.html.

39. Hill et al., "A Combined Analysis of Genetically Correlated Traits Identifies 187 Loci and a Role for Neurogenesis and Myelination in Intelligence."

40. Woodley et al., "The Relationship between Microcephalin, ASPM and Intelligence."

41. Hill et al., "A Combined Analysis of Genetically Correlated Traits Identifies 187 Loci and a Role for Neurogenesis and Myelination in Intelligence."

Chapter 11. Mean Genes: The (Un)Holy Grail of Race-Realism

1. "QuickStats: Age-Adjusted Rates for Homicides, by Race/Ethnicity—United States, 1999–2015," *MMWR: Morbidity and Mortality Weekly Report* 66, no. 31 (2017): 839, https://doi.org/10.15585/mmwr.mm6631a9.

2. "FBI Releases 2016 Report on Crime in the United States," *U.S. Department of Justice*, September 25, 2017, accessed December 18, 2018, https://www.justice.gov/opa/pr/fbi-releases-2016-report-crime-united-states.

3. "FBI Releases 2016 Report on Crime in the United States."

4. Jesse Singal, "The White and Black Murder Rates Are Both Rising But for Different Reasons," *Intelligencer*, January 1, 2018, accessed December 18, 2018, http://nymag.com/intelligencer/2018/01/the-white-and-black-murder-rates-are-both-rising.html.

5. Singal, "The White and Black Murder Rates Are Both Rising But for Different Reasons."

6. Amy L. Byrd and Stephen B. Manuck, "MAOA, Childhood Maltreatment, and Antisocial Behavior: Meta-Analysis of a Gene–Environment Interaction," *Biological Psychiatry* 75, no. 1 (January 2014): 9–17, https://doi.org/10.1016/j.biopsych.2013.05.004.

7. Kim Veroude et al., "Genetics of Aggressive Behavior: An Overview," *American Journal of Medical Genetics Part B: Neuropsychiatric Genetics* 171, no. 1 (January 2016): 3–43, https://doi.org/10.1002/ajmg.b.32364.

8. Byrd and Manuck, "MAOA, Childhood Maltreatment, and Antisocial Behavior."

9. Catrina M. Schilling, Anthony Walsh, and Ilhong Yun, "ADHD and Criminality: A Primer on the Genetic, Neurobiological, Evolutionary, and Treatment Literature for Criminologists," *Journal of Criminal Justice* 39, no. 1 (January 2011): 3–11, https://doi.org/10.1016/j.jcrimjus.2010.11.001.

10. Chittaranjan Andrade and N. Sanjay Kumar Rao, "How Antidepressant Drugs Act: A Primer on Neuroplasticity as the Eventual Mediator of Antidepressant Efficacy," *Indian Journal of Psychiatry* 52, no. 4 (October 2010): 378–86, https://doi.org/10.4103/0019-5545.74318; Dongju Seo, Christopher J. Patrick, and Patrick J. Kennealy, "Role of Serotonin

and Dopamine System Interactions in the Neurobiology of Impulsive Aggression and Its Comorbidity with Other Clinical Disorders," *Aggression and Violent Behavior* 13, no. 5 (October 2008): 383–95, https://doi.org/10.1016/j.avb.2008.06.003; Larry R. Squire, *Fundamental Neuroscience*, 3rd ed. (Amsterdam and Boston: Elsevier/Academic Press, 2008).

11. Seo, Patrick, and Kennealy, "Role of Serotonin and Dopamine System Interactions in the Neurobiology of Impulsive Aggression and Its Comorbidity with Other Clinical Disorders."

12. Philip Kitcher, *The Lives to Come: The Genetic Revolution and Human Possibilities* (New York: Simon & Schuster, 1996); Squire, *Fundamental Neuroscience*.

13. J. Tiihonen et al., "Genetic Background of Extreme Violent Behavior," *Molecular Psychiatry* 20, no. 6 (June 2015): 786–92, https://doi.org/10.1038/mp.2014.130.

14. Guang Guo et al., "The Dopamine Transporter Gene, a Spectrum of Most Common Risky Behaviors, and the Legal Status of the Behaviors," *PLoS ONE* 5, no. 2 (February 22, 2010): e9352, https://doi.org/10.1371/journal.pone.0009352; F. E. Lotrich, B. G. Pollock, and R. E. Ferrell, "Polymorphism of the Serotonin Transporter: Implications for the Use of Selective Serotonin Reuptake Inhibitors," *American Journal of Pharmacogenomics: Genomics–Related Research in Drug Development and Clinical Practice* 1, no. 3 (2001): 153–64; Peter McGuffin, Shaza Alsabban, and Rudolf Uher, "The Truth about Genetic Variation in the Serotonin Transporter Gene and Response to Stress and Medication," *British Journal of Psychiatry* 198, no. 06 (June 2011): 424–27, https://doi.org/10.1192/bjp.bp.110.085225.

15. Tiihonen et al., "Genetic Background of Extreme Violent Behavior."

16. Avshalom Caspi et al., "Role of Genotype in the Cycle of Violence in Maltreated Children," *Science* 297, no. 5,582 (August 2, 2002): 851–54, https://doi.org/10.1126/science.1072290; Mary-Ane Enoch, "Genetic and Environmental Influences on the Development of Alcoholism: Resilience vs. Risk," *Annals of the New York Academy of Sciences* 1,094 (December 2006): 193–201, https://doi.org/10.1196/annals.1376.019; Camilla Gokturk et al., "Serotonin Transporter (5-HTTLPR) and Monoamine Oxidase (MAOA) Promoter Polymorphisms in Women with Severe Alcoholism," *Archives of Women's Mental Health* 11, no. 5–6 (December 2008): 347–55, https://doi.org/10.1007/s00737-008-0033-6.

17. Rod Lea and Geoff Chambers, "Monoamine Oxidase, Addiction, and the 'Warrior' Gene Hypothesis," *New Zealand Medical Journal* 120, no. 1,250 (2007): U2441, https://www.ncbi.nlm.nih.gov/pubmed/17339897; J. Kim-Cohen et al., "MAOA, Maltreatment, and Gene–Environment Interaction Predicting Children's Mental Health: New Evidence and a Meta-Analysis," *Molecular Psychiatry* 11, no. 10 (October 2006): 903–13, https://doi.org/10.1038/sj.mp.4001851.

18. Oscar Arias-Carrión and Ernst Pöppel, "Dopamine, Learning, and Reward-Seeking Behavior," *Acta Neurobiologiae Experimentalis* 67, no. 4 (2007): 481–88; Erik B. Oleson and Jonté B. Roberts, "The Power of Price Compels You: Behavioral Economic Insights into Dopamine-Based Valuation of Rewarding and Aversively Motivated Behavior," *Brain Research* 1,713 (December 10, 2018): 32–41, https://doi.org/10.1016/j.brainres.2018.11.043.

19. "Attention Deficit Hyperactivity Disorder (ADHD)," *Centers for Disease Control/National Center for Health Statistics*, May 3, 2017, accessed December 18, 2018, https://www.cdc.gov/nchs/fastats/adhd.htm.

20. K. Blum, "Dopamine Genetics and Function in Food and Substance Abuse," *Journal of Genetic Syndromes and Gene Therapy* 4, no. 1 (2013), https://doi.org/10.4172/2157-7412.1000121; Francesca Ducci and David Goldman, "The Genetic Basis of Addictive

Disorders," *Psychiatric Clinics of North America* 35, no. 2 (June 2012): 495–519, https://doi.org/10.1016/j.psc.2012.03.010.

21. Richard P. Ebstein and R. H. Belmaker, "Saga of an Adventure Gene: Novelty-Seeking, Substance Abuse, and the Dopamine D4 Receptor (D4DR) Exon III Repeat Polymorphism," *Molecular Psychiatry* 2, no. 5 (September 1997): 381–84, https://doi.org/10.1038/sj.mp.4000315; Richard P. Ebstein et al., "Dopamine D4 Receptor (D4DR) Exon III Polymorphism Associated with the Human Personality Trait of Novelty-Seeking," *Nature Genetics* 12, no. 1 (January 1996): 78–80, https://doi.org/10.1038/ng0196-78.

22. Taylor Wingo et al., "Novelty-Seeking and Drug Addiction in Humans and Animals: From Behavior to Molecules," *Journal of Neuroimmune Pharmacology: The Official Journal of the Society on Neuroimmune Pharmacology* 11, no. 3 (2016): 456–70, https://doi.org/10.1007/s11481-015-9636-7.

23. Ebstein et al., "Dopamine D4 Receptor (D4DR) Exon III Polymorphism Associated with the Human Personality Trait of Novelty-Seeking."

24. Schilling, Walsh, and Yun, "ADHD and Criminality."

25. Jolanta Chmielowiec et al., "Associations between the Dopamine D4 Receptor and DAT1 Dopamine Transporter Genes Polymorphisms and Personality Traits in Addicted Patients," *International Journal of Environmental Research and Public Health* 15, no. 10 (September 21, 2018), https://doi.org/10.3390/ijerph15102076.

26. Radek Ptácek, Hana Kuzelová, and George B. Stefano, "Dopamine D4 Receptor Gene DRD4 and Its Association with Psychiatric Disorders," *Medical Science Monitor: International Medical Journal of Experimental and Clinical Research* 17, no. 9 (September 2011): RA215–220.

27. Dan T. A. Eisenberg et al., "Dopamine Receptor Genetic Polymorphisms and Body Composition in Undernourished Pastoralists: An Exploration of Nutrition Indices among Nomadic and Recently Settled Ariaal Men of Northern Kenya," *BMC Evolutionary Biology* 8, no. 1 (2008): 173, https://doi.org/10.1186/1471-2148-8-173.

28. Eisenberg et al., "Dopamine Receptor Genetic Polymorphisms and Body Composition in Undernourished Pastoralists."

29. Eisenberg et al., "Dopamine Receptor Genetic Polymorphisms and Body Composition in Undernourished Pastoralists."

30. Marina L. Butovskaya et al., "Aggression and Polymorphisms in AR, DAT1, DRD2, and COMT Genes in Datoga Pastoralists of Tanzania," *Scientific Reports* 3, no. 1 (December 2013), https://doi.org/10.1038/srep03148.

31. Butovskaya et al., "Aggression and Polymorphisms in AR, DAT1, DRD2, and COMT Genes in Datoga Pastoralists of Tanzania"; Michael G. Vaughn et al., "DAT1 and 5HTT Are Associated with Pathological Criminal Behavior in a Nationally Representative Sample of Youth," *Criminal Justice and Behavior* 36, no. 11 (November 2009): 1,113–24, https://doi.org/10.1177/0093854809342839.

32. Xavier Caldú et al., "Impact of the COMT Val108/158 Met and DAT Genotypes on Prefrontal Function in Healthy Subjects," *NeuroImage* 37, no. 4 (October 1, 2007): 1,437–44, https://doi.org/10.1016/j.neuroimage.2007.06.021; Samuel R. Chamberlain and Jon E. Grant, "Selective Role for the COMT Polymorphism in a Trans-Diagnostic Compulsivity Phenotype," *Psychological Medicine* 48, no. 15 (November 2018): 2,626–27, https://doi.org/10.1017/S0033291718002180; Annelise Júlio-Costa et al., "Count on Dopamine: Influences

of COMT Polymorphisms on Numerical Cognition," *Frontiers in Psychology* 4 (2013): 531, https://doi.org/10.3389/fpsyg.2013.00531.

33. Butovskaya et al., "Aggression and Polymorphisms in AR, DAT1, DRD2, and COMT Genes in Datoga Pastoralists of Tanzania."

34. Butovskaya et al., "Aggression and Polymorphisms in AR, DAT1, DRD2, and COMT Genes in Datoga Pastoralists of Tanzania."

35. Butovskaya et al., "Aggression and Polymorphisms in AR, DAT1, DRD2, and COMT Genes in Datoga Pastoralists of Tanzania."

36. Eric L. Thibodeau, Dante Cicchetti, and Fred A. Rogosch, "Child Maltreatment, Impulsivity, and Antisocial Behavior in African American Children: Moderation Effects from a Cumulative Dopaminergic Gene Index," *Development and Psychopathology* 27, no. 4, pt. 2 (November 2015): 1,621–36, https://doi.org/10.1017/S095457941500098X.

37. Patrick Davies, Dante Cicchetti, and Rochelle F. Hentges, "Maternal Unresponsiveness and Child Disruptive Problems: The Interplay of Uninhibited Temperament and Dopamine Transporter Genes," *Child Development* 86, no. 1 (January 2015): 63–79, https://doi.org/10.1111/cdev.12281.

38. Thibodeau, Cicchetti, and Rogosch, "Child Maltreatment, Impulsivity, and Antisocial Behavior in African American Children."

39. Vaughn et al., "DAT1 and 5HTT Are Associated with Pathological Criminal Behavior in a Nationally Representative Sample of Youth."

40. Thibodeau, Cicchetti, and Rogosch, "Child Maltreatment, Impulsivity, and Antisocial Behavior in African American Children."

41. Thibodeau, Cicchetti, and Rogosch, "Child Maltreatment, Impulsivity, and Antisocial Behavior in African American Children."

42. V. E. Golimbet et al., "Relationship between Dopamine System Genes and Extraversion and Novelty-Seeking," *Neuroscience and Behavioral Physiology* 37, no. 6 (July 2007): 601–6, https://doi.org/10.1007/s11055-007-0058-8.

43. S. Teodorovic and B. Uzelac, "Genetic Basis of Aggression: Overview and Implications for Legal Proceedings," *Romanian Journal of Legal Medicine* 23, no. 3 (September 2015): 193–202, https://doi.org/10.4323/rjlm.2015.193.

44. Guo et al., "The Dopamine Transporter Gene, a Spectrum of Most Common Risky Behaviors, and the Legal Status of the Behaviors."

45. Guo et al., "The Dopamine Transporter Gene, a Spectrum of Most Common Risky Behaviors, and the Legal Status of the Behaviors."

46. Marissa A. Gorlick et al., "DRD4 Long Allele Carriers Show Heightened Attention to High-Priority Items Relative to Low-Priority Items," *Journal of Cognitive Neuroscience* 27, no. 3 (March 2015): 509–21, https://doi.org/10.1162/jocn_a_00724.

47. Gorlick et al., "DRD4 Long Allele Carriers Show Heightened Attention to High-Priority Items Relative to Low-Priority Items."

48. Miles Berger, John A. Gray, and Bryan L. Roth, "The Expanded Biology of Serotonin," *Annual Review of Medicine* 60, no. 1 (February 2009): 355–66, https://doi.org/10.1146/annurev.med.60.042307.110802; Seo, Patrick, and Kennealy, "Role of Serotonin and Dopamine System Interactions in the Neurobiology of Impulsive Aggression and Its Comorbidity with Other Clinical Disorders."

49. P. R. Albert, C. Benkelfat, and L. Descarries, "The Neurobiology of Depression: Revisiting the Serotonin Hypothesis. I. Cellular and Molecular Mechanisms," *Philosophical*

Transactions of the Royal Society B: Biological Sciences 367, no. 1,601 (September 5, 2012): 2,378–81, https://doi.org/10.1098/rstb.2012.0190; S. Ramamoorthy et al., "Antidepressant- and Cocaine-Sensitive Human Serotonin Transporter: Molecular Cloning, Expression, and Chromosomal Localization.," *Proceedings of the National Academy of Sciences* 90, no. 6 (March 15, 1993): 2,542–46, https://doi.org/10.1073/pnas.90.6.2542.

50. K. P. Lesch et al., "Association of Anxiety-Related Traits with a Polymorphism in the Serotonin Transporter Gene Regulatory Region," *Science* 274, no. 5,292 (November 29, 1996): 1,527–31.

51. J. Gelernter, H. Kranzler, and J. F. Cubells, "Serotonin Transporter Protein (SLC6A4) Allele and Haplotype Frequencies and Linkage Disequilibria in African- and Europe-an-American and Japanese Populations and in Alcohol-Dependent Subjects," *Human Genetics* 101, no. 2 (December 1997): 243–46; Brett C. Haberstick et al., "Population Frequencies of the Triallelic 5HTTLPR in Six Ethnically Diverse Samples from North America, Southeast Asia, and Africa," *Behavior Genetics* 45, no. 2 (March 2015): 255–61, https://doi.org/10.1007/s10519-014-9703-5.

52. Noreen Goldman et al., "The Serotonin Transporter Polymorphism (5-HTTLPR): Allelic Variation and Links with Depressive Symptoms," *Depression and Anxiety* 27, no. 3 (March 2010): 260–69, https://doi.org/10.1002/da.20660; Edna Grünblatt et al., "Association Study of the 5-HTTLPR Polymorphism and Depression in 75-Year-Old Nonde-mented Subjects from the Vienna Transdanube Aging (VITA) Study," *Journal of Clinical Psychiatry* 67, no. 9 (September 2006): 1,373–78; Lesch et al., "Association of Anxiety-Related Traits with a Polymorphism in the Serotonin Transporter Gene Regulatory Region."

53. Teodorovic and Uzelac, "Genetic Basis of Aggression."

54. Teodorovic and Uzelac, "Genetic Basis of Aggression."

55. Vernon A. Barnes et al., "Impact of Williams LifeSkills Training on Anger, Anxiety, and Ambulatory Blood Pressure in Adolescents," *Translational Behavioral Medicine* 2, no. 4 (December 2012): 401–10, https://doi.org/10.1007/s13142-012-0162-3.

56. "Genetic Variant May Increase Risk of Stress-Induced Cardiovascular Damage," *Duke Health*, March 22, 2001, accessed December 18, 2018, https://corporate.dukehealth.org/news-listing/genetic-variant-may-increase-risk-stress-induced-cardiovascular-damage?h=nl; R. B. Williams et al., "Central Nervous System Serotonin Function and Cardiovascular Responses to Stress," *Psychosomatic Medicine* 63, no. 2 (April 2001): 300–305.

57. Anthony Walsh and Ilhong Yun, "Evoked Culture and Evoked Nature: The Promise of Gene-Culture Co-Evolution Theory for Sociology," *Frontiers in Sociology* 1 (July 19, 2016), https://doi.org/10.3389/fsoc.2016.00008.

58. J. Gelernter et al., "Population Studies of Polymorphisms of the Serotonin Transporter Protein Gene," *American Journal of Medical Genetics* 88, no. 1 (February 5, 1999): 61–66, https://doi.org/10.1002/(SICI)1096-8628(19990205)88:1<61::AID-AJMG11>3.0.CO;2-K; Redford B. Williams et al., "Serotonin-Related Gene Polymorphisms and Central Nervous System Serotonin Function," *Neuropsychopharmacology* 28, no. 3 (March 2003): 533–41, https://doi.org/10.1038/sj.npp.1300054.

59. Williams et al., "Serotonin-Related Gene Polymorphisms and Central Nervous System Serotonin Function."

60. "Genetic Variant May Increase Risk of Stress-Induced Cardiovascular Damage."

61. "Genetic Variant May Increase Risk of Stress-Induced Cardiovascular Damage."

62. Redford B. Williams et al., "Population Differences in Associations of Serotonin Transporter Promoter Polymorphism (5HTTLPR) Di- and Triallelic Genotypes with Blood Pressure and Hypertension Prevalence," *American Heart Journal* 185 (March 2017): 110–22, https://doi.org/10.1016/j.ahj.2016.12.013.

63. David Guwatudde et al., "The Burden of Hypertension in Sub-Saharan Africa: A Four-Country Cross-Sectional Study," *BMC Public Health* 15, no. 1 (December 2015), https://doi.org/10.1186/s12889-015-2546-z.

64. Williams et al., "Population Differences in Associations of Serotonin Transporter Promoter Polymorphism (5HTTLPR) Di- and Triallelic Genotypes with Blood Pressure and Hypertension Prevalence."

65. A. Córdova-Palomera et al., "Genome-Wide Methylation Study on Depression: Differential Methylation and Variable Methylation in Monozygotic Twins," *Translational Psychiatry* 5 (April 28, 2015): e557, https://doi.org/10.1038/tp.2015.49.

66. Córdova-Palomera et al., "Genome-Wide Methylation Study on Depression."

67. Caspi et al., "Role of Genotype in the Cycle of Violence in Maltreated Children."

68. Debra L. Foley et al., "Childhood Adversity, Monoamine Oxidase a Genotype, and Risk for Conduct Disorder," *Archives of General Psychiatry* 61, no. 7 (July 2004): 738–44, https://doi.org/10.1001/archpsyc.61.7.738; Brett C. Haberstick et al., "Monoamine Oxidase A (MAOA) and Antisocial Behaviors in the Presence of Childhood and Adolescent Maltreatment," *American Journal of Medical Genetics: Part B, Neuropsychiatric Genetics: The Official Publication of the International Society of Psychiatric Genetics* 135B, no. 1 (May 5, 2005): 59–64, https://doi.org/10.1002/ajmg.b.30176; Kim-Cohen et al., "MAOA, Maltreatment, and Gene–Environment Interaction Predicting Children's Mental Health"; S. B. Manuck et al., "A Regulatory Polymorphism of the Monoamine Oxidase A Gene May Be Associated with Variability in Aggression, Impulsivity, and Central Nervous System Serotonergic Responsivity," *Psychiatry Research* 95, no. 1 (July 24, 2000): 9–23; Andreas Meyer-Lindenberg et al., "Neural Mechanisms of Genetic Risk for Impulsivity and Violence in Humans," *Proceedings of the National Academy of Sciences of the United States of America* 103, no. 16 (April 18, 2006): 6,269–74, https://doi.org/10.1073/pnas.0511311103.

69. Paul Chapman, "Violence Is Blamed on 'Warrior Gene' in the Maoris," *Telegraph*, August 10, 2006, accessed December 18, 2018, https://www.telegraph.co.uk/news/1526042/Violence-is-blamed-on-warrior-gene-in-the-Maoris.html.

70. Meyer-Lindenberg et al., "Neural Mechanisms of Genetic Risk for Impulsivity and Violence in Humans."

71. Joan Arehart-Treichel, "Multiple Factors at Root of Antisocial Behavior," *Psychiatric News* 43, no. 15 (August 2008): 4–18, https://doi.org/10.1176/pn.43.15.0004; Meyer-Lindenberg et al., "Neural Mechanisms of Genetic Risk for Impulsivity and Violence in Humans."

72. J Arehart-Treichel, "Psychosocial Factors Outweigh Genes in Development of Antisocial Behavior," *Psychiatr News* 40, no. 21 (2005): 30; Kent W. Nilsson et al., "Role of Monoamine Oxidase A Genotype and Psychosocial Factors in Male Adolescent Criminal Activity," *Biological Psychiatry* 59, no. 2 (January 15, 2006): 121–27, https://doi.org/10.1016/j.biopsych.2005.06.024.

73. Nilsson et al., "Role of Monoamine Oxidase A Genotype and Psychosocial Factors in Male Adolescent Criminal Activity."

74. Brett C. Haberstick et al., "MAOA Genotype, Childhood Maltreatment, and Their Interaction in the Etiology of Adult Antisocial Behaviors," *Biological Psychiatry* 75, no. 1 (January 2014): 25–30, https://doi.org/10.1016/j.biopsych.2013.03.028.

75. Haberstick et al., "MAOA Genotype, Childhood Maltreatment, and Their Interaction in the Etiology of Adult Antisocial Behaviors."

76. Sarah Richardson, "A Violence in the Blood," *Discover Magazine*, October 1, 1993, accessed December 18, 2018, http://discovermagazine.com/1993/oct/aviolenceinthebl293.

77. Sean C. Godar et al., "The Role of Monoamine Oxidase A in Aggression: Current Translational Developments and Future Challenges," *Progress in Neuro-Psychopharmacology and Biological Psychiatry* 69 (January 2016): 90–100, https://doi.org/10.1016/j.pnpbp.2016.01.001; Richardson, "A Violence in the Blood."

78. Godar et al., "The Role of Monoamine Oxidase A in Aggression."

79. Byrd and Manuck, "MAOA, Childhood Maltreatment, and Antisocial Behavior."

80. Dean A. Stetler et al., "Association of Low-Activity MAOA Allelic Variants with Violent Crime in Incarcerated Offenders," *Journal of Psychiatric Research* 58 (November 2014): 69–75, https://doi.org/10.1016/j.jpsychires.2014.07.006.

81. Stetler et al., "Association of Low-Activity MAOA Allelic Variants with Violent Crime in Incarcerated Offenders."

82. Stetler et al., "Association of Low-Activity MAOA Allelic Variants with Violent Crime in Incarcerated Offenders."

83. Irving M. Reti et al., "Monoamine Oxidase A Regulates Antisocial Personality in Whites with No History of Physical Abuse," *Comprehensive Psychiatry* 52, no. 2 (March 2011): 188–94, https://doi.org/10.1016/j.comppsych.2010.05.005.

84. Reti et al., "Monoamine Oxidase A Regulates Antisocial Personality in Whites with No History of Physical Abuse."

85. Reti et al., "Monoamine Oxidase A Regulates Antisocial Personality in Whites with No History of Physical Abuse."

86. Caspi et al., "Role of Genotype in the Cycle of Violence in Maltreated Children."

87. Guang Guo et al., "The VNTR 2 Repeat in MAOA and Delinquent Behavior in Adolescence and Young Adulthood: Associations and MAOA Promoter Activity," *European Journal of Human Genetics* 16, no. 5 (May 2008): 626–34, https://doi.org/10.1038/sj.ejhg.5201999.

88. Kevin M. Beaver et al., "Exploring the Association between the 2-Repeat Allele of the MAOA Gene Promoter Polymorphism and Psychopathic Personality Traits, Arrests, Incarceration, and Lifetime Antisocial Behavior," *Personality and Individual Differences* 54, no. 2 (January 2013): 164–68, https://doi.org/10.1016/j.paid.2012.08.014; Kevin M. Beaver, J. C. Barnes, and Brian B. Boutwell, "The 2-Repeat Allele of the MAOA Gene Confers an Increased Risk for Shooting and Stabbing Behaviors," *Psychiatric Quarterly* 85, no. 3 (September 2014): 257–65, https://doi.org/10.1007/s11126-013-9287-x.

89. Beaver, Barnes, and Boutwell, "The 2-Repeat Allele of the MAOA Gene Confers an Increased Risk for Shooting and Stabbing Behaviors."

90. Beaver, Barnes, and Boutwell, "The 2-Repeat Allele of the MAOA Gene Confers an Increased Risk for Shooting and Stabbing Behaviors."

91. Alondra Oubré, "The Extreme Warrior Gene: A Reality Check," *Scientia Salon* (blog), July 31, 2014, accessed December 18, 2018, https://scientiasalon.wordpress.com/2014/07/31/the-extreme-warrior-gene-a-reality-check/.

92. Oubré, "The Extreme Warrior Gene."

93. Daniel Ewon Choe et al., "Interactions between Monoamine Oxidase A and Punitive Discipline in African American and Caucasian Men's Antisocial Behavior," *Clinical Psychological Science* 2, no. 5 (September 2014): 591–601, https://doi.org/10.1177/2167702613518046.

94. Choe et al., "Interactions between Monoamine Oxidase A and Punitive Discipline in African American and Caucasian Men's Antisocial Behavior."

95. Choe et al., "Interactions between Monoamine Oxidase A and Punitive Discipline in African American and Caucasian Men's Antisocial Behavior."

96. Choe et al., "Interactions between Monoamine Oxidase A and Punitive Discipline in African American and Caucasian Men's Antisocial Behavior"; Oubré, "The Extreme Warrior Gene."

97. Michael E. Roettger et al., "The Association between the MAOA 2R Genotype and Delinquency Over Time among Men: The Interactive Role of Parental Closeness and Parental Incarceration," *Criminal Justice and Behavior* 43, no. 8 (August 2016): 1,076–94, https://doi.org/10.1177/0093854816629184.

98. Roettger et al., "The Association between the MAOA 2R Genotype and Delinquency Over Time among Men."

99. Roettger et al., "The Association between the MAOA 2R Genotype and Delinquency Over Time among Men."

100. Roettger et al., "The Association between the MAOA 2R Genotype and Delinquency Over Time among Men."

101. Scott A. Kelly, Tami M. Panhuis, and Andrew M. Stoehr, "Phenotypic Plasticity: Molecular Mechanisms and Adaptive Significance," *Comprehensive Physiology* 2, no. 2 (April 2012): 1,417–39, https://doi.org/10.1002/cphy.c110008.

102. Elena Shumay et al., "Evidence That the Methylation State of the Monoamine Oxidase A (MAOA) Gene Predicts Brain Activity of MAO A Enzyme in Healthy Men," *Epigenetics* 7, no. 10 (October 13, 2012): 1,151–60, https://doi.org/10.4161/epi.21976.

103. Christiane Ziegler et al., "Monoamine Oxidase A Gene Methylation and Its Role in Posttraumatic Stress Disorder: First Evidence from the South Eastern Europe (SEE)-PTSD Study," *International Journal of Neuropsychopharmacology* 21, no. 5 (May 1, 2018): 423–32, https://doi.org/10.1093/ijnp/pyx111.

104. Nathan J. Kolla et al., "Lower Monoamine Oxidase-A Total Distribution Volume in Impulsive and Violent Male Offenders with Antisocial Personality Disorder and High Psychopathic Traits: An [11C] Harmine Positron Emission Tomography Study," *Neuropsychopharmacology* 40, no. 11 (October 2015): 2,596–603, https://doi.org/10.1038/npp.2015.106.

105. Haberstick et al., "MAOA Genotype, Childhood Maltreatment, and Their Interaction in the Etiology of Adult Antisocial Behaviors."

106. Haberstick et al., "MAOA Genotype, Childhood Maltreatment, and Their Interaction in the Etiology of Adult Antisocial Behaviors."

107. Haberstick et al., "MAOA Genotype, Childhood Maltreatment, and Their Interaction in the Etiology of Adult Antisocial Behaviors."

108. Kent W. Nilsson et al., "Genotypes Do Not Confer Risk for Delinquency but Rather Alter Susceptibility to Positive and Negative Environmental Factors: Gene–Environment Interactions of BDNF Val66Met, 5-HTTLPR, and MAOA-UVNTR," *International Journal of Neuropsychopharmacology* 18, no. 5 (March 2015): 107, https://doi.org/10.1093/ijnp/pyu107.

109. Shumay et al., "Evidence That the Methylation State of the Monoamine Oxidase A (MAOA) Gene Predicts Brain Activity of MAO A Enzyme in Healthy Men."

110. Kelly, Panhuis, and Stoehr, "Phenotypic Plasticity"; Shumay et al., "Evidence That the Methylation State of the Monoamine Oxidase A (MAOA) Gene Predicts Brain Activity of MAO A Enzyme in Healthy Men."

111. Kelly, Panhuis, and Stoehr, "Phenotypic Plasticity."

112. Nilsson et al., "Genotypes Do Not Confer Risk for Delinquency but Rather Alter Susceptibility to Positive and Negative Environmental Factors."

113. J. C. Barnes, Kevin M. Beaver, and Brian B. Boutwell, "Examining the Genetic Underpinnings to Moffitt's Developmental Taxonomy: A Behavioral Genetic Analysis," *Criminology* 49, no. 4 (November 2011): 923–54, https://doi.org/10.1111/j.1745-9125 .2011.00243.x; "Genes Influence Criminal Behavior, Research Suggests," *ScienceDaily*, January 26, 2012, accessed December 18, 2018, https://www.sciencedaily.com/ releases/2012/01/120125151841.htm.

114. Guo et al., "The VNTR 2 Repeat in MAOA and Delinquent Behavior in Adolescence and Young Adulthood"; Vaughn et al., "DAT1 and 5HTT Are Associated with Pathological Criminal Behavior in a Nationally Representative Sample of Youth."

115. Butovskaya et al., "Aggression and Polymorphisms in AR, DAT1, DRD2, and COMT Genes in Datoga Pastoralists of Tanzania."

116. Sytske Besemer, David P. Farrington, and Catrien C. J. H. Bijleveld, "Labeling and Intergenerational Transmission of Crime: The Interaction between Criminal Justice Intervention and a Convicted Parent," *PLoS ONE* 12, no. 3 (March 8, 2015): e0172419, https:// doi.org/10.3886/E100376V2.

117. Alison J. Douglas, "Baby Love? Oxytocin–Dopamine Interactions in Mother–Infant Bonding," *Endocrinology* 151, no. 5 (May 2010): 1,978–80, https://doi.org/10.1210/en.2010 -0259.

118. Douglas, "Baby Love?"; Andrew H. Kemp and Adam J. Guastella, "The Role of Oxytocin in Human Affect: A Novel Hypothesis," *Current Directions in Psychological Science* 20, no. 4 (August 2011): 222–31, https://doi.org/10.1177/0963721411417547.

119. Kemp and Guastella, "The Role of Oxytocin in Human Affect."

120. Sarina M. Rodrigues et al., "Oxytocin Receptor Genetic Variation Relates to Empathy and Stress Reactivity in Humans," *Proceedings of the National Academy of Sciences of the United States of America* 106, no. 50 (December 15, 2009): 21,437–441, https://doi.org/ 10.1073/pnas.0909579106.

121. Catherine Maud et al., "The Role of Oxytocin Receptor Gene (OXTR) DNA Methylation (DNAm) in Human Social and Emotional Functioning: A Systematic Narrative Review," *BMC Psychiatry* 18, no. 1 (May 29, 2018): 154, https://doi.org/10.1186/s12888 -018-1740-9.

122. Erica L. Smearman et al., "Oxytocin Receptor Genetic and Epigenetic Variations: Association with Child Abuse and Adult Psychiatric Symptoms," *Child Development* 87, no. 1 (February 2016): 122–34, https://doi.org/10.1111/cdev.12493.

123. Smearman et al., "Oxytocin Receptor Genetic and Epigenetic Variations."

124. Rodrigues et al., "Oxytocin Receptor Genetic Variation Relates to Empathy and Stress Reactivity in Humans."

125. Bekh Bradley et al., "Association between Childhood Maltreatment and Adult Emotional Dysregulation in a Low-Income, Urban, African American Sample: Moderation

by Oxytocin Receptor Gene," *Development and Psychopathology* 23, no. 2 (May 2011): 439–52, https://doi.org/10.1017/S0954579411000162.

126. Bradley et al., "Association between Childhood Maltreatment and Adult Emotional Dysregulation in a Low-Income, Urban, African American Sample."

127. Douglas, "Baby Love?"

128. Maud et al., "The Role of Oxytocin Receptor Gene (OXTR) DNA Methylation (DNAm) in Human Social and Emotional Functioning."

Chapter 12. Girl Growth: Sexual Maturation in Black and Brown Bodies

1. Marcia E. Herman-Giddens et al., "Secondary Sexual Characteristics and Menses in Young Girls Seen in Office Practice: A Study from the Pediatric Research in Office Settings Network," *Pediatrics* 99, no. 4 (April 1997): 505–12.

2. William Cameron Chumlea et al., "Age at Menarche and Racial Comparisons in U.S. Girls," *Pediatrics* 111, no. 1 (January 2003): 110–13; Herman-Giddens et al., "Secondary Sexual Characteristics and Menses in Young Girls Seen in Office Practice"; Mary Lewis, "Children Aren't Starting Puberty Younger, Medieval Skeletons Reveal," *Conversation*, February 12, 2018, accessed December 18, 2018, http://theconversation.com/children-arent-starting-puberty-younger-medieval-skeletons-reveal-91095; Mary Pierce and Rebecca Hardy, "Commentary: The Decreasing Age of Puberty—as Much a Psychosocial as Biological Problem?" *International Journal of Epidemiology* 41, no. 1 (February 2012): 300–302, https://doi.org/10.1093/ije/dyr227.

3. M. E. Herman-Giddens et al., "Secondary Sexual Characteristics in Boys: Data from the Pediatric Research in Office Settings Network," *Pediatrics* 130, no. 5 (November 1, 2012): e1058–68, https://doi.org/10.1542/peds.2011-3291.

4. Herman-Giddens et al., "Secondary Sexual Characteristics in Boys."

5. Herman-Giddens et al., "Secondary Sexual Characteristics and Menses in Young Girls Seen in Office Practice."

6. Phyllis B. Eveleth and J. M. Tanner, *Worldwide Variation in Human Growth*, 2nd ed. (Cambridge, UK, and New York: Cambridge University Press, 1990); Christopher W. Kuzawa and Jared M. Bragg, "Plasticity in Human Life History Strategy: Implications for Contemporary Human Variation and the Evolution of Genus *Homo*," *Current Anthropology* 53, no. S6 (December 2012): S369–82, https://doi.org/10.1086/667410.

7. Eveleth and Tanner, *Worldwide Variation in Human Growth*; Kuzawa and Bragg, "Plasticity in Human Life History Strategy."

8. F. Thomas et al., "International Variability of Ages at Menarche and Menopause: Patterns and Main Determinants," *Human Biology* 73, no. 2 (April 2001): 271–90.

9. Eveleth and Tanner, *Worldwide Variation in Human Growth*; Kuzawa and Bragg, "Plasticity in Human Life History Strategy"; Thomas et al., "International Variability of Ages at Menarche and Menopause."

10. Eveleth and Tanner, *Worldwide Variation in Human Growth*.

11. Z. M. Ofuya, "The Age at Menarche in Nigerian Adolescents from Two Different Socioeconomic Classes," *Online Journal of Health and Allied Sciences* 4, no. 3 (2007): 1–4.

12. P. Pasquet et al., "Age at Menarche and Urbanization in Cameroon: Current Status and Secular Trends," *Annals of Human Biology* 26, no. 1 (February 1999): 89–97.

13. Thomas et al., "International Variability of Ages at Menarche and Menopause."

14. K. O. Rogo, R. K. Oniang'o, and L. A. Muruli, "Menarche in African Girls in Some Postsecondary Institutions in Kenya," *East African Medical Journal* 64, no. 11 (November 1987): 745–50.

15. A. D. Groos and T. A. Smith, "Age at Menarche and Associated Nutritional Status Variables in Karimui and Daribi Census Divisions of Simbu Province," *Papua and New Guinea Medical Journal* 35, no. 2 (June 1992): 84–94.

16. Eveleth and Tanner, *Worldwide Variation in Human Growth.*

17. Eveleth and Tanner, *Worldwide Variation in Human Growth*; Thomas et al., "International Variability of Ages at Menarche and Menopause."

18. Chumlea et al., "Age at Menarche and Racial Comparisons in U.S. Girls."

19. Lewis, "Children Aren't Starting Puberty Younger, Medieval Skeletons Reveal."

20. Shumei S. Sun et al., "National Estimates of the Timing of Sexual Maturation and Racial Differences among U.S. Children," *Pediatrics* 110, no. 5 (November 2002): 911–19.

21. Mickey Emmanuel and Brooke R. Bokor, "Tanner Stages," in *StatPearls* (Treasure Island, FL: StatPearls, 2018), http://www.ncbi.nlm.nih.gov/books/NBK470280/.

22. Herman-Giddens et al., "Secondary Sexual Characteristics and Menses in Young Girls Seen in Office Practice."

23. A. C. Hergenroeder et al., "Validity of Self-Assessment of Pubertal Maturation in African American and European American Adolescents," *Journal of Adolescent Health: Official Publication of the Society for Adolescent Medicine* 24, no. 3 (March 1999): 201–5.

24. Chang Chen et al., "Investigating the Relationship between Precocious Puberty and Obesity: A Cross-Sectional Study in Shanghai, China," *BMJ Open* 7, no. 4 (August 23, 2017): e014004, https://doi.org/10.1136/bmjopen-2016-014004.

25. Folakemi O. Irewole-Ojo et al., "Age of Pubertal Events among School Girls in Lagos, Nigeria," *Journal of Pediatric Endocrinology and Metabolism* 31, no. 3 (March 28, 2018): 313–21, https://doi.org/10.1515/jpem-2017-0344.

26. K. L. Howdeshell et al., "Exposure to Bisphenol A Advances Puberty," *Nature* 401, no. 6,755 (October 21, 1999): 763–64, https://doi.org/10.1038/44517; Alberto Leonardi et al., "The Effect of Bisphenol A on Puberty: A Critical Review of the Medical Literature," *International Journal of Environmental Research and Public Health* 14, no. 9 (September 10, 2017): 1,044, https://doi.org/10.3390/ijerph14091044.

27. Leonardi et al., "The Effect of Bisphenol A on Puberty."

28. Ellen W. Demerath et al., "Genome-Wide Association Study of Age at Menarche in African American Women," *Human Molecular Genetics* 22, no. 16 (August 15, 2013): 3,329–346, https://doi.org/10.1093/hmg/ddt181.

29. B. J. Ellis et al., "Quality of Early Family Relationships and Individual Differences in the Timing of Pubertal Maturation in Girls: A Longitudinal Test of an Evolutionary Model," *Journal of Personality and Social Psychology* 77, no. 2 (August 1999): 387–401.

30. Sylvie Mrug et al., "Early Puberty, Negative Peer Influence, and Problem Behaviors in Adolescent Girls," *Pediatrics* 133, no. 1 (January 2014): 7–14, https://doi.org/10.1542/peds.2013-0628.

31. Wanzhu Tu et al., "Associations between Menarche-Related Genetic Variants and Pubertal Growth in Male and Female Adolescents," *Journal of Adolescent Health: Official Publication of the Society for Adolescent Medicine* 56, no. 1 (January 2015): 66–72, https://doi.org/10.1016/j.jadohealth.2014.07.020.

32. Demerath et al., "Genome-Wide Association Study of Age at Menarche in African American Women"; Tu et al., "Associations between Menarche-Related Genetic Variants and Pubertal Growth in Male and Female Adolescents."

33. K. Paige Harden, "Genetic Influences on Adolescent Sexual Behavior: Why Genes Matter for Environmentally Oriented Researchers," *Psychological Bulletin* 140, no. 2 (March 2014): 434–65, https://doi.org/10.1037/a0033564.

34. Tu et al., "Associations between Menarche-Related Genetic Variants and Pubertal Growth in Male and Female Adolescents."

35. Demerath et al., "Genome-Wide Association Study of Age at Menarche in African American Women."

36. Demerath et al., "Genome-Wide Association Study of Age at Menarche in African American Women."

37. Demerath et al., "Genome-Wide Association Study of Age at Menarche in African American Women."

38. Demerath et al., "Genome-Wide Association Study of Age at Menarche in African American Women."

39. Chris Fradkin et al., "Associations between Socioeconomic Status and Obesity in Diverse, Young Adolescents: Variation across Race/Ethnicity and Gender," *Health Psychology* 34, no. 1 (2015): 1–9, https://doi.org/10.1037/hea0000099.

40. John Cline, "The Mysterious Benefits of Deep Sleep," *Psychology Today*, October 11, 2010, accessed December 18, 2019, https://www.psychologytoday.com/us/blog/sleepless -in-america/201010/the-mysterious-benefits-deep-sleep.

41. N. D. Shaw et al., "Insights into Puberty: The Relationship between Sleep Stages and Pulsatile LH Secretion," *Journal of Clinical Endocrinology and Metabolism* 97, no. 11 (November 2012): E2055–62, https://doi.org/10.1210/jc.2012-2692.

42. L. A. Proos, Y. Hofvander, and T. Tuvemo, "Menarcheal Age and Growth Pattern of Indian Girls Adopted in Sweden: I. Menarcheal Age," *Acta Paediatrica Scandinavica* 80, no. 8–9 (September 1991): 852–58.

43. Catherine E. Aiken and Susan E. Ozanne, "Transgenerational Developmental Programming," *Human Reproduction Update* 20, no. 1 (February 2014): 63–75, https://doi.org/ 10.1093/humupd/dmt043; Kuzawa and Bragg, "Plasticity in Human Life History Strategy."

44. Kuzawa and Bragg, "Plasticity in Human Life History Strategy."

45. Sarah Prentice et al., "Evidence for a Downward Secular Trend in Age of Menarche in a Rural Gambian Population," *Annals of Human Biology* 37, no. 5 (October 2010): 717–21, https://doi.org/10.3109/03014461003727606.

46. B. Rey DeCastro et al., "Estrogenic Activity of Polychlorinated Biphenyls Present in Human Tissue and the Environment," *Environmental Science and Technology* 40, no. 8 (April 15, 2006): 2,819–825; Francesco Massart et al., "High Growth Rate of Girls with Precocious Puberty Exposed to Estrogenic Mycotoxins," *Journal of Pediatrics* 152, no. 5 (May 2008): 690–695.e1, https://doi.org/10.1016/j.jpeds.2007.10.020.

47. Massart et al., "High Growth Rate of Girls with Precocious Puberty Exposed to Estrogenic Mycotoxins."

48. Chandra M. Tiwary, "A Survey of Use of Hormone/Placenta-Containing Hair Preparations by Parents and/or Children Attending Pediatric Clinics," *Military Medicine* 162, no. 4 (April 1997): 252–56.

49. Chandra M. Tiwary, "Premature Sexual Development in Children Following the Use of Estrogen- or Placenta-Containing Hair Products," *Clinical Pediatrics* 37, no. 12 (December 1998): 733–39, https://doi.org/10.1177/000992289803701204.

50. S. Pagan, "Early Puberty Linked to Shampoos," *New Scientist* (blog), April 3, 2002, accessed December 18, 2018, https://www.newscientist.com/article/dn2119-early-puberty-linked-to-shampoos/.

51. E. A. Graham and N. Sugar, "Secondary Sexual Characteristics and Menses in Young Girls.," *New Scientist* 101, no. 5 (1998): 949–50.

52. Tiwary, "A Survey of Use of Hormone/Placenta-Containing Hair Preparations by Parents and/or Children Attending Pediatric Clinics"; Tiwary, "Premature Sexual Development in Children Following the Use of Estrogen- or Placenta-Containing Hair Products."

53. Tiwary, "Premature Sexual Development in Children Following the Use of Estrogen- or Placenta-Containing Hair Products."

54. Tiwary, "Premature Sexual Development in Children Following the Use of Estrogen- or Placenta-Containing Hair Products."

55. Tiwary, "Premature Sexual Development in Children Following the Use of Estrogen- or Placenta-Containing Hair Products."

56. Massart et al., "High Growth Rate of Girls with Precocious Puberty Exposed to Estrogenic Mycotoxins."

57. Joyce A. Martin et al., "Births: Final Data for 2016," *National Vital Statistics Reports: From the Centers for Disease Control and Prevention, National Center for Health Statistics, National Vital Statistics System* 67, no. 1 (January 2018): 1–55.

58. Harden, "Genetic Influences on Adolescent Sexual Behavior."

59. Ellis et al., "Quality of Early Family Relationships and Individual Differences in the Timing of Pubertal Maturation in Girls."

60. Ellis et al., "Quality of Early Family Relationships and Individual Differences in the Timing of Pubertal Maturation in Girls."

61. Mrug et al., "Early Puberty, Negative Peer Influence, and Problem Behaviors in Adolescent Girls."

62. Mrug et al., "Early Puberty, Negative Peer Influence, and Problem Behaviors in Adolescent Girls."

63. Harden, "Genetic Influences on Adolescent Sexual Behavior."

64. Harden, "Genetic Influences on Adolescent Sexual Behavior."

65. Harden, "Genetic Influences on Adolescent Sexual Behavior."

66. Harden, "Genetic Influences on Adolescent Sexual Behavior."

67. Harden, "Genetic Influences on Adolescent Sexual Behavior."

68. Riya R. Kanherkar, Naina Bhatia-Dey, and Antonei B. Csoka, "Epigenetics across the Human Lifespan," *Frontiers in Cell and Developmental Biology* 2 (2014): 49, https://doi.org/10.3389/fcell.2014.00049.

69. Kristian Almstrup et al., "Pubertal Development in Healthy Children Is Mirrored by DNA Methylation Patterns in Peripheral Blood," *Scientific Reports* 6, no. 1 (September 2016), https://doi.org/10.1038/srep28657.

70. Emma E. Thompson et al., "Global DNA Methylation Changes Spanning Puberty Are Near Predicted Estrogen-Responsive Genes and Enriched for Genes Involved in Endocrine and Immune Processes," *Clinical Epigenetics* 10 (2018): 62, https://doi.org/10.1186/s13148-018-0491-2.

71. Thompson et al., "Global DNA Methylation Changes Spanning Puberty Are Near Predicted Estrogen-Responsive Genes and Enriched for Genes Involved in Endocrine and Immune Processes."

Chapter 13. Testosterone Tales of Race and Aggression

1. P. B. Gray, "The Descent of a Man's Testosterone," *Proceedings of the National Academy of Sciences* 108, no. 39 (September 27, 2011): 16,141–42, https://doi.org/10.1073/pnas.1113323108; Hui Hu et al., "Racial Differences in Age-Related Variations of Testosterone Levels among U.S. Males: Potential Implications for Prostate Cancer and Personalized Medication," *Journal of Racial and Ethnic Health Disparities* 2, no. 1 (March 2015): 69–76, https://doi.org/10.1007/s40615-014-0049-8; David S. Lopez et al., "Racial/Ethnic Differences in Serum Sex Steroid Hormone Concentrations in U.S. Adolescent Males," *Cancer Causes and Control* 24, no. 4 (April 2013): 817–26, https://doi.org/10.1007/s10552-013-0154-8; A. Mazur and A. Booth, "Testosterone and Dominance in Men," *Behavioral and Brain Sciences* 21, no. 3 (June 1998): 353–63, http://dx.doi.org/10.1017/S0140525X98001228.

2. Coren L. Apicella, Justin M. Carré, and Anna Dreber, "Testosterone and Economic Risk Taking: A Review," *Adaptive Human Behavior and Physiology* 1, no. 3 (September 2015): 358–85, https://doi.org/10.1007/s40750-014-0020-2; Mazur and Booth, "Testosterone and Dominance in Men"; J. T. Trevors and M. H. Saier, "The Cure for Our Global Problems," *Water, Air, and Soil Pollution* 201, no. 1–4 (July 2009): 1–2, https://doi.org/10.1007/s11270-009-0101-1.

3. L. Ellis and H. Nyborg, "Racial/Ethnic Variations in Male Testosterone Levels: A Probable Contributor to Group Differences in Health," *Steroids* 57, no. 2 (February 1992): 72–75; Satoshi Kanazawa, "The Evolution of General Intelligence," *Personality and Individual Differences* 53, no. 2 (July 2012): 90–93, https://doi.org/10.1016/j.paid.2011.05.015; Richard Lynn, "Rushton's r–K Life History Theory of Race Differences in Penis Length and Circumference Examined in 113 Populations," *Personality and Individual Differences* 55, no. 3 (July 2013): 261–66, https://doi.org/10.1016/j.paid.2012.02.016; J. Philippe Rushton, *Race, Evolution, and Behavior: A Life History Perspective*, 3rd ed. (Puerto Ayora: Charles Darwin Research Institute, 2000); J. Philippe Rushton and Donald I. Templer, "Do Pigmentation and the Melanocortin System Modulate Aggression and Sexuality in Humans as They Do in Other Animals?" *Personality and Individual Differences* 53, no. 1 (July 2012): 4–8, https://doi.org/10.1016/j.paid.2012.02.015.

4. Carleton Coon, *Racial Adaptations* (Chicago: Nelson-Hall, 1982).

5. Alexander Alland, *Race in Mind: Race, IQ, and Other Racisms* (New York: Palgrave Macmillan, 2006); C. Leslie, "Scientific Racism: Reflections on Peer Review, Science, and Ideology," *Social Science and Medicine* 31, no. 8 (1990): 891–905.

6. Carleton Stevens Coon, *Origin of Races* (New York: Random House, 1962).

7. Coon, *Racial Adaptations*.

8. Coon, *Racial Adaptations*.

9. Ellis and Nyborg, "Racial/Ethnic Variations in Male Testosterone Levels"; Edward M. Miller, "Paternal Provisioning versus Mate Seeking in Human Populations," *Personality and Individual Differences* 17, no. 2 (August 1994): 227–55, https://doi.org/10.1016/0191-8869(94)90029-9; Rushton, *Race, Evolution, and Behavior*.

10. Ellis and Nyborg, "Racial/Ethnic Variations in Male Testosterone Levels"; Michael E. Levin, *Why Race Matters: Race Differences and What They Mean*, Human Evolution, Behavior, and Intelligence (Westport, CT: Praeger, 1997); Miller, "Paternal Provisioning versus Mate Seeking in Human Populations"; Rushton, *Race, Evolution, and Behavior*; Anthony Walsh, *Race and Crime: A Biosocial Analysis* (New York: Nova Science Publishers, 2009).

11. Rushton, *Race, Evolution, and Behavior*.

12. Rushton, *Race, Evolution, and Behavior*.

13. David G. Gardner, Dolores M. Shoback, and Francis S. Greenspan, eds., *Greenspan's Basic and Clinical Endocrinology*, 8th ed. (New York and London: McGraw-Hill Medical, 2007).

14. Gardner, Shoback, and Greenspan, eds., *Greenspan's Basic and Clinical Endocrinology*.

15. E. David Crawford, "Epidemiology of Prostate Cancer," *Urology* 62, no. 6, suppl 1 (December 22, 2003): 3–12; R. Ross et al., "Serum Testosterone Levels in Healthy Young Black and White Men," *Journal of the National Cancer Institute* 76, no. 1 (January 1986): 45–48.

16. Ross et al., "Serum Testosterone Levels in Healthy Young Black and White Men."

17. Trevors and Saier, "The Cure for Our Global Problems."

18. Apicella, Carré, and Dreber, "Testosterone and Economic Risk Taking."

19. Stephen J. Winters et al., "Testosterone, Sex Hormone-Binding Globulin, and Body Composition in Young Adult African American and Caucasian Men," *Metabolism* 50, no. 10 (October 2001): 1,242–47, https://doi.org/10.1053/meta.2001.26714.

20. Heather J. Litman et al., "Serum Androgen Levels in Black, Hispanic, and White Men," *Journal of Clinical Endocrinology and Metabolism* 91, no. 11 (November 2006): 4,326–34, https://doi.org/10.1210/jc.2006-0037.

21. Mazur and Booth, "Testosterone and Dominance in Men."

22. J. R. Udry et al., "Serum Androgenic Hormones Motivate Sexual Behavior in Adolescent Boys," *Fertility and Sterility* 43, no. 1 (January 1985): 90–94.

23. Udry et al., "Serum Androgenic Hormones Motivate Sexual Behavior in Adolescent Boys."

24. James M. Dabbs et al., "Testosterone, Crime, and Misbehavior among 692 Male Prison Inmates," *Personality and Individual Differences* 18, no. 5 (May 1995): 627–33, https://doi.org/10.1016/0191-8869(94)00177-T.

25. Jerald Bain et al., "Sex Hormones in Murderers and Assaulters," *Behavioral Sciences and the Law* 5, no. 1 (1987): 95–101, https://doi.org/10.1002/bsl.2370050109.

26. Mazur and Booth, "Testosterone and Dominance in Men"; A. Richard et al., "Racial Variation in Sex Steroid Hormone Concentration in Black and White Men: A Meta-Analysis," *Andrology* 2, no. 3 (May 2014): 428–35, https://doi.org/10.1111/j.2047-2927.2014.00206.x.

27. Ellis and Nyborg, "Racial/Ethnic Variations in Male Testosterone Levels."

28. Mazur and Booth, "Testosterone and Dominance in Men," 360.

29. Mazur and Booth, "Testosterone and Dominance in Men."

30. Mazur and Booth, "Testosterone and Dominance in Men."

31. Mazur and Booth, "Testosterone and Dominance in Men," 360.

32. Mazur and Booth, "Testosterone and Dominance in Men."

33. A. B. Losecaat Vermeer, I. Riečanský, and C. Eisenegger, "Competition, Testosterone, and Adult Neurobehavioral Plasticity," in *Progress in Brain Research*, vol. 229, ed. Stephen

Waxman, Donald G. Stein, Dick Swaab, and Howard Fields (Amsterdam: Elsevier, 2016), 213–38, https://doi.org/10.1016/bs.pbr.2016.05.004.

Chapter 14. The AR Gene: The "New Testosterone"

1. Nancy Ferrari, "Prostate Cancer Risk in African Americans," *Harvard Health Blog*, November 3, 2009, accessed December 18, 2018, https://www.health.harvard.edu/blog/prostate-cancer-risk-in-african-americans-20091103209.

2. Ferrari, "Prostate Cancer Risk in African Americans"; Emanuela Taioli et al., "Polymorphisms in CYP17 and CYP3A4 and Prostate Cancer in Men of African Descent," *Prostate* 73, no. 6 (May 2013): 668–76, https://doi.org/10.1002/pros.22612; Heinric Williams and Isaac J. Powell, "Epidemiology, Pathology, and Genetics of Prostate Cancer among African Americans Compared with Other Ethnicities," *Methods in Molecular Biology* 472 (2009): 439–53, https://doi.org/10.1007/978-1-60327-492-0_21.

3. Peter Boyle et al., "Endogenous and Exogenous Testosterone and the Risk of Prostate Cancer and Increased Prostate-Specific Antigen (PSA) Level: A Meta-Analysis," *BJU International* 118, no. 5 (November 2016): 731–41, https://doi.org/10.1111/bju.13417.

4. Xueying Mao et al., "Involvement of Different Mechanisms for the Association of CAG Repeat Length Polymorphism in Androgen Receptor Gene with Prostate Cancer," *American Journal of Cancer Research* 4, no. 6 (2014): 886–96; J. L. Stanford et al., "Polymorphic Repeats in the Androgen Receptor Gene: Molecular Markers of Prostate Cancer Risk," *Cancer Research* 57, no. 6 (March 15, 1997): 1,194–98; Williams and Powell, "Epidemiology, Pathology, and Genetics of Prostate Cancer among African Americans Compared with Other Ethnicities."

5. C. M. Ackerman et al., "Ethnic Variation in Allele Distribution of the Androgen Receptor (AR) (CAG)n Repeat," *Journal of Andrology* 33, no. 2 (March 1, 2012): 210–15, https://doi.org/10.2164/jandrol.111.013391; R. A. Irvine et al., "The CAG and GGC Microsatellites of the Androgen Receptor Gene Are in Linkage Disequilibrium in Men with Prostate Cancer," *Cancer Research* 55, no. 9 (May 1, 1995): 1,937–40; O. Sartor, Q. Zheng, and J. A. Eastham, "Androgen Receptor Gene CAG Repeat Length Varies in a Race-Specific Fashion in Men without Prostate Cancer," *Urology* 53, no. 2 (February 1999): 378–80.

6. Marina L. Butovskaya et al., "Androgen Receptor Gene Polymorphism, Aggression, and Reproduction in Tanzanian Foragers and Pastoralists," *PLoS ONE* 10, no. 8 (August 20, 2015): e0136208, https://doi.org/10.1371/journal.pone.0136208; Daniel Cheng et al., "Association Study of Androgen Receptor CAG Repeat Polymorphism and Male Violent Criminal Activity," *Psychoneuroendocrinology* 31, no. 4 (May 2006): 548–52, https://doi.org/10.1016/j.psyneuen.2005.11.004; Peter Frost, "Polygyny Makes Men Bigger, Tougher . . . and Meaner," *Unz Review: An Alternative Media Selection* (blog), October 24, 2015, accessed December 18, 2018, http://www.unz.com/pfrost/polygyny-makes-men-bigger-tougher-and-meaner/; Singh Rajender et al., "Reduced CAG Repeats Length in Androgen Receptor Gene Is Associated with Violent Criminal Behavior," *International Journal of Legal Medicine* 122, no. 5 (September 2008): 367–72, https://doi.org/10.1007/s00414-008-0225-7.

7. Genetics Home Reference, "AR Gene," *National Institutes of Health U.S. National Library of Medicine*, accessed December 18, 2018, https://ghr.nlm.nih.gov/gene/AR.

8. Giacomo Tirabassi et al., "Influence of CAG Repeat Polymorphism on the Targets of Testosterone Action," *International Journal of Endocrinology* 2015 (2015): 1–12, https://doi.org/10.1155/2015/298107; Benjamin C. Trumble, Adrian V. Jaeggi, and Michael Gurven,

"Evolving the Neuroendocrine Physiology of Human and Primate Cooperation and Collective Action," *Philosophical Transactions of the Royal Society B: Biological Sciences* 370, no. 1,683 (December 5, 2015), https://doi.org/10.1098/rstb.2015.0014.

9. Charles L. Bennett et al., "Racial Variation in CAG Repeat Lengths within the Androgen Receptor Gene among Prostate Cancer Patients of Lower Socioeconomic Status," *Journal of Clinical Oncology* 20, no. 17 (September 2002): 3,599–604, https://doi.org/10.1200/JCO.2002.11.085; E. Giovannucci et al., "The CAG Repeat within the Androgen Receptor Gene and Its Relationship to Prostate Cancer," *Proceedings of the National Academy of Sciences* 94, no. 7 (April 1, 1997): 3,320–23, https://doi.org/10.1073/pnas.94.7.3320; Irvine et al., "The CAG and GGC Microsatellites of the Androgen Receptor Gene Are in Linkage Disequilibrium in Men with Prostate Cancer"; Sartor, Zheng, and Eastham, "Androgen Receptor Gene CAG Repeat Length Varies in a Race-Specific Fashion in Men without Prostate Cancer."

10. Giovannucci et al., "The CAG Repeat within the Androgen Receptor Gene and Its Relationship to Prostate Cancer."

11. Giovannucci et al., "The CAG Repeat within the Androgen Receptor Gene and Its Relationship to Prostate Cancer."

12. Sartor, Zheng, and Eastham, "Androgen Receptor Gene CAG Repeat Length Varies in a Race-Specific Fashion in Men without Prostate Cancer."

13. Giovannucci et al., "The CAG Repeat within the Androgen Receptor Gene and Its Relationship to Prostate Cancer."

14. Zhiqiang Qin et al., "Association between Polymorphic CAG Repeat Lengths in the Androgen Receptor Gene and Susceptibility to Prostate Cancer: A Systematic Review and Meta-Analysis," *Medicine* 96, no. 25 (June 2017): e7258, https://doi.org/10.1097/MD.0000000000007258.

15. W. D. Aiken, "Historical Determinants of Contemporary Attributes of African Descendants in the Americas: The Androgen Receptor Holds the Key," *Medical Hypotheses* 77, no. 6 (December 2011): 1,121–24, https://doi.org/10.1016/j.mehy.2011.09.017.

16. Aiken, "Historical Determinants of Contemporary Attributes of African Descendants in the Americas."

17. Aiken, "Historical Determinants of Contemporary Attributes of African Descendants in the Americas"; R. A. Kittles et al., "Extent of Linkage Disequilibrium between the Androgen Receptor Gene CAG and GGC Repeats in Human Populations: Implications for Prostate Cancer Risk," *Human Genetics* 109, no. 3 (September 2001): 253–61, https://doi.org/10.1007/s004390100576.

18. Matthew L. Freedman et al., "Systematic Evaluation of Genetic Variation at the Androgen Receptor Locus and Risk of Prostate Cancer in a Multiethnic Cohort Study," *American Journal of Human Genetics* 76, no. 1 (January 2005): 82–90, https://doi.org/10.1086/427224; Timothy Gilligan et al., "Absence of a Correlation of Androgen Receptor Gene CAG Repeat Length and Prostate Cancer Risk in an African American Population," *Clinical Prostate Cancer* 3, no. 2 (September 2004): 98–103; Vanessa R. Panz et al., "Tandem CAG Repeats of the Androgen Receptor Gene and Prostate Cancer Risk in Black and White Men," *Endocrine* 15, no. 2 (2001): 213–16, https://doi.org/10.1385/ENDO:15:2:213.

19. Ethan M. Lange et al., "The Androgen Receptor CAG and GGN Repeat Polymorphisms and Prostate Cancer Susceptibility in African American Men: Results from the

Flint Men's Health Study," *Journal of Human Genetics* 53, no. 3 (2008): 220–26, https://doi.org/10.1007/s10038-007-0240-4.

20. Lange et al., "The Androgen Receptor CAG and GGN Repeat Polymorphisms and Prostate Cancer Susceptibility in African American Men."

21. T. V. Rajan et al., "Evidence for Low Androgenicity among Indian (South Asian) Men," *Aging Male: The Official Journal of the International Society for the Study of the Aging Male* 17, no. 1 (March 2014): 30–34, https://doi.org/10.3109/13685538.2013.832192.

22. Annette Mouritsen et al., "Androgen Receptor CAG Repeat Length Is Associated with Body Fat and Serum SHBG in Boys: A Prospective Cohort Study," *Journal of Clinical Endocrinology and Metabolism* 98, no. 3 (March 2013): E605–9, https://doi.org/10.1210/jc.2012-3778.

23. D. E. Comings et al., "Association of the Androgen Receptor Gene (AR) with ADHD and Conduct Disorder," *Neuroreport* 10, no. 7 (May 14, 1999): 1,589–92.

24. Rajender et al., "Reduced CAG Repeats Length in Androgen Receptor Gene Is Associated with Violent Criminal Behavior."

25. Marina L. Butovskaya et al., "Aggression and Polymorphisms in AR, DAT1, DRD2, and COMT Genes in Datoga Pastoralists of Tanzania," *Scientific Reports* 3, no. 1 (December 2013), https://doi.org/10.1038/srep03148; Rajender et al., "Reduced CAG Repeats Length in Androgen Receptor Gene Is Associated with Violent Criminal Behavior."

26. Butovskaya et al., "Aggression and Polymorphisms in AR, DAT1, DRD2, and COMT Genes in Datoga Pastoralists of Tanzania"; Peter L. Hurd, Kathryn L. Vaillancourt, and Natalie L. Dinsdale, "Aggression, Digit Ratio, and Variation in Androgen Receptor and Monoamine Oxidase a Genes in Men," *Behavior Genetics* 41, no. 4 (July 2011): 543–56, https://doi.org/10.1007/s10519-010-9404-7.

27. Cheng et al., "Association Study of Androgen Receptor CAG Repeat Polymorphism and Male Violent Criminal Activity."

28. Cheng et al., "Association Study of Androgen Receptor CAG Repeat Polymorphism and Male Violent Criminal Activity."

29. Joseph L. Graves, "What a Tangled Web He Weaves: Race, Reproductive Strategies, and Rushton's Life History Theory," *Anthropological Theory* 2, no. 2 (June 2002): 131–54, https://doi.org/10.1177/1469962002002002627.

30. J. Philippe Rushton, *Race, Evolution, and Behavior: A Life History Perspective*, 3rd ed. (Puerto Ayora: Charles Darwin Research Institute, 2000).

31. Rushton, *Race, Evolution, and Behavior*.

32. Rushton, *Race, Evolution, and Behavior*.

33. C. Loring Brace, *"Race" Is a Four-Letter Word: The Genesis of the Concept* (New York: Oxford University Press, 2005); Leonard Lieberman, "How 'Caucasoids' Got Such Big Crania and Why They Shrank: From Morton to Rushton," *Current Anthropology* 42, no. 1 (February 2001): 69–95; Audrey Smedley and Brian D. Smedley, *Race in North America: Origin and Evolution of a Worldview*, 4th ed. (Boulder, CO: Westview Press, 2012).

34. Heather J. Litman et al., "Serum Androgen Levels in Black, Hispanic, and White Men," *Journal of Clinical Endocrinology and Metabolism* 91, no. 11 (November 2006): 4,326–34, https://doi.org/10.1210/jc.2006-0037.

35. A. Mazur and A. Booth, "Testosterone and Dominance in Men," *Behavioral and Brain Sciences* 21, no. 3 (June 1998): 353–63.

36. Trumble, Jaeggi, and Gurven, "Evolving the Neuroendocrine Physiology of Human and Primate Cooperation and Collective Action."

37. D. V. Shibalev et al., "Molecular Genetic Polymorphism of Androgen Receptor Gene (AR) in African Populations of Hadza and Datoga," *Russian Journal of Genetics* 49, no. 12 (December 2013): 1,258–60, https://doi.org/10.1134/S1022795413120119.

38. Trumble, Jaeggi, and Gurven, "Evolving the Neuroendocrine Physiology of Human and Primate Cooperation and Collective Action."

39. J. S. Mascaro, P. D. Hackett, and J. K. Rilling, "Testicular Volume Is Inversely Correlated with Nurturing-Related Brain Activity in Human Fathers," *Proceedings of the National Academy of Sciences* 110, no. 39 (September 24, 2013): 15,746–51, https://doi.org/10.1073/pnas.1305579110; Trumble, Jaeggi, and Gurven, "Evolving the Neuroendocrine Physiology of Human and Primate Cooperation and Collective Action."

40. Butovskaya et al., "Androgen Receptor Gene Polymorphism, Aggression, and Reproduction in Tanzanian Foragers and Pastoralists"; Peter B. Gray, Peter T. Ellison, and Benjamin C. Campbell, "Testosterone and Marriage among Ariaal Men of Northern Kenya," *Current Anthropology* 48, no. 5 (October 2007): 750–55, https://doi.org/10.1086/522061; Peter B. Gray, Timothy S. McHale, and Justin M. Carré, "A Review of Human Male Field Studies of Hormones and Behavioral Reproductive Effort," *Hormones and Behavior* 91 (May 2017): 52–67, https://doi.org/10.1016/j.yhbeh.2016.07.004; Shibalev et al., "Molecular Genetic Polymorphism of Androgen Receptor Gene (AR) in African Populations of Hadza and Datoga."

41. Edward Dutton, Dimitri van der Linden, and Richard Lynn, "Population Differences in Androgen Levels: A Test of the Differential K Theory," *Personality and Individual Differences* 90 (February 2016): 289–95, https://doi.org/10.1016/j.paid.2015.11.030.

42. Dutton, van der Linden, and Lynn, "Population Differences in Androgen Levels."

43. Zachary L. Simmons and James R. Roney, "Variation in CAG Repeat Length of the Androgen Receptor Gene Predicts Variables Associated with Intrasexual Competitiveness in Human Males," *Hormones and Behavior* 60, no. 3 (August 2011): 306–12, https://doi.org/10.1016/j.yhbeh.2011.06.006.

44. Simmons and Roney, "Variation in CAG Repeat Length of the Androgen Receptor Gene Predicts Variables Associated with Intrasexual Competitiveness in Human Males."

45. C. Loring Brace, "Racialism and Racist Agendas: *Race, Evolution, and Behavior: A Life History Perspective*," J. Philippe Rushton," *American Anthropologist* 98, no. 1 (March 1996): 176–77, https://doi.org/10.1525/aa.1996.98.1.02a00250.

46. Simmons and Roney, "Variation in CAG Repeat Length of the Androgen Receptor Gene Predicts Variables Associated with Intrasexual Competitiveness in Human Males."

47. Simmons and Roney, "Variation in CAG Repeat Length of the Androgen Receptor Gene Predicts Variables Associated with Intrasexual Competitiveness in Human Males."

48. Simmons and Roney, "Variation in CAG Repeat Length of the Androgen Receptor Gene Predicts Variables Associated with Intrasexual Competitiveness in Human Males."

49. Simmons and Roney, "Variation in CAG Repeat Length of the Androgen Receptor Gene Predicts Variables Associated with Intrasexual Competitiveness in Human Males."

50. O. Akinloye, J. Gromoll, and M. Simoni, "Variation in CAG and GGN Repeat Lengths and CAG/GGN Haplotype in Androgen Receptor Gene Polymorphism and Prostate Carcinoma in Nigerians," *British Journal of Biomedical Science* 68, no. 3 (January 2011): 138–42, https://doi.org/10.1080/09674845.2011.11730341.

51. Akinloye, Gromoll, and Simoni, "Variation in CAG and GGN Repeat Lengths and CAG/GGN Haplotype in Androgen Receptor Gene Polymorphism and Prostate Carcinoma in Nigerians."

52. Simmons and Roney, "Variation in CAG Repeat Length of the Androgen Receptor Gene Predicts Variables Associated with Intrasexual Competitiveness in Human Males."

53. Simmons and Roney, "Variation in CAG Repeat Length of the Androgen Receptor Gene Predicts Variables Associated with Intrasexual Competitiveness in Human Males."

54. Gray, McHale, and Carré, "A Review of Human Male Field Studies of Hormones and Behavioral Reproductive Effort."

55. Butovskaya et al., "Aggression and Polymorphisms in AR, DAT1, DRD2, and COMT Genes in Datoga Pastoralists of Tanzania."

56. Shibalev et al., "Molecular Genetic Polymorphism of Androgen Receptor Gene (AR) in African Populations of Hadza and Datoga."

57. Shibalev et al., "Molecular Genetic Polymorphism of Androgen Receptor Gene (AR) in African Populations of Hadza and Datoga."

58. Butovskaya et al., "Androgen Receptor Gene Polymorphism, Aggression, and Reproduction in Tanzanian Foragers and Pastoralists"; Shibalev et al., "Molecular Genetic Polymorphism of Androgen Receptor Gene (AR) in African Populations of Hadza and Datoga"; V. A. Vasilyev et al., "Polymorphism of the Dopamine D4 Receptor (DRD4) and Serotonin Transporter (5-HTTL) Gene Promoter Regions in African Tribes of Hadza and Datoga," *Russian Journal of Genetics* 47, no. 2 (February 2011): 226–29, https://doi.org/10.1134/S1022795411020190.

59. Vasilyev et al., "Polymorphism of the Dopamine D4 Receptor (DRD4) and Serotonin Transporter (5-HTTL) Gene Promoter Regions in African Tribes of Hadza and Datoga."

60. Butovskaya et al., "Androgen Receptor Gene Polymorphism, Aggression, and Reproduction in Tanzanian Foragers and Pastoralists"; Shibalev et al., "Molecular Genetic Polymorphism of Androgen Receptor Gene (AR) in African Populations of Hadza and Datoga"; Vasilyev et al., "Polymorphism of the Dopamine D4 Receptor (DRD4) and Serotonin Transporter (5-HTTL) Gene Promoter Regions in African Tribes of Hadza and Datoga."

61. Butovskaya et al., "Aggression and Polymorphisms in AR, DAT1, DRD2, and COMT Genes in Datoga Pastoralists of Tanzania."

62. Shibalev et al., "Molecular Genetic Polymorphism of Androgen Receptor Gene (AR) in African Populations of Hadza and Datoga."

63. Butovskaya et al., "Aggression and Polymorphisms in AR, DAT1, DRD2, and COMT Genes in Datoga Pastoralists of Tanzania."

64. J. M. Carré and N. A. Olmstead, "Social Neuroendocrinology of Human Aggression: Examining the Role of Competition-Induced Testosterone Dynamics," *Neuroscience* 286 (February 2015): 171–86, https://doi.org/10.1016/j.neuroscience.2014.11.029.

65. Carré and Olmstead, "Social Neuroendocrinology of Human Aggression."

66. Butovskaya et al., "Aggression and Polymorphisms in AR, DAT1, DRD2, and COMT Genes in Datoga Pastoralists of Tanzania."

67. Carré and Olmstead, "Social Neuroendocrinology of Human Aggression."

68. Elliot M. Fratkin, *Ariaal Pastoralists of Kenya: Studying Pastoralism, Drought, and Development in Africa's Arid Lands*, 2nd ed., Cultural Survival Studies in Ethnicity and Change (Boston: Pearson/Allyn and Bacon, 2004).

69. Dan T. A. Eisenberg et al., "Dopamine Receptor Genetic Polymorphisms and Body Composition in Undernourished Pastoralists: An Exploration of Nutrition Indices among Nomadic and Recently Settled Ariaal Men of Northern Kenya," *BMC Evolutionary Biology* 8, no. 1 (2008): 173, https://doi.org/10.1186/1471-2148-8-173.

70. Benjamin C. Campbell et al., "Androgen Receptor CAG Repeats and Body Composition among Ariaal Men," *International Journal of Andrology* 32, no. 2 (April 2009): 140–48, https://doi.org/10.1111/j.1365-2605.2007.00825.x.

71. Esther Esteban et al., "Androgen Receptor CAG and GGC Polymorphisms in Mediterraneans: Repeat Dynamics and Population Relationships," *Journal of Human Genetics* 51, no. 2 (February 2006): 129–36, https://doi.org/10.1007/s10038-005-0336-7; Kittles et al., "Extent of Linkage Disequilibrium between the Androgen Receptor Gene CAG and GGC Repeats in Human Populations."

72. Esteban et al., "Androgen Receptor CAG and GGC Polymorphisms in Mediterraneans."

73. Gray, Ellison, and Campbell, "Testosterone and Marriage among Ariaal Men of Northern Kenya."

74. Gray, Ellison, and Campbell, "Testosterone and Marriage among Ariaal Men of Northern Kenya."

75. Butovskaya et al., "Aggression and Polymorphisms in AR, DAT1, DRD2, and COMT Genes in Datoga Pastoralists of Tanzania"; Louis Calistro Alvarado, "Population Differences in the Testosterone Levels of Young Men Are Associated with Prostate Cancer Disparities in Older Men," *American Journal of Human Biology: The Official Journal of the Human Biology Council* 22, no. 4 (August 2010): 449–55, https://doi.org/10.1002/ajhb.21016; Kesson Magid et al., "Childhood Ecology Influences Salivary Testosterone, Pubertal Age, and Stature of Bangladeshi UK Migrant Men," *Nature Ecology and Evolution* 2, no. 7 (July 2018): 1,146–54, https://doi.org/10.1038/s41559-018-0567-6.

76. Lee T. Gettler et al., "The Role of Testosterone in Coordinating Male Life History Strategies: The Moderating Effects of the Androgen Receptor CAG Repeat Polymorphism," *Hormones and Behavior* 87 (January 2017): 164–75, https://doi.org/10.1016/j.yhbeh.2016.10.012.

77. Gettler et al., "The Role of Testosterone in Coordinating Male Life History Strategies."

78. Butovskaya et al., "Aggression and Polymorphisms in AR, DAT1, DRD2, and COMT Genes in Datoga Pastoralists of Tanzania."

79. John Archer, Nicola Graham-Kevan, and Michelle Davies, "Testosterone and Aggression: A Reanalysis of Book, Starzyk, and Quinsey's (2001) Study," *Aggression and Violent Behavior* 10, no. 2 (January 2005): 241–61, https://doi.org/10.1016/j.avb.2004.01.001.

80. Stephen B. Manuck et al., "Salivary Testosterone and a Trinucleotide (CAG) Length Polymorphism in the Androgen Receptor Gene Predict Amygdala Reactivity in Men," *Psychoneuroendocrinology* 35, no. 1 (January 2010): 94–104, https://doi.org/10.1016/j.psyneuen.2009.04.013.

81. Akinloye, Gromoll, and Simoni, "Variation in CAG and GGN Repeat Lengths and CAG/GGN Haplotype in Androgen Receptor Gene Polymorphism and Prostate Carcinoma in Nigerians."

82. Akinloye, Gromoll, and Simoni, "Variation in CAG and GGN Repeat Lengths and CAG/GGN Haplotype in Androgen Receptor Gene Polymorphism and Prostate Carcinoma in Nigerians."

83. Menelaos L. Batrinos, "Testosterone and Aggressive Behavior in Man," *International Journal of Endocrinology and Metabolism* 10, no. 3 (June 1, 2012): 563–68, https://doi.org/10.5812/ijem.3661.

84. Marco Del Giudice et al., "Individual Differences in Developmental Plasticity: A Role for Early Androgens?" *Psychoneuroendocrinology* 90 (April 2018): 165–73, https://doi.org/10.1016/j.psyneuen.2018.02.025.

85. Batrinos, "Testosterone and Aggressive Behavior in Man."

86. James R. Roney, Zachary L. Simmons, and Aaron W. Lukaszewski, "Androgen Receptor Gene Sequence and Basal Cortisol Concentrations Predict Men's Hormonal Responses to Potential Mates," *Proceedings: Biological Sciences* 277, no. 1,678 (January 7, 2010): 57–63, https://doi.org/10.1098/rspb.2009.1538.

87. Roney, Simmons, and Lukaszewski, "Androgen Receptor Gene Sequence and Basal Cortisol Concentrations Predict Men's Hormonal Responses to Potential Mates."

88. Gray, McHale, and Carré, "A Review of Human Male Field Studies of Hormones and Behavioral Reproductive Effort."

89. Batrinos, "Testosterone and Aggressive Behavior in Man."

90. Gray, McHale, and Carré, "A Review of Human Male Field Studies of Hormones and Behavioral Reproductive Effort."

91. Erica L. Smearman et al., "Oxytocin Receptor Genetic and Epigenetic Variations: Association with Child Abuse and Adult Psychiatric Symptoms," *Child Development* 87, no. 1 (February 2016): 122–34, https://doi.org/10.1111/cdev.12493.

92. Bekh Bradley et al., "Association between Childhood Maltreatment and Adult Emotional Dysregulation in a Low-Income, Urban, African American Sample: Moderation by Oxytocin Receptor Gene," *Development and Psychopathology* 23, no. 2 (May 2011): 439–52, https://doi.org/10.1017/S0954579411000162; Smearman et al., "Oxytocin Receptor Genetic and Epigenetic Variations."

93. Butovskaya et al., "Aggression and Polymorphisms in AR, DAT1, DRD2, and COMT Genes in Datoga Pastoralists of Tanzania."

94. Butovskaya et al., "Aggression and Polymorphisms in AR, DAT1, DRD2, and COMT Genes in Datoga Pastoralists of Tanzania."

95. Butovskaya et al., "Aggression and Polymorphisms in AR, DAT1, DRD2, and COMT Genes in Datoga Pastoralists of Tanzania."

96. Butovskaya et al., "Aggression and Polymorphisms in AR, DAT1, DRD2, and COMT Genes in Datoga Pastoralists of Tanzania."

97. Butovskaya et al., "Aggression and Polymorphisms in AR, DAT1, DRD2, and COMT Genes in Datoga Pastoralists of Tanzania."

98. Jie Li et al., "Ethnical Disparities of Prostate Cancer Predisposition: Genetic Polymorphisms in Androgen-Related Genes," *American Journal of Cancer Research* 3, no. 2 (2013): 127–51.

99. Vito Cucchiara et al., "Epigenomic Regulation of Androgen Receptor Signaling: Potential Role in Prostate Cancer Therapy," *Cancers* 9, no. 1 (January 16, 2017), https://doi.org/10.3390/cancers9010009.

100. Changmeng Cai, Xin Yuan, and Steven P. Balk, "Androgen Receptor Epigenetics," *Translational Andrology and Urology* 2, no. 3 (September 1, 2013): 148–57, https://doi.org/10.3978/j.issn.2223-4683.2013.09.02.

101. Joshua M. Galanter et al., "Differential Methylation between Ethnic Subgroups Reflects the Effect of Genetic Ancestry and Environmental Exposures," *ELife* 6 (January 3, 2017): e20532, https://doi.org/10.7554/eLife.20532.

Epilogue

1. Aravinda Chakravarti, "Perspectives on Human Variation through the Lens of Diversity and Race," *Cold Spring Harbor Perspectives in Biology* 7, no. 9 (September 1, 2015): a023358, https://doi.org/10.1101/cshperspect.a023358.

2. Jonathan Michael Kaplan, "Race, IQ, and the Search for Statistical Signals Associated with So-Called 'X' Factors: Environments, Racism, and the 'Hereditarian Hypothesis,'" *Biology and Philosophy* 30, no. 1 (January 2015): 1–17, https://doi.org/10.1007/s10539-014 -9428-0.

3. Kaplan, "Race, IQ, and the Search for Statistical Signals Associated with So-Called 'X' Factors."

4. Marina L. Butovskaya et al., "Serotonergic Gene Polymorphisms (5-HTTLPR, 5HTR1A, 5HTR2A) and Population Differences in Aggression: Traditional (Hadza and Datoga) and Industrial (Russians) Populations Compared," *Journal of Physiological Anthropology* 37, no. 1 (April 16, 2018): 10, https://doi.org/10.1186/s40101-018-0171-0.

5. Becky Mansfield and Julie Guthman, "Epigenetic Life: Biological Plasticity, Abnormality, and New Configurations of Race and Reproduction," *Cultural Geographies* 22, no. 1 (January 2015): 3–20, https://doi.org/10.1177/1474474014555659.

6. Mansfield and Guthman, "Epigenetic Life."

7. Theresa Phillips, "The Role of Methylation in Gene Expression," *Nature Education* 1, no. 1 (2008): 16.

8. James R. Lupski et al., "Clan Genomics and the Complex Architecture of Human Disease," *Cell* 147, no. 1 (September 30, 2011): 32–43, https://doi.org/10.1016/j.cell.2011.09.008; "New Genetic Roots for Intelligence Discovered," *ScienceDaily*, May 23, 2017, accessed October 16, 2018, https://www.sciencedaily.com/releases/2017/05/170523083324.htm.

9. Lupski et al., "Clan Genomics and the Complex Architecture of Human Disease."

10. Lupski et al., "Clan Genomics and the Complex Architecture of Human Disease."

Index